TURBULENT FLOW COMPUTATION

FLUID MECHANICS AND ITS APPLICATIONS
Volume 66

Series Editor: R. MOREAU
MADYLAM
Ecole Nationale Supérieure d'Hydraulique de Grenoble
Boîte Postale 95
38402 Saint Martin d'Hères Cedex, France

Aims and Scope of the Series

The purpose of this series is to focus on subjects in which fluid mechanics plays a funda
mental role.

As well as the more traditional applications of aeronautics, hydraulics, heat and mas:
transfer etc., books will be published dealing with topics which are currently in a state o
rapid development, such as turbulence, suspensions and multiphase fluids, super and hy
personic flows and numerical modelling techniques.

It is a widely held view that it is the interdisciplinary subjects that will receive intens
scientific attention, bringing them to the forefront of technological advancement. Fluid:
have the ability to transport matter and its properties as well as transmit force, therefor
fluid mechanics is a subject that is particulary open to cross fertilisation with other scien
ces and disciplines of engineering. The subject of fluid mechanics will be highly relevan
in domains such as chemical, metallurgical, biological and ecological engineering. Thi:
series is particularly open to such new multidisciplinary domains.

The median level of presentation is the first year graduate student. Some texts are mono
graphs defining the current state of a field; others are accessible to final year undergradu
ates; but essentially the emphasis is on readability and clarity.

For a list of related mechanics titles, see final pages.

Turbulent Flow Computation

Edited by

D. DRIKAKIS

Queen Mary, University of London, Department of Engineering,
London, United Kingdom

and

B.J. GEURTS

University of Twente, Faculty of Mathematical Sciences,
Enschede, The Netherlands

KLUWER ACADEMIC PUBLISHERS

DORDRECHT / BOSTON / LONDON

A C.I.P. Catalogue record for this book is available from the Library of Congress.

ISBN 978-90-481-5981-9

Published by Kluwer Academic Publishers,
P.O. Box 17, 3300 AA Dordrecht, The Netherlands.

Sold and distributed in North, Central and South America
by Kluwer Academic Publishers,
101 Philip Drive, Norwell, MA 02061, U.S.A.

In all other countries, sold and distributed
by Kluwer Academic Publishers,
P.O. Box 322, 3300 AH Dordrecht, The Netherlands.

Printed on acid-free paper

Contents

Preface

In various branches of fluid mechanics, our understanding is inhibited by the presence of turbulence. Although many experimental and theoretical studies have significantly helped to increase our physical understanding, a comprehensive and predictive theory of turbulent flows has not yet been established. Therefore, the prediction of turbulent flow relies heavily on simulation strategies. The development of reliable methods for turbulent flow computation will have a significant impact on a variety of technological advancements. These range from aircraft and car design, to turbomachinery, combustors, and process engineering. Moreover, simulation approaches are important in materials design, prediction of biologically relevant flows, and also significantly contribute to the understanding of environmental processes including weather and climate forecasting.

The material that is compiled in this book presents a coherent account of contemporary computational approaches for turbulent flows. It aims to provide the reader with information about the current state of the art as well as to stimulate directions for future research and development. The book puts particular emphasis on computational methods for incompressible and compressible turbulent flows as well as on methods for analysing and quantifying numerical errors in turbulent flow computations. In addition, it presents turbulence modelling approaches in the context of large eddy simulation, and unfolds the challenges in the field of simulations for multiphase flows and computational fluid dynamics (CFD) of engineering flows in complex geometries. Apart from reviewing main research developments, new material is also included in many of the chapters.

This book aims to appeal to a broad audience of applied scientists and engineers who are involved in the development and application of CFD methods for turbulent flows. The readership includes academic researchers as well as CFD practitioners in industry. The chapters were composed to constitute an advanced textbook for PhD candidates in the field of CFD and turbulence. The book can also be used as a complementary textbook at the level of MSc (or MEng) studies in engineering, applied mathematics and physics departments.

In the rest of this preface we will briefly outline the content of the contributing chapters.

The first chapter provides a review of spectral element methods on unstructured grids and focuses primarily on the numerical uncertainties in under-resolved simulations using spectral methods. The authors discuss the effect of under-resolved discretization of the advective (nonlinear) terms and dealiasing on non-uniform grids. They present several flow examples and diagnostic approaches that can be employed to detect erroneous predictions.

Chapter 2 outlines the basic structure of high resolution methods and proposes their use as an effective turbulence model in the context of large eddy simulation (LES). Theoretical arguments as well as computational evidence, based on comparisons of multimaterial mixing simulations and experiments, are provided. These concern the ability of high resolution methods for hyperbolic conservation laws to achieve many of the properties of subgrid scale models.

Issues related to symmetry-preserving discretization of the Navier-Stokes equations and their effects on turbulent flow simulations, are discussed in Chapter 3. The authors draw the attention of the reader to difference operators that have the same symmetry properties as the corresponding differential operators. These are illustrated for model equations such as the convection-diffusion equation, as well as for the full Navier-Stokes equations.

The fourth chapter presents methods for analysing and quantifying numerical errors in direct numerical simulations (DNS) and LES of turbulent flow. The author discusses all types of errors arising in these simulations, i.e., aliasing, truncation, time-stepping and commutation errors, and comments on their analysis.

Chapter 5 presents the design of adaptive low-dissipative high order schemes for turbulent flow computations. A summary of linear and nonlinear stability, and an outline of the rationale and criteria for designing such adaptive schemes, is presented. The control of adaptive numerical dissipation for high order schemes is also covered.

The sixth chapter discusses issues pertinent to the predictability and reliability of numerical simulations of multiscale complex nonlinear problems in computational fluid dynamics. The author outlines the sources of nonlinearities, the knowledge gained by studying the dynamics of numerics for nonlinear model problems, as well as specific approaches for ensuring confidence in the numerical simulations.

The Lagrangian-Averaged Navier-Stokes$-\alpha$ (LANS$-\alpha$) approach for modelling turbulence is presented in Chapter 7. This approach aims to eliminate some of the heuristic elements that would otherwise be involved in the modelling of subgrid scale stresses. The authors investigate the physical and numerical properties of the LANS$-\alpha$ subgrid parameterization through simulations of a turbulent mixing layer.

Developments and challenges in the computation of geophysical turbulence are discussed in Chapter 8. It is demonstrated that in the absence of an explicit subgrid-scale turbulence model, nonoscillatory forward-in-time (NFT) methods offer means of implicit subgrid-scale modelling. These methods can be quite effective in large-eddy simulations of high Reynolds number flows. Theoretical discussions are illustrated with examples of meteorological flows that address the range of applications from micro-turbulence to change in climate.

The ninth chapter discusses developments in the field of turbulent multiphase flows. A method based on solving the Navier-Stokes equations by a finite difference/front tracking technique is presented. The inclusion of fully deformable interfaces and surface tension, in addition to inertial and viscous effects, is described. The chapter also highlights new areas where computations are still in their infancy.

Chapter 10 demonstrates the applicability of modern computational methods and engineering turbulence models to a number of turbulent flows of engineering interest. The techniques discussed in this chapter include traditional turbulence closures as well as hybrid RANS/LES models for unsteady flows. The generation of high quality grids and the development of efficient methods for unsteady flows are identified as major challenges in future CFD research.

We would like to gratefully acknowledge the contributing authors for their substantial effort in preparing the chapters. We hope that the book will provide a good introduction to the various facets of computational fluid dynamics as well as stimulate further research in the fields of numerical methods and turbulence modelling, and their application in technology and natural sciences.

DIMITRIS DRIKAKIS, LONDON, UNITED KINGDOM.

BERNARD GEURTS, ENSCHEDE, THE NETHERLANDS.

Contributing Authors

Paul Batten obtained his Ph.D. in parallel computing at the University of Southampton, UK, in 1991. He subsequently held positions at the Manchester Metropolitan University and UMIST, Manchester, performing research into blast-wave simulation and turbulence modeling, respectively. Dr. Batten moved to California in January 1999 to work for Metacomp Technologies, Inc., where his current research activities include acoustics and turbulence modeling for unsteady flows. His research interests include computational methods, numerical algorithms and turbulence modeling.

Bernard Bunner is an engineer in the Microfluidics and Biotechnology group at Coventor, a MEMS consulting company. He obtained his PhD in Mechanical Engineering and an MSc in Electrical Engineering from the University of Michigan. His research interests include multiphase flows, microfluidics and numerical methods.

Sukumar Chakravarty is the founder and president of Metacomp Technologies, Inc., since 1994. From 1979 till then he was Manager of CFD at the Rockwell Science Center, earning the Rockwell Engineer of the Year award in 1989. He is an Adjunct Professor at UCLA, teaching numerical analysis and computational aerodynamics. He is a pioneer in numerical algorithms and several other computational areas. He was co-author of a paper that won NASA's H. Julian Allen Award for 1985.

Dimitris Drikakis is a Professor of Fluid Mechanics at Queen Mary, University of London (QMUL). Previous academic appointments include a visiting Professorship at the Laboratory of Aerodynamics and Biomechanics of the Universite de la Méditerranée, Marseille, France; a Readership in Computational Fluid Dynamics at QMUL; a Lectureship at the University of Manchester Institute of Science and Technology (UMIST); the posts of senior scientist and group leader at the Chair of Fluid Mechanics of the University of Erlangen-Nuremberg,

Germany. His research interests include computational fluid dynamics, computational methods, turbulence, aerodynamics, biofluid mechanics, microflows, nano-engineering and scientific computing.

Asghar Esmaeeli is an assistant research professor at the Worcester Polytechnic Institute. Before his current position he spent six years as a senior research fellow at the University of Michigan, where he received a PhD in Mechanical Engineering and Scientific Computing in 1995. His research interests include multiphase flows, boiling and bubbly flows, microgravity flows, and numerical modeling.

Arturo Fernandez is a research scientist at the Worcester Polytechnic Institute. He received his PhD degree from the "Universidad Politecnica de Madrid" in 2000. His research interests focus on multiphase flows, parallel computing and its application to fluid mechanics, vortex methods and electrohydrodynamics.

Bernard J. Geurts teaches Computational Mathematical Physics at the University of Twente since 1991. He is a visiting professor at Queen Mary, University of London since 1999. He coordinates the special interest group of ERCOFTAC (European Research Council On Flow, Turbulence And Combustion) on computational fluid dynamics (LESig) and is author of the book *Elements of Direct and Large-Eddy Simulation*. His research interests center around flows with complex physics, e.g., turbulence, multiphase flows, polymer-physics, combustion and biology. Moreover, he works in nonlinear dynamics, numerical methods and simulation.

Sandip Ghosal is currently an Associate Professor of Mechanical Engineering at Northwestern University. He obtained a Ph.D. in Physics from Columbia University in the area of Astrophysical Fluid Dynamics and was employed as a researcher at Stanford University, Los Alamos National Laboratories and Sandia National Laboratories in the United States. He is most well known in the turbulence community through his contributions to LES, in particular in the mathematical development of the "dynamic model" and the systematic methods of error analysis that he introduced in LES. In addition to turbulence, his current research interests include combustion and the fluid mechanics of microfluidic systems.

Uriel Goldberg received his Ph.D.in physics of fluids from Case Institute of Technology, 1984, was Member of the Technical Staff, Rockwell Science Center, 1984-1995 and research associate, UMIST, Manchester, England, 1995-

1996. Dr. Goldberg joined Metacomp Technologies, Inc. as Senior Scientist in 1996. His activities include development of turbulence models for use in CFD.

Darryl D. Holm works in the Mathematical Modeling and Analysis Group in the Theoretical Division at Los Alamos National Laboratory. He is co-founder and past Acting Director of the Los Alamos Center for Nonlinear Studies and is now a Laboratory Fellow. He currently leads the Turbulence Working Group at Los Alamos. Holm's primary scientific interest is nonlinear science – ranging from integrable to chaotic behavior – especially nonlinear dynamics of optical pulses and fluids. In fluid dynamics, Holm is applying Lagrangian averaging, asymptotics and other dynamical systems methods to study the mean effects of subgrid scales on nonlinear wave structures and momentum transport as a basis for turbulence closure models.

George Em Karniadakis received his M.Sc. and Ph.D. from Massachusetts Institute of Technology in 1984 and 1987, respectively, both in Mechanical Engineering. He did his postdoc at Stanford University, and he previously taught at Princeton University. He is currently Professor of Applied Mathematics at Brown University. He has pioneered spectral methods on unstructured grids, parallel simulations of turbulence in complex geometries, and microfluidics simulations.

Robert M. Kirby received his B.Sc. in Applied Mathematics and Computer and Information Sciences from the Florida State University in 1997. He received his Sc.M. in Applied Mathematics from Brown University in 1999, and is currently working on a Sc.M. in Computer Science and Ph.D. in Applied Mathematics at Brown University. His research interests are in software design, parallel computing and direct numerical simulation of flow-structure interactions.

Joseph M. Prusa is a Collaborating Associate Professor at Iowa State University, Ames, IA. He has also been a visiting scientist at the National Center for Atmospheric Research, Boulder, CO. His research interests include atmospheric physics and the development of generalized coordinate models for analysis and computation of thermo-fluid dynamics problems.

William Rider is a Technical Staff Member in the Computer and Computational Science Division at Los Alamos National Laboratory. He has worked at Los Alamos since 1989. His principal research activities include the development of high resolution shock capturing methods, solution methods for multi-material incompressible flows, numerical methods for mixing flow, Newton-

Krylov methods, interface tracking and verification and validation of simulation codes.

Björn Sjögreen is associate professor in numerical analysis at the Royal Institute of Technology in Stockholm, Sweden. He obtained a Ph.D. in numerical analysis from Uppsala University, Sweden, in 1988. Research interests are in the design of numerical methods for high speed fluid flows, and fluid flows with combustion.

Piotr Smolarkiewicz is a Senior Scientist in the Microscale Mesoscale Meteorology Division at National Center for Atmospheric Research (NCAR), Boulder, Colorado. Before arriving to NCAR in the early eighties, he has held a faculty appointment at the Department of Physics of the University of Warsaw, Poland. His current interests include geophysical flows of all scales (viz, low Mach number, high Reynolds number flows), non-Newtonian fluids, and scientific computing (in general).

Gretar Tryggvason is a Professor and Head of Mechanical Engineering at the Worcester Polytechnic Institute. Previously, he was Professor of Mechanical Engineering at the University of Michigan in Ann Arbor. He has published papers on multiphase and free surface flows, vortex dynamics and combustion, boiling, solidification, and numerical methods. He is a fellow of the American Physical Society.

Arthur Veldman obtained a Ph.D. in Applied Mathematics from the University of Groningen. In 1977 he joined the National Aerospace Laboratory NLR in Amsterdam, where he was involved in various projects in the area of computational aero- and hydrodynamics. Additionally, between 1984 and 1990 he was part-time professor of CFD at Delft University of Technology. In 1990 he returned to Groningen, where he occupies the chair in Computational Mechanics.

Roel Verstappen obtained his Ph.D. in Applied Mathematics at the University of Twente in 1989. Since 1990, he is an assistant professor of Computational Mechanics at the University of Groningen (The Netherlands). His main area of reseach concerns the development of simulation methods for DNS of turbulent flow, and the mathematical analysis of DNS results.

Helen C. Yee obtained her MS in Applied Mathematics and Ph.D. in Applied Mechanics (majoring in nonlinear dynamics) from the University of California at Berkeley, California, USA. She is a Senior Research Scientist at the NASA

Ames Research Center, Moffett Field, California, USA. Her research interests include dynamical systems and chaos, dynamics of numerics, nonlinear instability, numerical analysis and algorithm development for complex multiscale nonlinear CFD problems in air and space transportation systems, planetary and atmospheric sciences, and numerical combustion.

Chapter 1

UNDER-RESOLUTION AND DIAGNOSTICS IN SPECTRAL SIMULATIONS OF COMPLEX-GEOMETRY FLOWS

Robert M. Kirby
kirby@cfm.brown.edu

George Em Karniadakis
Division of Applied Mathematics, Brown University, 182 George St., Box F, Providence, RI 02912, USA
gk@cfm.brown.edu

Abstract Large-scale simulations are often under-resolved at some level, but they are still useful in extracting both qualitative and quantitative information about the flow. In order to use such results effectively we need to characterize the numerical uncertainty of under-resolved simulations. However, different numerical methods exhibit different behavior, and spectral-based methods in particular may over-predict fluctuations both in amplitude and frequency due to their very low artificial dissipation in contrast with finite differences. In this chapter, we provide insight into under-resolved spectral simulations and document several diagnostic signs of under-resolution for spectral/hp element methods. We first review the state-of-the art in direct numerical simulation and present a new class of spectral methods on unstructured grids for handling complex-geometry compressible and incompressible flows. We then focus on the effects of under-resolving the nonlinear contributions, and finally we present prototype cases for both transitional and turbulent flows.

Keywords: Spectral methods, complex-geometry, under-resolution, unstructured grids, turbulence

1. Introduction

Under-resolved simulations are perhaps the rule rather than the exception! This can be understood, as in practice users attempt high Reynolds number

D. Drikakis and B.J. Geurts (eds.), Turbulent Flow Computation, 1–42.

simulations in problems with new physics and thus unknown resolution requirements. Verification and validation of the solution is a very tedious process ([1]), and at present there are no established efficient methods to assess numerical accuracy. Also, for large-scale turbulence simulations, existing computational resources may often be inadequate for the attempted simulations, so additional error checking simulations would be prohibitively expensive.

However, an under-resolved simulation is not useless but, in fact, it can provide a lot of information if proper characterization is established combined with experience for the specific discretization used in such a simulation. One such example is a relatively early direct numerical simulation of turbulent channel flow by ([2]) which has remained largely unnoticed. In figure 1.1 we plot one of his results, i.e. the Reynolds stress distribution across the channel. It is in good agreement with high resolution DNS even for the lowest resolution employed in Zores's simulation, corresponding to four Fourier modes in the streamwise, 16 Chebyshev modes in the normal, and six Fourier modes in the spanwise direction. The Reynolds number based on the wall shear velocity is $R_* \approx 120$. In order to achieve smooth profiles a very long- time averaging was employed. Clearly, this is an example of an under-resolved simulation, which however sustains the turbulence fluctuations and leads to better than 10% accuracy in second-order statistics. In contrast, a low resolution simulation based on finite differences would typically converge to the laminar steady state solution.

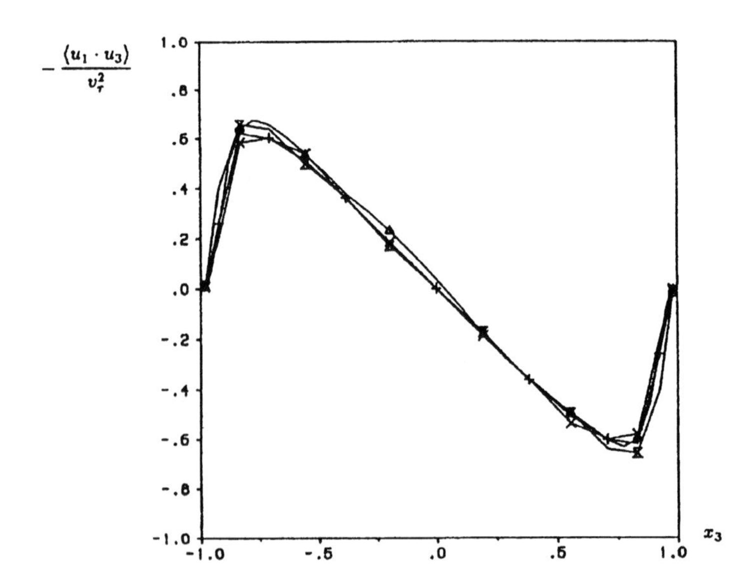

Figure 1.1. Reynolds stress distribution of a low-resolution simulation of a turbulent channel flow by ([2]). The different symbols correspond to different resolution in x (stream), y (normal) and z (span) as follows: $\triangle$: $8 \times 32 \times 8$; $\bowtie$: $8 \times 16 \times 8$; $+$: $6 \times 16 \times 6$; $\times$: $4 \times 16 \times 6$.

For under-resolved simulations to be useful we need to characterize both numerical and physical uncertainty, creating appropriate composite error bars similar to experiments. This is a very difficult task, and work on uncertainty associated with the data, i.e. input, is still at an early stage. On the numerical side, there are still many classical issues which are unresolved today, e.g. skew-symmetry of advection operators in the discrete sense, time-integration algorithms with large time step, efficient treatment of geometric complexity, efficient adaptivity, etc.

There are two major challenges today in direct numerical simulations (DNS) of turbulence following the successes of the last two decades ([3]): The first is that the maximum Reynolds number possible in simulations is still much lower compared to turbulent flows of practical interest. For example, at present or in the near future, the maximum Re_λ (based on the Taylor micro-scale) for homogeneous turbulence that can be accurately simulated is less than 500. However, in geophysical flows the typical Reynolds number Re_λ may be orders of magnitude higher. The second challenge we face is that complex-geometry flows are still largely untackled; geometries beyond the standard channel flow with flat walls have only recently been considered, and most of them involve at least one homogeneous direction.

A summary of the range of Reynolds number and geometries for which direct numerical simulations have been successfully completed is plotted in the sketch of figure 1.2. It shows that accurate direct numerical simulations of turbulence in simple-geometry domains can handle much higher Reynolds number flows than in complex-geometry domains. This, in essence, reflects the additional computational complexity associated with discretization of the Navier-Stokes equations in complex-geometry domains. Clearly, the Fourier discretization which is employed for all three directions in homogeneous turbulence cannot be used in inhomogeneous directions, where Chebyshev or Legendre spectral discretizations (or alternatively some high-order finite difference variant) are used. More specifically, on non-separable or multiply-connected domains (e.g. flow past a circular cylinder) these classical methods are also inappropriate, and thus new domain-decomposition based methods need to be used effectively.

As regards the type of discretization, it was evident even from the early attempts to perform DNS of turbulence in the seventies, that high-order discretization was not only computationally advantageous but also a necessity. Simulating turbulence requires long-time integration, however non-negligible dispersion errors associated with low-order discretization could eventually render the computational results erroneous. There is plenty of anecdotal evidence about such results from the early practitioners, and modern numerical analysis can rigorously document this as well. The importance of high-order discretization has also been recognized for large eddy simulations (LES) ([4]), as

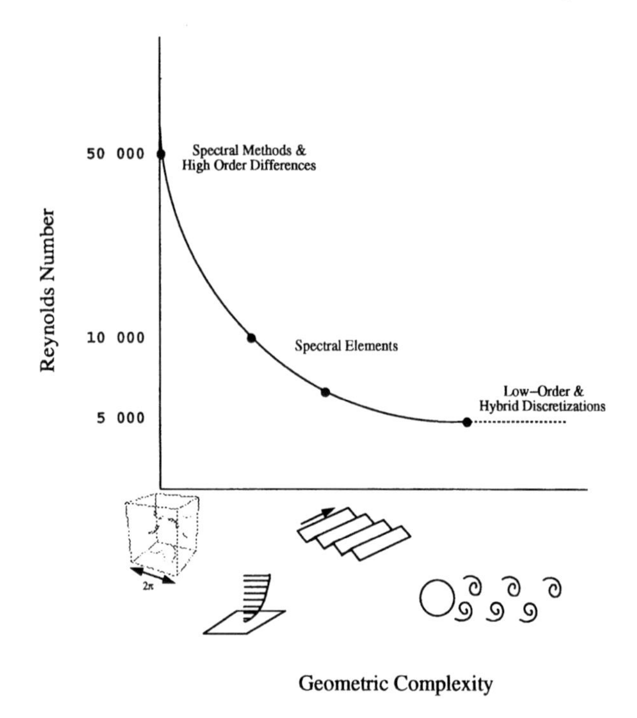

Figure 1.2. Conceptual overview of DNS of turbulent flows: Maximum Reynolds number versus geometric complexity.

discretization errors seem to interact with the subgrid modeling errors in an adverse way.

As we go through the multi-Teraflop (and beyond) computing era and are capable of performing simulations of 1 billion points or more at reasonable turn-around time, high-order numerical methods will play a key role in simulating high Reynolds number and complex-geometry turbulence. They provide

- fast convergence,
- small diffusion and dispersion errors,
- easier implementation of the *inf-sup* condition for incompressible Navier-Stokes,
- better data volume-over-surface ratio for efficient parallel processing, and
- better input/output handling due to the smaller volume of data.

For many engineering applications where accuracy of the order of 10% is acceptable, quadratic convergence is usually sufficient for stationary problems. However, this is not true in time-dependent flow simulations where long-time integration is required. Also, in DNS a 10% inaccuracy in phase errors may lead to flow re-laminarization. Therefore, we must ask how *long-time integration* relates to the formal order of accuracy of a numerical scheme, and what is the corresponding computational cost? To this end, let us consider the convection of a waveform at a constant speed. Let us now assume that there are $N^{(k)}$ grid points required per wavelength to reduce the error to a level ϵ, where k denotes the formal order of the scheme. In addition, let us assume that we integrate for M *time periods*. We can neglect temporal errors $\mathcal{O}(\Delta t)^J$ (where J is the order of the time integration) by assuming a sufficiently small time step Δt. We wish to estimate the phase error in this simulation for second- $N^{(2)}$, fourth- $N^{(4)}$, and sixth- $N^{(6)}$ order finite difference schemes. The complete analysis is presented in ([5]), and here we present the results for the computational work. In figure 1.3 we compare the efficiency of these three different discretizations for the *same phase error* by plotting the computational work required to maintain an "engineering" accuracy of 10% versus the number of time periods for the integration. This comparison favors the fourth-order scheme for short times ($M \propto \mathcal{O}(1)$) over both the second-order and the sixth-order schemes. However, for long-time integration ($M \propto \mathcal{O}(100)$), even for this engineering accuracy of 10%, the sixth-order scheme is superior as the corresponding operation count $W^{(6)}$ is about 6 times lower than the operation count of the second-order scheme $W^{(2)}$, and half the work of the fourth-order scheme $W^{(4)}$. For an accuracy of 1% in the solution of this convection problem, the sixth-order scheme is superior even for short-time integration.

High-order accuracy, however, does not automatically imply a resolved and thus accurate DNS or LES. In particular, spectral-based methods tend to behave differently when the number of grid points or modes is insufficient. For example, they tend to be more unstable, lead to over-prediction of amplitudes, and could even result in erroneous unsteady flow at subcritical conditions. This is the primary topic that we focus on in the present paper.

Specifically, we first review some key developments in extending spectral methods to unstructured grids for both incompressible and compressible flows. We then discuss in some detail the effect of under-resolving the discretization of nonlinear terms and how dealiasing can be handled on non-uniform grids presenting both one-dimensional examples but also full DNS of turbulent flow that may suffer from aliasing. We then proceed with several examples of internal and external flows and document diagnostics that can be employed to detect erroneous physics. We also include simulations of some turbulent flows which, although clearly under-resolved, lead to useful results in agreement with the experiments. Finally, we conclude with a perspective on simulating turbulence in fully three-dimensional domains where both numerical uncertainty and physical uncertainty are adequately characterized.

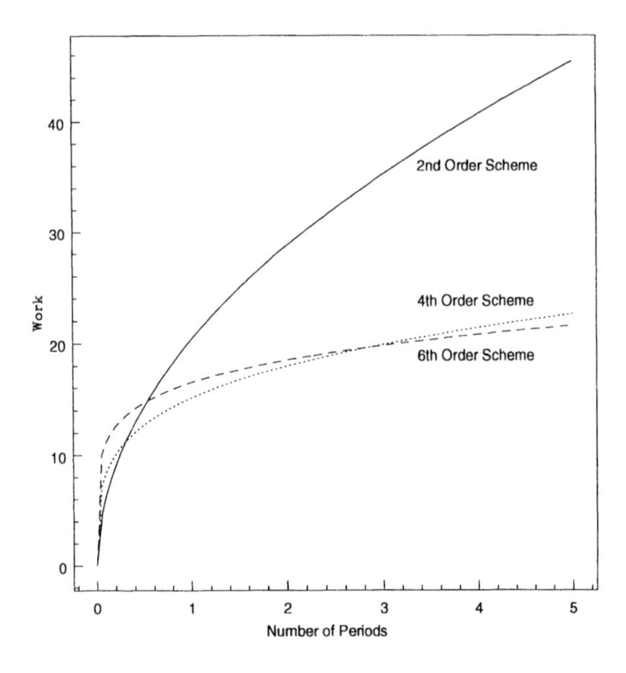

Figure 1.3. Computational work (number of floating-point operations) required to integrate a linear advection equation for M periods while maintaining a cumulative phase error of 10%.

2. Spectral Methods on Unstructured Grids

There have been more than fifteen years of developments in extending spectral methods to complex-geometry domains ([5]), starting with the pioneering work of ([6]), who developed spectral methods in the context of a multi-element variational formulation similar to finite element methods. This allowed the use of spectral (Chebyshev or Legendre) expansions as trial basis in general quadrilateral subdomains. Continuity of data and unknowns across subdomains is ensured via appropriate construction of the trial basis similar to finite element methods for second-order elliptic problems. Such methods were used to produce the first spectral DNS of turbulence in complex-geometry domain, flow over riblets, in ([7]). An extension to non-conforming discretizations for turbulent flows, which are more appropriate for local refinement, was presented in ([8]).

The new generation of spectral methods developed recently is more appropriate for discretizations on unstructured grids consisting of triangles and tetrahedra, similar to grids used in aerodynamics ([9, 10]). In many simulations, however, it is more efficient to employ hybrid discretizations, i.e. discretizations using a combination of structured and unstructured subdomains. This is a recent trend in computational mechanics involving complex three-dimensional computational domains ([11, 12]). Such an approach combines the simplicity and convenience of structured domains with the geometric flexibility of an unstructured discretization. In two-dimensions, hybrid discretization simply implies the use of triangular and rectangular subdomains, however in three-dimensions the hybrid strategy is more complex requiring the use of hexahedra, prisms, pyramids and tetrahedra.

We have developed a unified description in dealiasing with elements of different shape in two- and three-dimensions. This unified approach generates polynomial expansions which can be expressed in terms of a *generalized* product of the form

$$\phi_{pqr}(x, y, z) = \phi_p^a(x)\phi_{pq}^b(y)\phi_{pqr}^c(z).$$

Here we have used the Cartesian co-ordinates x, y and z but, in general, they can be any set of co-ordinates defining a specified region. The standard tensor product is simply a degenerate case of this product, where the second and third functions are only dependent on one index. The primary motivation in developing an expansion of this form is computational efficiency. Such expansions can be evaluated in three-dimensions in $O(P^4)$ operations as compared to $O(P^6)$ operations with non-tensor products (where P is the number of spectral modes per direction).

2.1 Local Co-ordinate Systems

We start by defining a convenient set of local co-ordinates upon which we can construct the expansions. Unlike the barycentric co-ordinates, which are typically applied to unstructured domains in linear finite elements, we define a set of *collapsed Cartesian* co-ordinates in non-rectangular domains. These co-ordinates will form the foundation of the polynomial expansions. The advantage of this system is that every domain can be bounded by constant limits of the new local co-ordinates; accordingly operations such as integration and differentiation can be performed using standard one-dimensional techniques.

The new co-ordinate systems are based upon the transformation of a triangular region to a rectangular domain (and vice versa) as shown in figure 1.4. The main effect of the transformation is to map the vertical lines in the rectangular domain (i.e. lines of constant ξ_1) onto lines radiating out of the point ($\eta_1 = -1, \eta_2 = 1$) in the triangular domain. The triangular region can now be described using the "ray" co-ordinate (ξ_1) and the standard horizontal co-ordinate ($\eta_2 = \xi_2$). The triangular domain is therefore defined by ($-1 \leq \xi_1, \xi_2 \leq 1$) rather than the Cartesian description ($-1 \leq \eta_1, \eta_2;\ \eta_1 + \eta_2 \leq 0$) where the upper bound couples the two co-ordinates. The "ray" co-ordinate (ξ_1) is multi-valued at ($\eta_1 = -1, \eta_2 = 1$). Nevertheless, we note that the use of singular co-ordinate systems is very common, arising in both cylindrical and spherical co-ordinate systems.

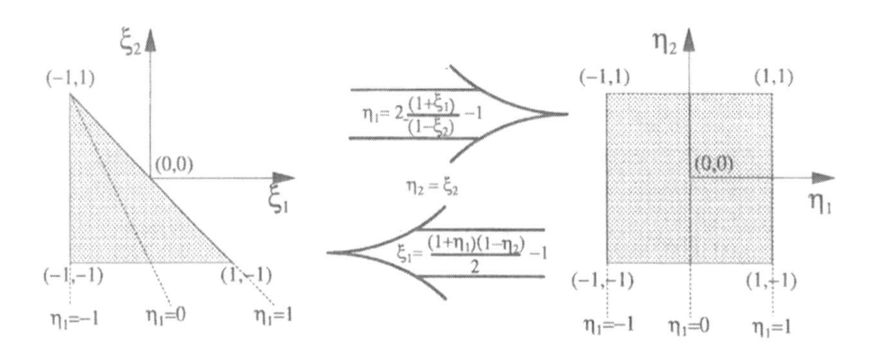

Figure 1.4. Triangle to rectangle transformation

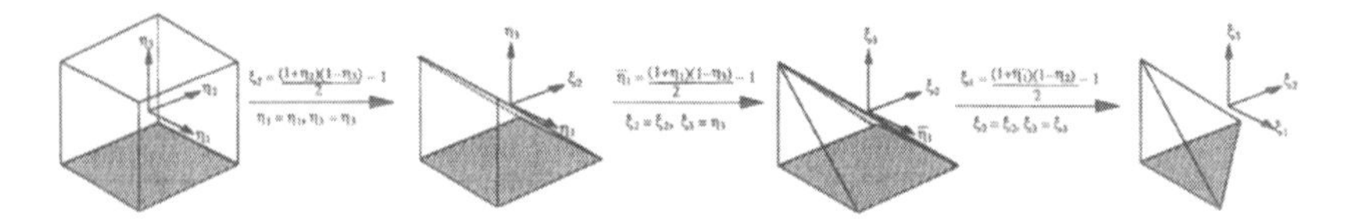

Figure 1.5. Hexahedron to tetrahedron transformation

As illustrated in figure 1.5, the same transformation can be repeatedly applied to generate new co-ordinate systems in three-dimensions. Here, we start from the bi-unit hexahedral domain and apply the triangle to rectangle transformation in the vertical plane to generate a prismatic region. The transformation is then used in the second vertical plane to generate the pyramidic region. Finally, the rectangle to triangle transformation is applied to every square cross section parallel to the base of the pyramidic region to arrive at the tetrahedral domain.

By determining the hexahedral co-ordinates (ξ_1, ξ_2, ξ_3) in terms of the Cartesian co-ordinates of the tetrahedral region (η_1, η_2, η_3) we can generate a new co-ordinate system for the tetrahedron. This new system and the planes described by fixing the local co-ordinates are shown in figure 1.6. Also shown are the new systems for the intermediate domains which are generated in the same fashion. Here we have assumed that the local Cartesian co-ordinates for every domain are (η_1, η_2, η_3).

2.2 Hierarchical Expansions

For each of the hybrid domains we can develop a polynomial expansion based upon the local co-ordinate system derived in section 2.1. These expansions will be polynomials in terms of the local co-ordinates as well as the Cartesian co-ordinates (η_1, η_2, η_3). This is a significant property as primary operations such as integration and differentiation can be performed with respect to the local co-

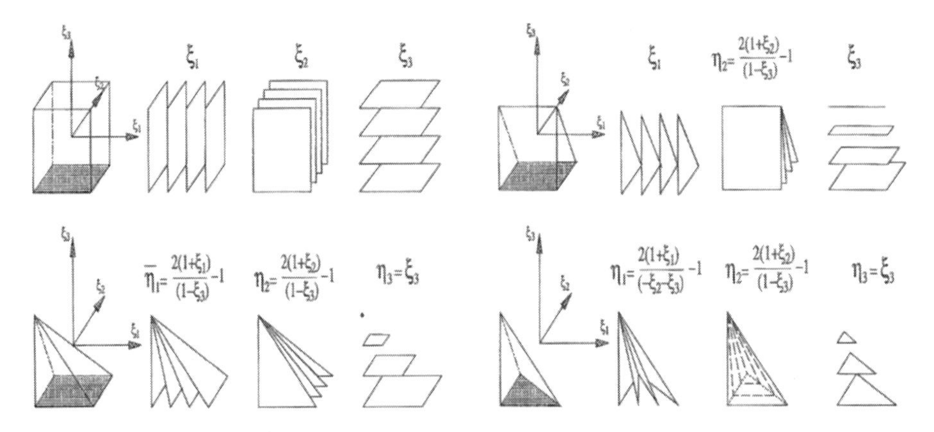

Figure 1.6. The local coordinate systems used in each of the hybrid elements and the planes described by fixing each local co-ordinate.

ordinates but the expansion may still be considered as a polynomial expansion in terms of the Cartesian system.

We shall initially consider expansions which are orthogonal in the Legendre inner product. We define three principle functions $\phi_i^a(z), \phi_{ij}^b(z)$ and $\phi_{ijk}^c(z)$, in terms of the Jacobi polynomial, $P_p^{\alpha,\beta}(z)$, as:

$$\phi_i^a(z) = P_i^{0,0}(z), \qquad \phi_{ij}^b(z) = \left(\tfrac{1-z}{2}\right)^i P_j^{2i+1,0}(z),$$

$$\phi_{ijk}^c(z) = \left(\tfrac{1-z}{2}\right)^{i+j} P_k^{2i+2j+2,0}(z).$$

Using these functions we can construct the orthogonal polynomial expansions:

Hexahedral expansion: $\phi_{pqr}(\eta_1,\eta_2,\eta_3) = \phi_p^a(\eta_1)\phi_q^a(\eta_2)\phi_r^a(\eta_3)$

Prismatic expansion: $\phi_{pqr}(\eta_1,\eta_2,\eta_3) = \phi_p^a(\eta_1)\phi_q^a(\xi_2)\phi_{qr}^b(\eta_3)$

Pyramidic expansion: $\phi_{pqr}(\eta_1,\eta_2,\eta_3) = \phi_p^a(\overline{\xi_1})\phi_q^a(\xi_2)\phi_{pqr}^c(\xi_3)$

Tetrahedral expansion: $\phi_{pqr}(\eta_1,\eta_2,\eta_3) = \phi_p^a(\xi_1)\phi_{pq}^b(\xi_2)\phi_{pqr}^c(\xi_3)$

where,

$$\xi_1 = \frac{2(1+\eta_1)}{(-\eta_2-\eta_3)} - 1, \quad \overline{\xi_1} = \frac{2(1+\eta_1)}{(1-\eta_3)} - 1, \quad \xi_2 = \frac{2(1+\eta_2)}{(1-\eta_3)} - 1, \quad \xi_3 = \eta_3,$$

are the local co-ordinates illustrated in figure 1.6.

The hexahedral expansion is simply a standard tensor product of Legendre polynomials (since $P_p^{0,0}(z) = L_p(z)$). In the other expansions the introduction of the degenerate local co-ordinate systems is linked to the use of the more unusual functions $\phi^b_{ij}(z)$ and $\phi^c_{ijk}(z)$. These functions both contain factors of the form $\left(\frac{1-z}{2}\right)^p$ which is necessary to keep the expansion as a polynomial of the Cartesian co-ordinates (η_1, η_2, η_3). For example, the co-ordinate ξ_2 in the prismatic expansion necessitates the use of the function $\phi^b_{qr}(\eta_3)$ which introduces a factor of $\left(\frac{1-\eta_3}{2}\right)^q$. The product of this factor with $\phi^a_q(\xi_2)$ is a polynomial function in η_2 and η_3. Since the remaining part of the prismatic expansion, $\phi^a_p(\eta_1)$, is already in terms of a Cartesian co-ordinate the whole expansion is a polynomial in terms of the Cartesian system.

The polynomial space, in Cartesian co-ordinates, for each expansion is:

$$\mathcal{P} = \text{Span}\{\eta_1^p \, \eta_2^q \, \eta_3^r\} \tag{1.1}$$

where pqr for each domain is

$$\begin{array}{llll} \text{Hexahedron} & 0 \le p \le P_1 & 0 \le q \le P_2 & 0 \le r \le P_3 \\ \text{Prism} & 0 \le p \le P_1 & 0 \le q \le P_2 & 0 \le q + r \le P_3 \\ \text{Pyramidic} & 0 \le p \le P_1 & 0 \le q \le P_2 & 0 \le p + q + r \le P_3 \\ \text{Tetrahedron} & 0 \le p \le P_1 & 0 \le p + q \le P_2 & 0 \le p + q + r \le P_3. \end{array} \tag{1.2}$$

The range of the p, q and r indices indicate how the expansions should be expanded to generate a complete polynomial space. We note that if $P_1 = P_2 = P_3$ then the tetrahedral and pyramidic expansions span the same space and are in a subspace of the prismatic expansion which is in turn a subspace of the hexahedral expansion.

2.3 Galerkin and Discontinuous Galerkin Projections

To obtain a system of nonlinear algebraic equations, we employ different projections and time integration algorithms. In particular, for *incompressible flows* we use a linear Galerkin projection in conjunction with the high-order fractional stepping scheme described in ([13, 14]). For *compressible flows*, we use a discontinuous Galerkin projection with multi-step explicit time integration ([15]).

We describe both approaches next with more emphasis on the latter which is a more recent development.

2.3.1 Incompressible Flows. The standard approach in treating the incompressible Navier-Stokes equations is to combine a semi-implicit scheme with a fractional procedure ([5]) following the Eulerian description. Here we consider a more recent development that takes advantage of semi-Lagrangian treatment for advection. This allows for large size time steps in simulations of turbulence, where at high Reynolds number the temporal scales are largely over-resolved. Following ([16]) we consider the Navier-Stokes equations in Lagrangian form

$$\frac{d\mathbf{u}}{dt} = -\nabla p + \nu \nabla^2 \mathbf{u}, \tag{1.3}$$

$$\nabla \cdot \mathbf{u} = 0, \tag{1.4}$$

where d/dt denotes a Lagrangian derivative. We employ a stiffly-stable second-order scheme to discretize the time derivative:

$$\frac{\frac{3}{2}u^{n+1} - 2u_d^n + \frac{1}{2}u_d^{n-1}}{\Delta t} = (-\nabla p + \nu \nabla^2 u)^{n+1}, \tag{1.5}$$

where u_d^n is the velocity u at the departure point x_d^n at time level t^n, and u_d^{n-1} is the velocity at the departure point x_d^{n-1} at time level t^{n-1}. The departure point x_d^n is obtained by solving

$$\frac{dx}{dt} = u^{n+1/2}(x,t), \quad x(t^{n+1}) = x_a$$

and also

$$u^{n+1/2} = 3/2u^n - 1/2u^{n-1}.$$

The point x_d^{n-1} is obtained by solving

$$\frac{dx}{dt} = u^n(x,t), \quad x(t^{n+1}) = x_a.$$

By using the above characteristic equations, the resulting scheme is second-order accurate in time.

Specifically, for computational convenience we use the following three sub-steps to solve equation (1.5)

$$\frac{\hat{u} - 2u_d^n + \frac{1}{2}u_d^{n-1}}{\Delta t} = 0, \tag{1.6}$$

$$\frac{\hat{\hat{u}} - \hat{u}}{\Delta t} = -\nabla p^{n+1}, \tag{1.7}$$

$$\frac{\frac{3}{2}u^{n+1} - \hat{\hat{u}}}{\Delta t} = \nu \nabla^2 u^{n+1}. \tag{1.8}$$

The discrete divergence-free condition results in a consistent Poisson equation for the pressure, i.e.

$$\nabla^2 p^{n+1} = \frac{1}{\Delta t} \nabla \cdot \hat{u},$$

with accurate pressure boundary conditions of the form ([5])

$$\frac{\partial p}{\partial n} = -\nu \cdot [\hat{u} + \nabla \times \omega^{n+1}],$$

where n is the unit normal, and ω is the vorticity ([14]).

The semi-Lagrangian approach is typically more expensive than the corresponding Eulerian approach, but in practice the larger size of time step allowed in the former leads to more efficient simulations. This was demonstrated for two- and three-dimensional flows in ([16]).

With regards to *spatial discretization*, in order to enforce the required C^0 continuity, the orthogonal expansion is modified by decomposing the expansion into an *interior* and *boundary contribution*. This results in partially sacrificing orthogonality. The interior modes (or bubble functions) are defined to be zero on the boundary of the local domain. The completeness of the expansion is then ensured by adding boundary modes which consist of

- Vertex, Edge, and Face contributions.

The *vertex modes* have unit value at one vertex and decay to zero at all other vertices; *edge modes* have local support along one edge and are zero on all other edges and vertices, and *face modes* have local support on one face and are zero on all other faces, edges and vertices. C^0 continuity between elements can then be enforced by matching similar shaped boundary modes. The local co-ordinate systems do impose some restrictions on the orientation in which triangular faces may connect. However, it has been shown in ([10]) that a C^0 tetrahedral expansion can be constructed for any tetrahedral mesh. A similar strategy could be applied to a hybrid discretization.

2.3.2 Compressible Flows.

We consider the non-dimensionalized compressible Navier-Stokes equations, which we write in compact form as

$$\vec{U}_t + \nabla \cdot \mathbf{F} = Re_\infty^{-1} \nabla \cdot \mathbf{F}^\nu \tag{1.9}$$

where $\mathbf{F}$ and $\mathbf{F}^\nu$ correspond to inviscid and viscous flux contributions, respectively. Here the vector $\vec{U} = [\rho, \rho u, \rho v, \rho w, E]^t$ with (u, v, w) the local fluid velocity, ρ the fluid density, and E the total internal energy. Splitting the Navier-Stokes operator in this form allows for the separate treatment of the inviscid and viscous contributions, which in general exhibit different mathematical properties. In the following, we review briefly the discontinuous Galerkin formulations employed in the proposed method. A systematic analysis of the advection operator was presented in ([17]), where a mixed formulation was used to treat the diffusion terms. No flux limiters are necessary as has been found before in ([18]) and has been justified theoretically in ([19]).

We first use a linear two-dimensional **advection equation** of a conserved quantity u in a region Ω, in order to illustrate the treatment of inviscid contributions:

$$\frac{\partial u}{\partial t} + \nabla \cdot \mathbf{F}(u) = 0, \tag{1.10}$$

where $\mathbf{F}(u) = (f(u), g(u), h(u))$ is the *flux* vector which defines the transport of $u(\mathbf{x}, t)$. We start with the variational statement of the standard Galerkin formulation of (1.10) by multiplying by a test function v and integrating by parts

$$\int_\Omega \frac{\partial u}{\partial t} v \, dx + \int_{\partial\Omega} v \, \hat{n} \cdot \mathbf{F}(u) \, ds - \int_\Omega \nabla v \cdot \mathbf{F}(u) \, dx = 0. \tag{1.11}$$

The solution $u \in \mathcal{X}$ (approximation space) satisfies this equation for all $v \in \mathcal{V}$ (test space). The requirement that $\mathcal{X}$ consist of continuous functions naturally leads to a basis consisting of functions with overlapping support, which implies equation (1.11) becomes a banded matrix equation. Solving the corresponding large system is not a trivial task for parallel implementations, and therefore a different type of formulation is desirable.

Another consideration from the point of view of advection is that continuous function spaces are not the natural place to pose the problem. Mathematically, hyperbolic problems of this type tend to have solutions in spaces of bounded variation. In physical problems, the best one can hope for in practice is that solutions will be piecewise continuous, that is, be smooth in regions separated by discontinuities (shocks). An additional consideration is that the formulation presented next preserves automatically conservativity in the element-wise sense.

These considerations suggest immediately a formulation where $\mathcal{X}$ may contain discontinuous functions. The discrete space $\mathcal{X}^\delta$ contains polynomials within each "element," but zero outside the element. Here the "element" is, for example, an individual triangular region T_i in the computational mesh applied to the problem. Thus the computational domain $\Omega = \bigcup_i T_i$, and T_i, T_j overlap only on edges.

Contending with the discontinuities requires a somewhat different approach to the variational formulation. Each element (E) is treated separately, giving a variational statement (after integrating by parts once more):

$$\frac{\partial}{\partial t}(u, v)_E + \int_{\partial T_E} v(\tilde{f}(u_i, u_e) - \mathbf{f}(u_i)) \cdot \mathbf{n} \, ds + (\nabla \cdot \mathbf{f}(u), v)_E = 0, \tag{1.12}$$

where $\mathbf{f}(u_i)$ is the flux of the interior values. Computations on each element are performed separately, and the connection between elements is a result of the way boundary conditions are applied. Here, boundary conditions are enforced

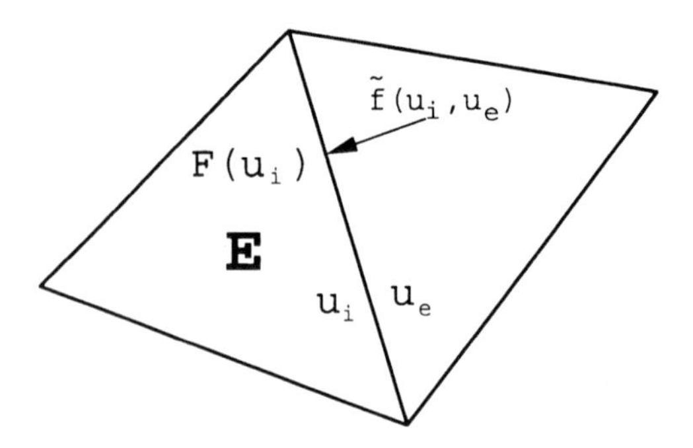

Figure 1.7. Interface conditions between two adjacent triangles.

via the numerical surface flux $\tilde{f}(u_i, u_e)$ that appears in equation (1.12). Because this value is computed at the boundary between adjacent elements, it may be computed from the value of u given at either element. These two possible values are denoted here as u_i in the interior of the element under consideration and and u_e in the exterior (see figure 1.7). Upwinding considerations dictate how this flux is computed. In the more complicated case of a hyperbolic system of equations, an approximate Riemann solver should be used to compute a value of f, g, h (in three-dimensions) based on u_i and u_e. Specifically, we compute the flux $\tilde{f}(u_i, u_e)$ using upwinding, i.e.

$$\tilde{f}(u) = R\Lambda^+ L u_i + R\Lambda^- L u_e$$

where A (the Jacobian matrix of F) is written in terms of the left and right eigenvectors, i.e. $A = R\Lambda L$ with Λ containing the corresponding eigenvalues in the diagonal; also, $\Lambda^{\pm} = (\Lambda \pm |\Lambda|)/2$. Alternatively, we can use a standard Lax-Friedrichs flux

$$\tilde{f}(u) = \frac{1}{2}(f(u_e) + f(u_i)) - \frac{1}{2} R|\Lambda|L(u_e - u_i).$$

This last form is what is used in the airfoil example presented in section 4.

Next, we consider as a model problem the **parabolic equation** with variable coefficient ν to demonstrate the treatment of the viscous contributions:

$$u_t = \nabla \cdot (\nu \nabla u) + f, \quad \text{in } \Omega, \quad u \in L^2(\Omega)$$
$$u = g(\mathbf{x}, t), \quad \text{on } \partial\Omega$$

We then introduce the flux variable

$$\mathbf{q} = -\nu \nabla u$$

with $\mathbf{q}(\mathbf{x}, t) \in \mathbf{L}^2(\Omega)$, and re-write the parabolic equation

$$u_t = -\nabla \cdot \mathbf{q} + f, \quad \text{in } \Omega$$

$$1/\nu \mathbf{q} = -\nabla u, \quad \text{in } \Omega$$
$$u = g(\mathbf{x}, t), \quad \text{on } \partial\Omega.$$

The weak formulation of the problem is then as follows: Find $(\mathbf{q}, u) \in \mathbf{L^2(\Omega)} \times L^2(\Omega)$ such that

$$(u_t, w)_E = (\mathbf{q}, \nabla w)_E - < w, \mathbf{q_b} \cdot \mathbf{n} >_E + (f, w)_E, \quad \forall w \in L^2(\Omega)$$
$$1/\nu(\mathbf{q}^m, \mathbf{v})_E = (u, \nabla \cdot \mathbf{v})_E - < u_b, \mathbf{v} \cdot \mathbf{n} >_E, \quad \forall \mathbf{v} \in \mathbf{L^2(\Omega)}$$
$$u = g(\mathbf{x}, t), \quad \text{on } \partial\Omega$$

where the parentheses denote standard inner product in an element (E) and the angle brackets denote boundary terms on each element, with $\mathbf{n}$ denoting the unit outwards normal. The surface terms contain weighted boundary values of v_b, q_b, which can be chosen as the arithmetic mean of values from the two sides of the boundary, i.e.

$$v_b = (v_i + v_e)/2,$$

and

$$q_b = (q_i + q_e)/2.$$

The consequences of choosing different numerical fluxes with regards to stability and accuracy have been investigated in ([20]).

By integrating by parts once more, we obtain an equivalent formulation which is easier to implement, and it is actually used in the computer code. The new variational problem is

$$(u_t, w)_E = (-\nabla \cdot \mathbf{q}, w)_E - < w, (\mathbf{q_b} - \mathbf{q_i}) \cdot \mathbf{n} >_E + (f, w)_E, \quad \forall w \in L^2(\Omega)$$
$$1/\nu(\mathbf{q}, \mathbf{v})_E = (-\nabla u, \mathbf{v})_E - < u_b - u_i, \mathbf{v} \cdot \mathbf{n} >_E, \quad \forall \mathbf{v} \in \mathbf{L^2(\Omega)}$$
$$u = g(\mathbf{x}, t), \quad \text{in } \partial\Omega$$

where the subscript (i) denotes contributions evaluated at the interior side of the boundary. We integrate the above system *explicitly* in time and employ the orthogonal Jacobi polynomials as trial and test basis.

3. Nonlinearities and Dealiasing

In spectral methods the quadratic nonlinearities in the incompressible Navier-Stokes equations or the cubic nonlinearities in the compressible Navier-Stokes are computed in the physical space. Specifically, the fields (velocity, pressure, energy) are first transformed into physical space and subsequently the products are obtained at all quadrature points in a collocation fashion. Another transform is then performed to bring the results back to modal space. More specifically, when the number of quadrature points Q is the same as the number of modes in the spectral expansion P we have a true collocation method, otherwise for $Q > P$ we have a super-collocation method.

The form in which we write the nonlinear terms, that is,

- in convective (flux), or
- skew-symmetric, or
- rotation form

is also important. In spectral DNS of boundary layers and channel flows, the rotation form is usually preferred over the convective form as it semi-conserves energy (in the inviscid limit). This, typically, makes it more stable, especially for the long-time integration required in DNS. In addition, it is more economical as it requires the evaluation of only six derivatives whereas the convective form requires nine derivative evaluations. The skew-symmetric form was found to be more "forgiving" in aliasing errors in under-resolved simulations of homogeneous turbulence compared to the rotation form. This is also true for finite difference methods; see the article of ([21]) in this volume where it is shown that skew-symmetry leads to symmetry preservation and enhanced stability. However, the skew-symmetric form requires the evaluation of 18 derivatives which is computationally more expensive.

There is not sufficient experience yet with spectral/*hp* element DNS to conclusively suggest one form or the other although there is some consensus that the convective form is quite accurate and leads to stable discretizations. A comparison of the different forms (convective, flux, and skew-symmetric) was performed in ([22]) for a constant advection velocity as well as for a spatially varying divergent-free velocity $(u, v) = (-\sin x_2 \cos x_1, \sin x_1 \cos x_2)$. The discretization was based on a nodal Gauss-Lobatto-Legendre basis. The result was that for the constant advection velocity all forms were the same in that they produced identical eigenspectrum with all imaginary eigenvalues. However, for the variable advection velocity, only the skew-symmetric form gave imaginary eigenvalues with the convective and conservative form producing complex eigenvalues with positive real parts. For purely convection equations, these spurious positive values may lead to instabilities if explicit time stepping is used. However, for Navier-Stokes computations at modest Reynolds number no such instabilities have been observed, presumably due to the stabilizing role of the viscous terms. Although the skew-symmetric form is usually considered the most accurate, problems may also be encountered with this form for Dirichlet and inflow/outflow conditions. This is presented in some detail in ([5]) using a numerical experiment performed by ([23]).

Finally, errors may be caused by insufficient quadrature used in the spectral/*hp* element discretization of the nonlinear terms, especially in complex-geometry flows. These errors can be eliminated effectively by employing over-integration, i.e. integrating the nonlinear terms in the variational statement with higher order quadrature than the one employed for the linear contributions, e.g. pressure and viscous terms. We will examine this issue in some detail next.

3.1 Accuracy, Stability and Over-Integration

To understand the ramifications of under-integration of nonlinear terms, we perform the following test:

1. Consider a single element in the space interval $[-1, 1]$ containing $P = 16$ Jacobi modes.

2. Initialize all the modal coefficients to one.

3. Evaluate the modal representation on a set of Q quadrature points.

4. Square (in a pointwise fashion) the values at the quadrature points.

5. Pre-multiply the set of points (as a vector) by the collocation derivative matrix of the appropriate size (rank $Q \times Q$).

6. Project back to modal coefficients by discrete inner products using Gaussian integration.

The procedure above mimics the "physical space" or pseudo-spectral evaluation of the term $\frac{\partial u^2}{\partial x}$ commonly used in spectral methods for evaluating nonlinear terms. This test was chosen because even in its simplicity it models the order of nonlinearity that occurs in the solution of the incompressible Navier-Stokes equations. All modes are set to *one* to mimic a case in which an element has under-resolved or marginally resolved the solution within the element. In the test above, the only unspecified parameter is the number of quadrature points Q to be used. In using Gauss-Lobattto points, the value of Q is taken to be one more than the number of modes P (in this case then $P = 16$ and $Q = 17$) ([24]), but this value is appropriate for the inner products corresponding to linear terms. For quadratic or cubic nonlinearities more quadrature points are required. The ramifications of under-integration of this form are shown in figure 1.8. The figure on the left was obtained for quadratic nonlinearity $(\frac{\partial}{\partial x}u^2)$ and the figure on the right was obtained for a cubic nonlinearity $(\frac{\partial}{\partial x}u^3)$. The difference in the modal coefficients at the conclusion of the algorithm above for different values of Q is provided. We observe that for the quadratic nonlinearity, once $\frac{3}{2}P$ quadrature points are used, the differences in the modal values do not change. Similarly for the cubic nonlinearity, once $2P$ quadrature points are used the differences in the modal values do not change.

In order to appreciate the effect of under-integration in the context of a numerical solution, we consider the inviscid Burgers equation, which we discretize

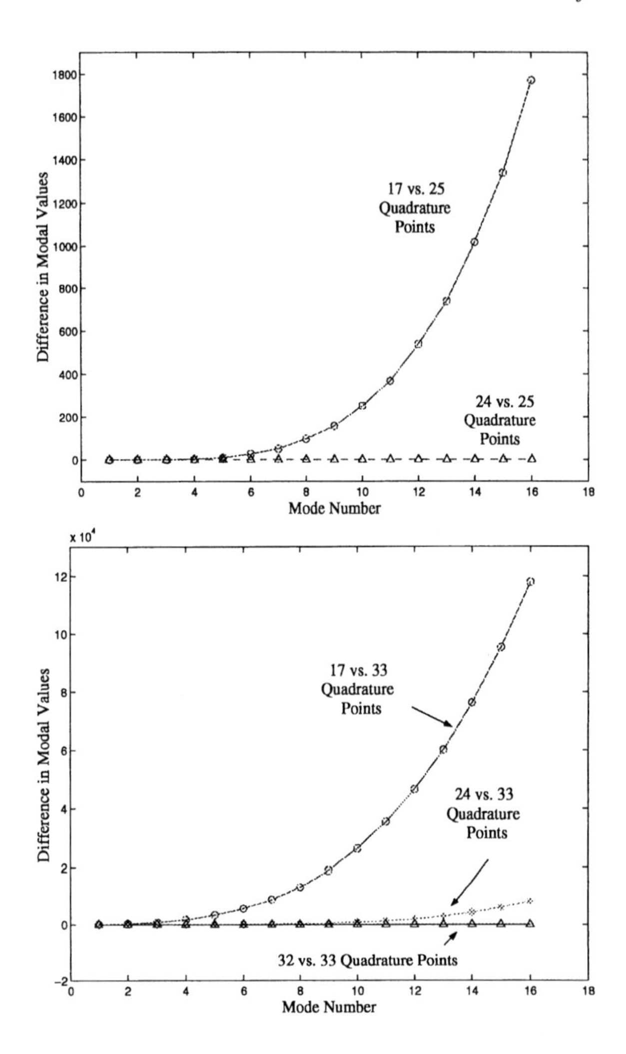

Figure 1.8. Comparison of the difference in modal coefficients when different numbers of quadrature points are used. Quadratic nonlinearity is shown on the left and cubic nonlinearity is shown on the right.

using the discontinuous Galerkin method. The initial condition is $-\sin(\pi x)$, and five equally spaced elements spanning $[-1, 1]$ were used, each one having $P = 16$ modes. In figure 1.9, we plot the L_2 norm of the solution versus the number of quadrature points used for numerical integration. When using $Q = 17, 19$ and $Q = 21$ points, the solution is unstable (denoted by the blue *). Once the number of quadrature points reaches $Q = 24$ ($\frac{3}{2}P$ where P is the number of modes), the L_2 norm of the solution does not change.

We can analyze this behavior by examining the energy in the modes (denoted by the square of the modal values) within the element that contains the jump in the inviscid Burgers solution. The modes were extracted at time $T = 0.35$, after

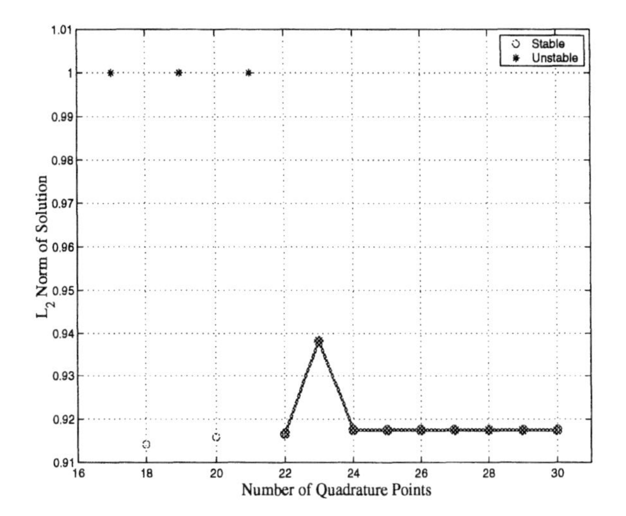

Figure 1.9. Solution of the inviscid Burgers equation evaluated at $T = 0.5$. Five equal spaced elements were used with 16 modes in each element. On the ordinate we plot the L_2 norm of the solution, and on the abscissa we plot the number of quadrature points used for numerical integration. Unstable solutions are denoted by blue *. Observe that after $Q = 24$ points, the L_2 norm of the solution does not change.

the shock has formed (at time $\frac{1}{\pi}$) and prior to the solution becoming unstable. In figure 1.10, we plot the square of the modal coefficients versus the mode number. Due to the symmetry of the element placement, only even number modes were excited.

This case corresponds to $Q = 17$ quadrature points being used, which from the figure 1.9, we know will become unstable by time $T = 0.5$. If a $\frac{3}{2}P$ rule is used, yielding $Q = 24$ points, the solution is stable, and the energy is much less than when the non-linear terms are under-integrated. This plot shows vividly the effects of aliasing when under-integration of the non-linear terms is performed.

An alternative way of handling instabilities associated with nonlinearities in hyperbolic conservation laws is through *monotonicity preserving schemes*. An approach suitable for high-order methods has been developed by ([25]) and was employed in large-eddy simulations in ([26]). It involves the addition of a *second-order convolution kernel* that acts on each mode separately and controls the high modes suppressing preferentially erroneous high-frequency oscillations. This type of nonlinear kernel has been termed as spectral vanishing viscosity (SVV). As an example, the inviscid Burgers equation with the SVV term added is

$$\frac{\partial u}{\partial t} + \frac{1}{2}\frac{\partial u^2}{\partial x} = \epsilon\frac{\partial}{\partial x}(Q_k * \frac{\partial u}{\partial x}). \tag{1.13}$$

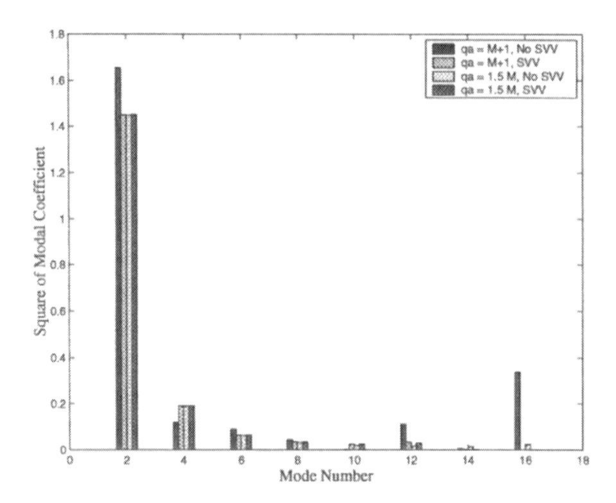

Figure 1.10. Modal coefficients of the inviscid Burgers solution before blow up. Both over-integration and SVV lead to a stable solution unlike the collocation approach.

Here $\epsilon \propto 1/P$, i.e. it is inversely proportional to the number of modes, and Q_k is a smooth kernel that facilitates a transition between the controlled high modes and the uncontrolled and more energetic low modes. It is given by

$$\hat{Q}_k = e^{-\frac{(k-P)^2}{(k-P_c)^2}}, \quad k > P_c.$$

The cut-off wave number P_c scales as $P_c \approx \sqrt{P}$, asymptotically for P large. Although the above can be considered as a viscosity regularization procedure there is a significant difference as has been demonstrated in ([27]), see also ([26]). The parameters ϵ and Q_k are chosen so that monotonicity is preserved while the spectral accuracy in the solution is also maintained. In figure 1.10 we demonstrate how the addition of SVV leads to effectively the same results as over-integration but operating with a collocation discretization, i.e. $Q = P+1$. The modal coefficients of the inviscid Burgers solution converge monotonically to zero leading to a stable simulation unlike the collocation untreated simulation.

3.2 Transition and Turbulence in a Triangular Duct

We demonstrate next the effect of under-integration and associated aliasing errors by simulating transition to turbulence of incompressible flow in a duct with its cross-section being an equilateral triangle. The laminar fully-developed solution is known analytically. We introduce some random disturbances in the flow and we integrate in time until these disturbances start decaying or growing in time. All simulations were performed in the domain shown in figure 1.11 with the cross-section discretized using one triangular element only and 16 Fourier modes (32 collocation points) in the streamwise (homogeneous) direction. The

Reynolds number is defined as $Re = UD_e/\nu$ where U is the average velocity and D_e is the equivalent (hydraulic) diameter. For $Re \leq 500$ all disturbances decay but for $Re = 1250$ the flow goes through transition and a turbulent state is sustained.

We have performed three simulations at $Re = 1250$ corresponding to three different combinations of polynomial and quadrature order. In the first one, shown in 1.11(a), we consider the case where $Q = P + 1$, where $P = 16$. The forces on the three walls of the duct are plotted as a function of time. From symmetry considerations, we expect that the statistical averages of the three forces are identical but obviously the symmetry is not preserved here. In figure 1.11(b) we plot the forces for the case with $Q = 2P$, and in figure 1.11(c) the case with $Q = 3P/2$. We have verified that in both cases the same statistical force average is obtained, consistent with the analysis presented above for handling under-integration induced errors.

Based on the above analysis and result as well as other similar results, we can state the following semi-empirical rule:

Dealiasing Rule: For quadratic nonlinearities employing super-collocation with $3/2\,P$ grid (quadrature) points per direction, where P is the polynomial order per direction, followed by a Galerkin projection leads to a dealiased turbulence simulation on non-uniform meshes.

4. Under-Resolution and Diagnostics

Spectral and spectral/hp element methods behave, in general, differently than low-order methods in under-resolved simulations. Spectral discretizations are more susceptible to numerical instabilities than other low-order discretizations. This could be frustrating for the users who seek robustness but it is actually safe-guarding against erroneous answers, as typically spectral codes blow up in seriously under-resolved simulations. In under-resolved spectral discretizations there are many more wiggles, and therefore it is easier to detect suspicious simulations before resorting to more rigorous error estimation. Also, spectral discretizations typically suffer from little numerical dissipation unlike finite-difference methods, which introduce an erroneous numerical viscosity in low resolution discretizations. This effectively lowers the nominal Reynolds number of the simulated flow and leads to stable simulations but with the incorrect physics. This is not true in spectral/hp discretizations where the nominal Reynolds number is also the effective Reynolds number. However, such behavior in conjunction also with the presence of high-wave number wiggles, may sometimes be the source of erroneous instabilities in under-resolved spectral flow simulations. For example, the resulting velocity profiles may not be monotonic and thus are susceptible to inviscid type instabilities, which in

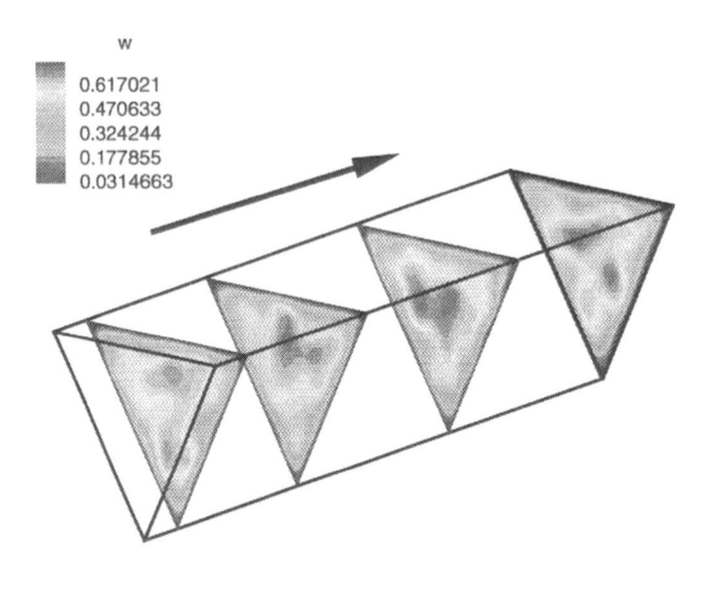

Figure 1.11. Duct flow domain: The cross-section is an equilateral triangle and the streamwise length is three times the triangle edge. Shown is a snapshot of streamwise velocity contours at $Re = 1250$.

turn promote transition from steady to unsteady flow or transition to higher bifurcations and eventually turbulence. For open unsteady flows, the amplitude of the oscillation in an under-resolved simulation is usually over-predicted.

In the following, we present a few examples of under-resolved flows that affect both transition to turbulence as well as turbulence statistics. We also include a case of transient compressible flow past an airfoil where under-resolution may seriously affect the lift. However, as we will see not all results from under-resolved simulations are inaccurate. Some flows exhibit low-dimensionality and the energetics of low modes dominate so even a very coarse grid turbulence simulation may predict the correct statistics – this is the case of the cylinder wake.

4.1 Erroneous Flow Transition

The first example is from the systematic spectral simulations presented by ([28]), in the study of bypass transition in a boundary layer. Using a Chebyshev discretization in the inhomogeneous direction and Fourier expansions in the other two directions, they demonstrated that with $P = 33$ Chebyshev modes

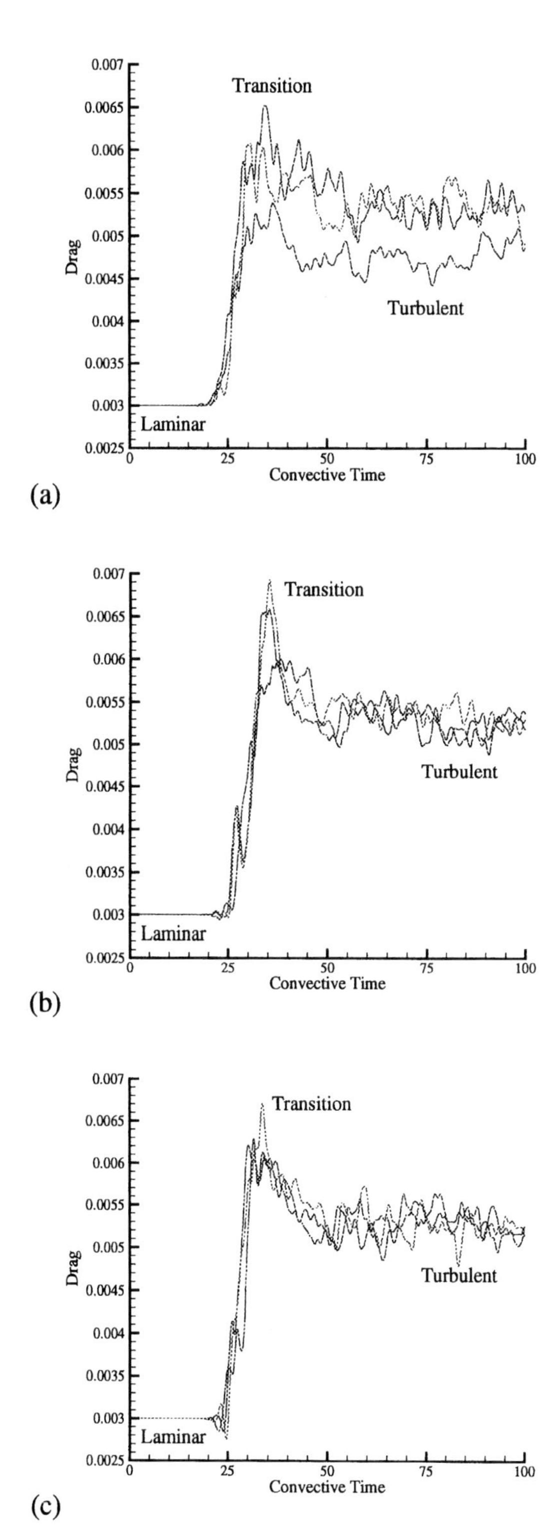

Figure 1.12. Wall shear forces on each wall as a function of time for (a) $(Q = M + 1)$; (b)

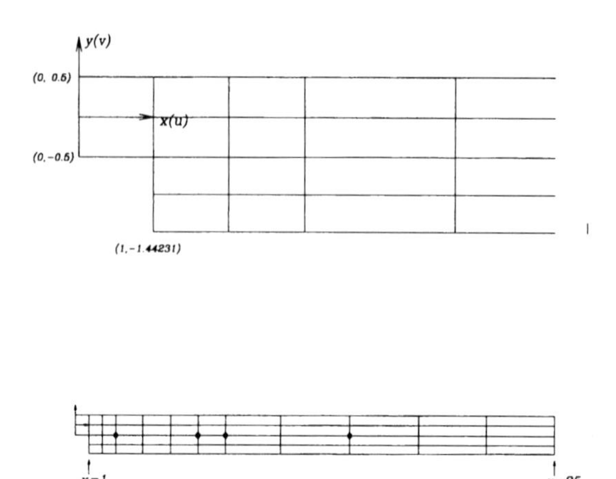

Figure 1.13. Low resolution mesh for flow over a backwards-facing step.

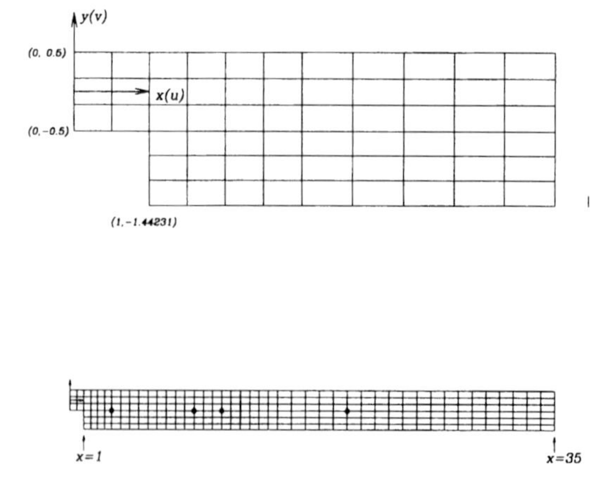

Figure 1.14. High resolution mesh for flow over a backwards-facing step.

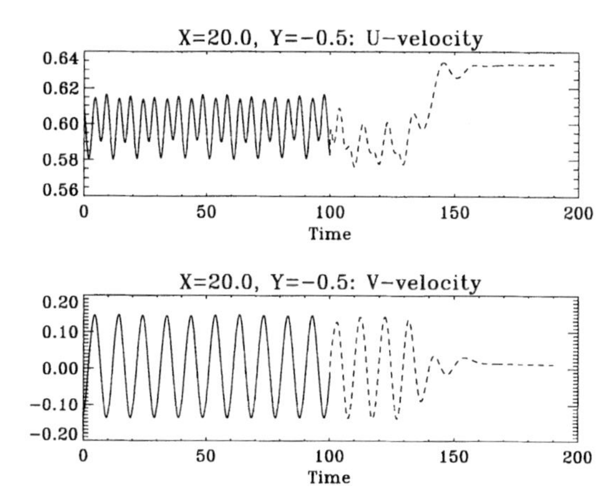

Figure 1.15. Time history at the third point (shown in the mesh in figures 13 and 14) at Re=700. Solid line: low resolution; Dash line: high resolution.

their simulation showed that a wave structure was present which would give rise to a secondary instability, as suggested earlier by other investigators. However, this structure changed, and the instability vanished completely when $P = 66$ Chebyshev modes were employed in the simulation. In fact, simulations with even higher resolution confirmed this explanation.

An example of similar behavior but with spectral element discretization is the simulation of flow over a backwards-facing step, which was first presented in ([29]). For the resolution shown in figure 1.13 the flow is unsteady as is evident in the plot that shows time history of velocity (figure 1.15). However, if higher resolution is used as shown in figure 1.14, then a steady state is reached (see corresponding figure 1.15), and this is true even at higher Reynolds numbers up to about $Re \approx 2,500$. Interestingly, the results of the under-resolved simulation are not totally irrelevant as they contain information about the actual flow albeit at a different set of parameters. For example, the two frequencies present in the unsteady case are the natural frequencies of the flow corresponding to the shear layer instability at the step corner and the Tollmien-Schlichting waves in the downstream channel portion. These modes are excited either by background noise, for example, some small turbulence level at the inflow, or spontaneously at a higher Reynolds number. Since no absolutely quiet wind tunnels exist, the results of the under-resolved "noisy" simulation, in this case, match the results of the experiment ([30]). We note that other inherently noisy discretizations employing vortex methods and lattice-Boltzmann methods also lead to an unsteady flow solution ([31, 32]).

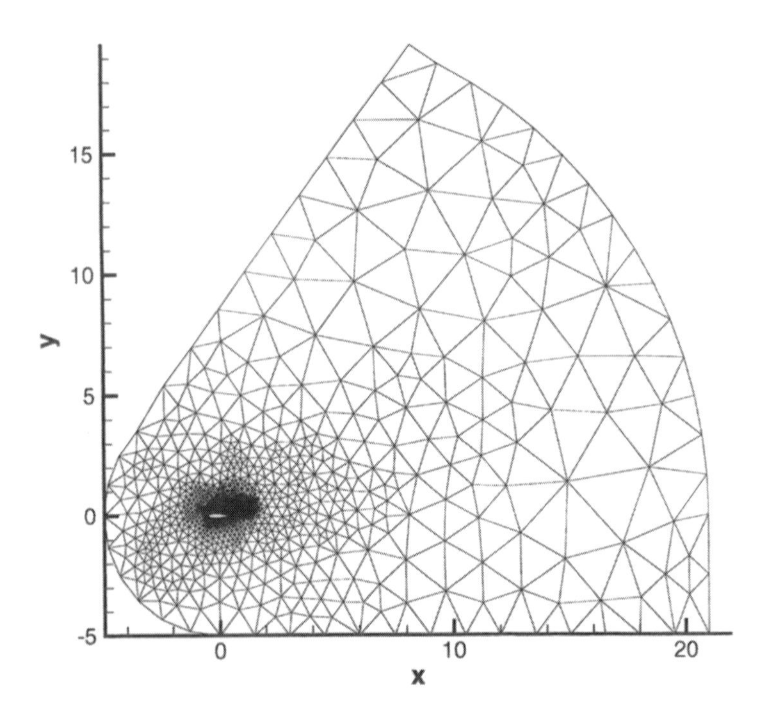

Figure 1.16. Domain and triangulization for the simulation around the pitching airfoil NACA 0015.

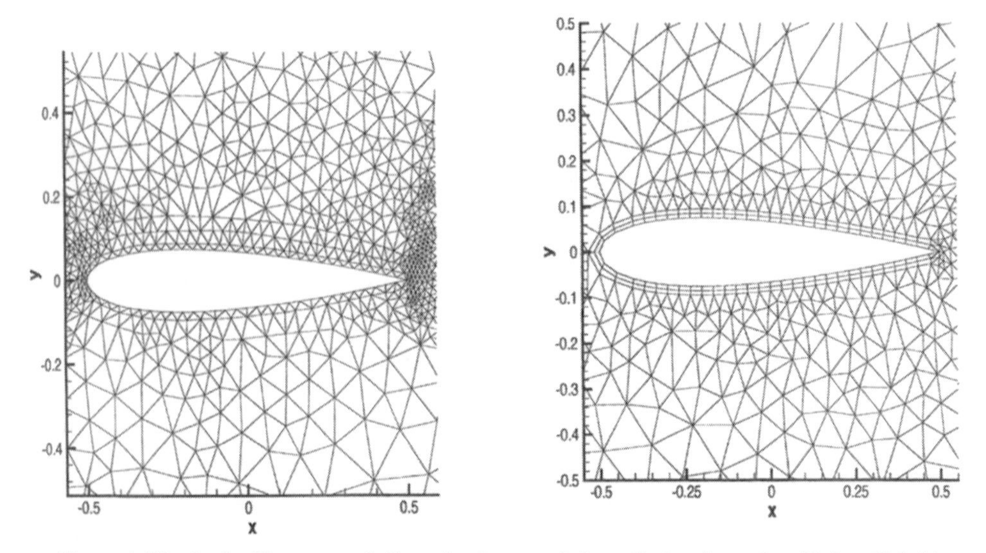

Figure 1.17. Left: Unstructured discretization consisting of triangles only. Right: Hybrid discretizations consisting of triangles and quadrilaterals. All dimensions are in units of chord length.

4.2 Fast-Pitching Airfoil

Next, we consider laminar flow around a rapidly pitching airfoil and compare discontinuous Galerkin spectral/hp element results against the finite volume results obtained in ([33]). In particular, we consider a NACA 0015 airfoil pitching upwards about a fixed axis at a constant rate from zero incidence to a maximum angle of attack of approximately 60 degrees. The pivot axis location is at $1/4$ of the chord measured from the leading edge. The temporal variation of the pitch given in ([33]) is

$$\Omega(t) = \Omega_0[1 - e^{-4.6t/t_0}], \quad t \geq 0$$

where t_0 denotes the time elapsed for the airfoil to reach 99% of its final pitch rate Ω_0. Here the non-dimensional values are $t_0^* = 1.0$ and $\Omega_0^* = 0.6$ based on the chord length and free stream velocity. As initial condition the computed field at 0 degrees angle of attack is used. The Mach number is $M = 0.2$ and the chord Reynolds number is $Re = 10,000$.

In ([33]) a similar simulation was obtained using a grid fixed to the airfoil by employing an appropriate transformation and discretizing the modified compressible Navier-Stokes equations using the implicit approximate factorization of ([34]). A typical grid used in ([33]) involved 203×101 points. In the present study, we employ the domain shown in figure 1.16. We performed two different sets of simulations, first with unstructured discretization around the airfoil (see figure 1.17; total of $3,888$ triangular elements), and subsequently with hybrid discretization with quadrilateral elements around the airfoil for better resolution of boundary layers (total of 116 quadrilateral and 2167 triangular elements). We demonstrate how the hybrid discretization combined with *variable* P-order per element allows accurate resolution of boundary elements without the need for re-meshing. We first performed simulations with constant P-order on all elements and subsequently with higher P-order in the inner layers of elements as shown in figure 1.18. We contrast the results in figure 1.19 for P-order $P = 3$ on the left, and P varying from 10 in the innermost layer to 2 in the far field. We see that the boundary layer is unresolved as indicated by the discontinuities at the element interfaces, but it is accurately resolved in the second simulation.

Returning now to the unstructured grid, we test convergence by also performing P-refinement on the same triangulization but with three different values of spectral order P corresponding to 2nd, 3rd and 4th order polynomial interpolation. In figure 1.20 we plot the computed lift and drag coefficients versus the angle of attack for grids corresponding to $P = 2, 3$ and $P = 4$. We also include (with symbols) the computational results of ([33]), and we see that in general there is very good agreement except at the large angles of attack close to 50 degrees. This difference is due to qualitative difference in flow structure at small scales, which are only resolved with the higher order simulations.

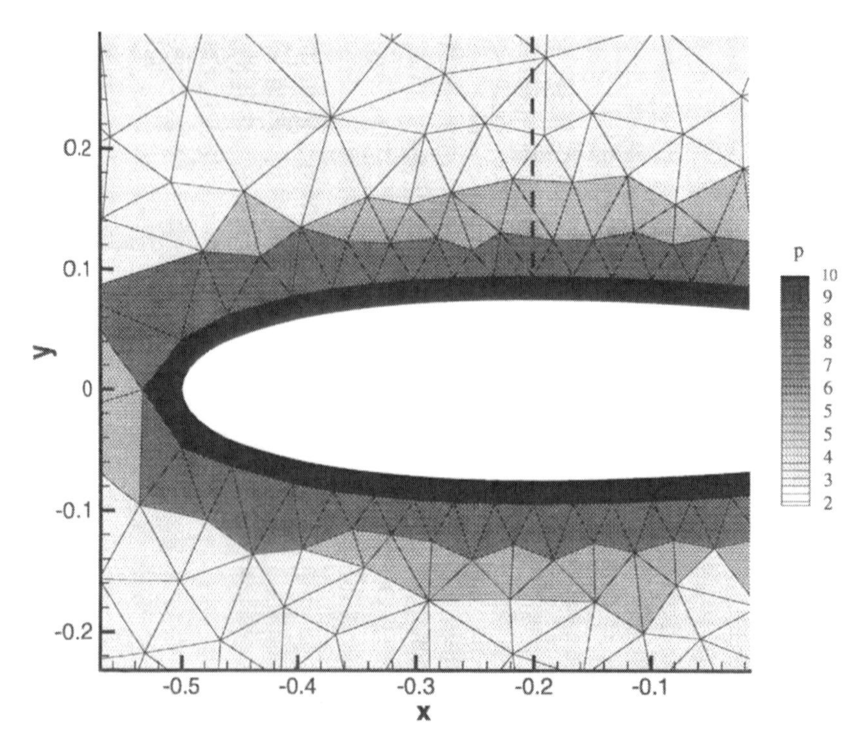

Figure 1.18. Hybrid discretization showing the variable p-order on a gray-scale map around the airfoil. The dash vertical line indicates the location where boundary layer profiles are taken (see figure 1.19).

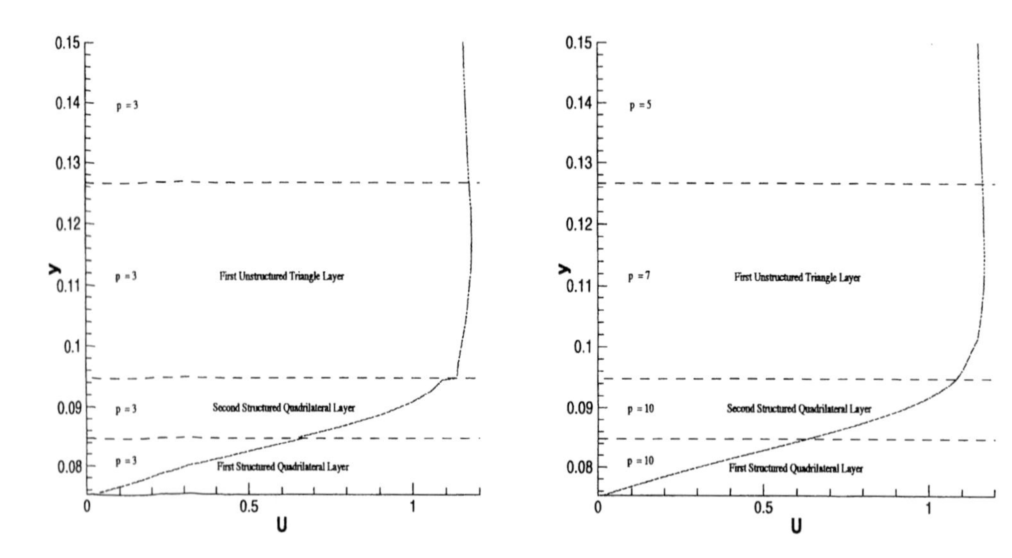

Figure 1.19. Boundary layer profiles for a simulation with uniform P-resolution (left) and variable P-resolution (right, as shown in figure 1.18).

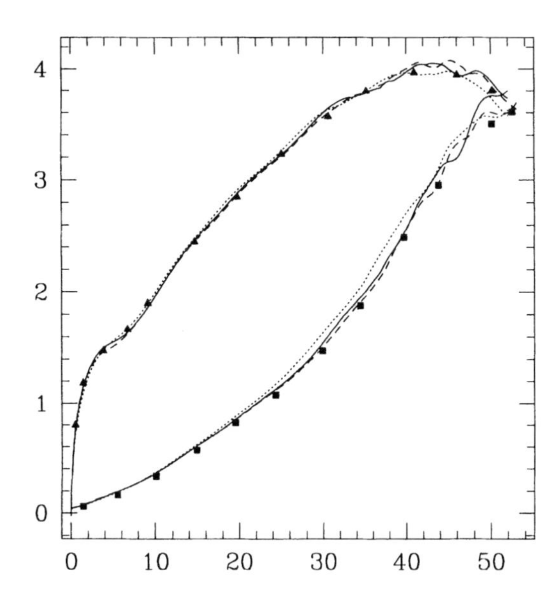

Figure 1.20. Lift (upper curve) and drag (lower curve) coefficients versus angle of attack in degrees. The symbols correspond to computations of ([33]), the dot line corresponds to our simulation at $P = 2$, the solid line to $P = 3$ and the dash line to $P = 4$.

To examine differences in the flow field due to spatial resolution we plot in figure 1.21 density contours for the cases $P = 2$ and $P = 3$ at non-dimensional time $t = 0.75$ corresponding to an angle of attack 18.55 degrees. We see that the higher resolution simulation provides a more detailed picture of the vortex shedding in the near-wake, but the contours around the airfoil are very similar. At a later time $t = 1.5$, corresponding to an angle of attack of 44.1 degrees, there are differences between the computations at resolution $P = 2$ and $P = 3$ and these differences are now extended to the upper surface of the airfoil where an interaction between the trailing edge vortex and the upstream propagating shed-vortex takes place, as shown in figure 1.22. These flow pattern differences are responsible for the aforementioned differences in the lift and drag coefficient at large angle of attack as shown in figure 1.20.

4.3 Turbulent Cylinder Wake

Numerical simulation of turbulent wakes has been computationally prohibitive and only preliminary results have been obtained in ([8]) using DNS. A more systematic study of the cylinder turbulent wake at $Re = 3,900$ was undertaken by ([35]) who used LES with an upwind discretization. A second LES study was performed by ([36]) with central-differencing in order to control the numerical damping reported in the first study, and more recently a high-order

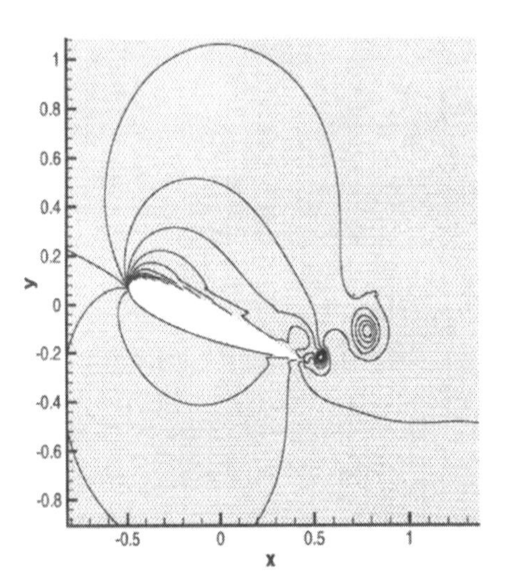

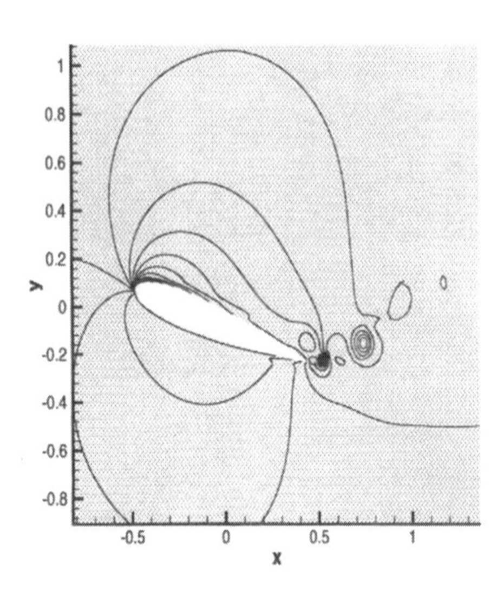

Figure 1.21. Density contours of the pitching airfoil at non-dimensional time $t = 0.75$ corresponding to 18.55 degrees angle of attack. Shown on the left are contours at spectral order $P = 2$ and on the right at $P = 3$.

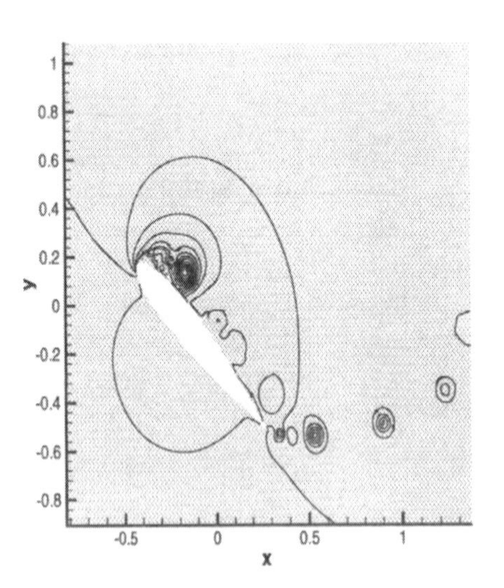

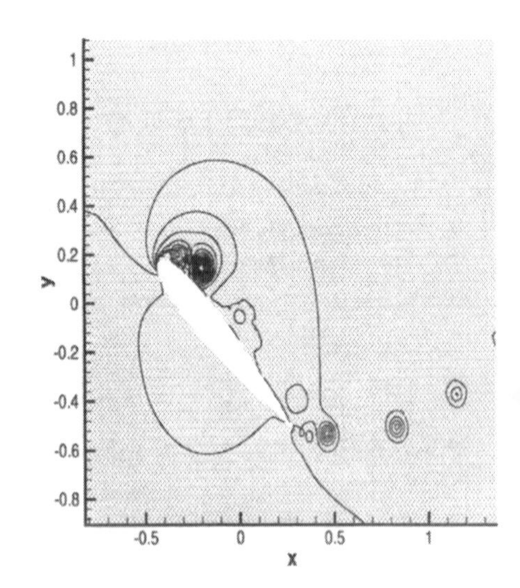

Figure 1.22. Density contours of the pitching airfoil at non-dimensional time $t = 1.5$ corresponding to 44.1 degrees angle of attack. Shown on the left are contours at spectral order $P = 2$ and on the right at $P = 3$.

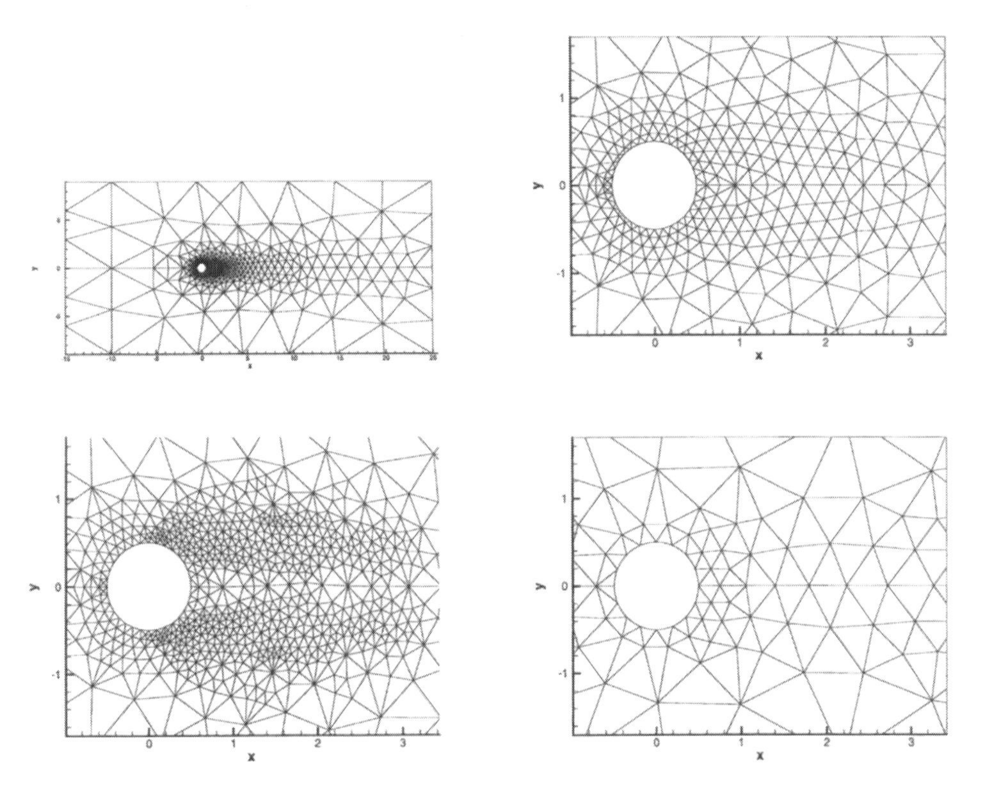

Figure 1.23. Two-dimensional "z-slice" of the entire domain (top left) and detail around the cylinder of the standard mesh ($K = 902$ elements - top right); refined ($K = 1,622$ elements - bottom left); and coarse mesh ($K = 412$ elements - bottom right) used in the spectral element simulations. The unstructured grid shown is the skeleton based on which hierarchical spectral expansions are constructed.

LES study was completed by ([37]). The results from the three studies are similar as far as the computed *mean* and *rms* velocities are concerned, i.e. LES predicts relatively accurately, although not uniformly, the experimental results in the region downstream of $x/D \geq 3$.

However, in the very-near-wake all simulations converge to a mean velocity profile in the *U-shape* unlike the experiments of ([43]) that show a *V-shape*. In contrast, an independent LES study by Rodi and co-workers ([38]) produced a V-shape velocity profile. Also, despite the higher fluctuations sustained in the central-differencing simulations by ([39]), no clear inertial range was obtained in either of the first two LES studies in contrast with the experiments. It is interesting to note that corresponding simulations with the subfilter model turned *off* produced an almost identical spectrum to the LES velocity spectrum. A systematic grid-refinement study performed in ([35]) also suggests that these results are resolution- independent for at least the first ten diameters in the near-wake. The high-order LES of Kravchenko & Moin, however, reproduced accurately the inertial range but predicted the same mean velocity field (i.e. U-shape) as the previous two simulations.

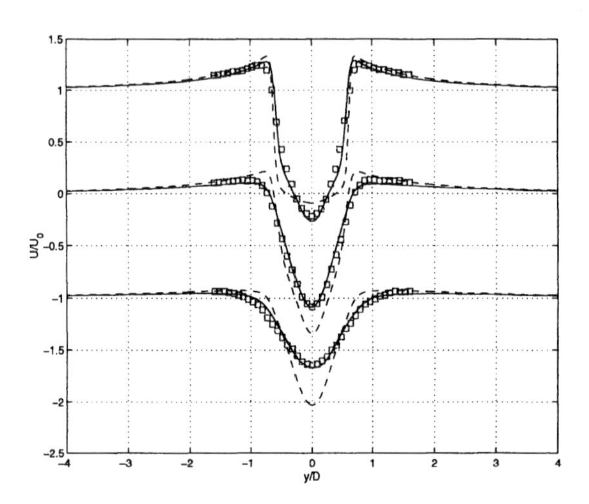

Figure 1.24. DNS *mean* streamwise velocity predictions at $x/D = 1.06; 1.54; 2.02$ (from top to bottom, respectively), (wide domain - solid line) and (narrow domain - dash line). Squares are data of ([43]).

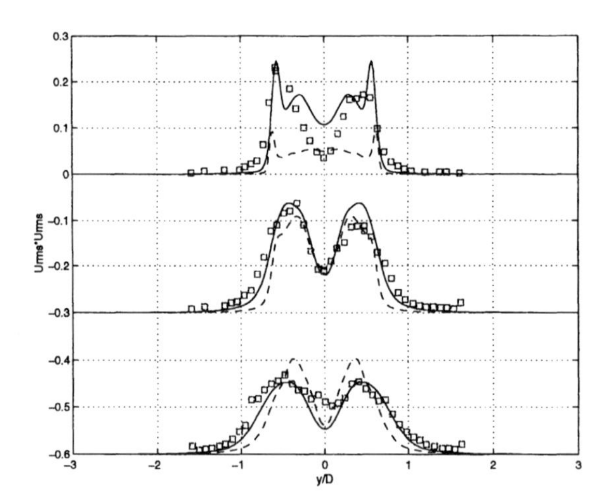

Figure 1.25. DNS *rms* streamwise velocity predictions at $x/D = 1.06; 1.54; 2.02$ (from top to bottom, respectively). (wide domain - solid line) and (narrow domain - dash line). Squares are data of ([43]).

For the simulations presented here a continuous Galerkin spectral/*hp* element method was employed in x- and y-directions while a Fourier expansion was employed along the homogeneous direction (cylinder-axis) with appropriate dealiasing. Specifically, triangular elements are used, filled with Jacobi polynomial modes of order P. We performed several simulations corresponding to h-refinement (i.e. with respect to number of elements K) and P-refinement (with respect to polynomial order P) ([40]). In figure 1.23 we show a "z-slice" of the computational domain in the $x - y$ plane with three different discretizations. The top plot shows a grid with $K = 902$ triangular prismatic elements, which has been the standard grid we have used for most cases. In the bottom plot we also show a grid with finer resolution around the shear layers corresponding to $K = 1,622$ elements, and also a grid with coarser resolution corresponding to $K = 412$ elements. The polynomial order per element varied from $P = 4$ to 10, and the number of Fourier modes varied from $N = 2$ to 128 (the corresponding number of physical points is twice the number of modes). The finest resolution simulation employed $K = 902$ elements of order $P = 10$ and 256 points ($N = 128$ Fourier modes) in the spanwise direction. The lowest resolution simulation employed $K = 412$ elements with $P = 6$ and only $N = 2$, i.e. a severe truncation of Fourier modes in the spanwise direction.

The domain extends from $-15D$ at the inflow to $25D$ at the outflow, and from $-9D$ to $9D$ in the cross-flow direction. Neumann boundary conditions (i.e. zero flux) were used at the outflow and on the sides of the domain to minimize the effect of normal boundary layers at the truncated domain. The spanwise length was varied as $L_z/D = \pi/2, \pi, 1.5\pi, 2\pi$. For reference, the spanwise length used in all simulations of ([35, 36, 37]) was $L_z/D = \pi$.

The experimental results of ([41]) suggest a value of correlation length less than $1.5D$ at three diameters downstream; this was obtained using the streamwise velocity only. However, from plots of the autocorrelation function for all three velocity components and the pressure we have seen that at a centerline point R_{uu} drops to zero at about $1.5D$ but that, in general, at points off-centerline R_{vv} and R_{uu} do not decay as fast ([40]). Such results indicate that values of R_{uu} obtained in experiments at centerline points may under-predict the spanwise correlation length. Therefore, it may be inadequate to use R_{uu} as the only criterion in deciding on the domain size. Indeed, we have found that the span length is very important in determining the *rms* values in the very-near-wake and correspondingly the mean velocity profiles.

In ([40]) high resolution results can be found for many different quantities. Typical velocity profiles for the mean and the variance and for different spans are shown in figures 1.24 and 1.25. Here, we examine how such results are affected by *substantially* reducing the grid resolution and without using any subfilter model. In particular, we present here results obtained on the grids shown in figure 1.23 (bottom right) consisting of $K = 412$ triangular elements and only $P = 6$ and the equivalent $K = 902$ and $P = 4$, both cases corresponding to approximately the same number of degrees of freedom. More comparisons with the subfilter model on the same grid can be found in ([42]). We will first use only *two Fourier modes* in the span, i.e. the mean mode and one perturbation ($N = 2$ or 4 points). We also choose a small value for the spanwise length $L_z/D = \pi/2$.

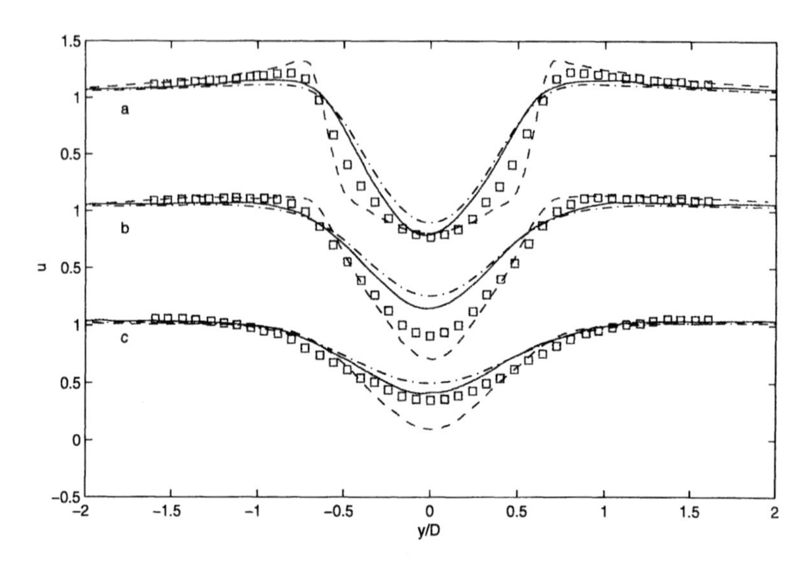

Figure 1.26. Streamwise mean velocity profile at $x/D = 1.06, 1.54, 2.02$. Squares denote experimental data of ([43]), solid line DNS ($K = 412$; $P = 6$), dash-dot line DNS ($K = 902$; $P = 4$) and dash line LES of ([35]).

We compare first with the experiments of ([43]) in the very near-wake and subsequently with the experiments of ([41]) farther downstream. In figure 1.26 we plot the mean streamwise velocity profile at locations $x/D = 1.06, 1.54$ and 2.02. We also include the experimental data of ([43]) taken from ([35]), and the LES data of ([35]). We see that the predictions from both low-resolution simulations without subfiltering are comparable to the LES predictions. In figures 1.27, 1.28 we plot the mean streamwise velocity and turbulent fluctuations, respectively, at locations $x/D = 4, 7, 10$ and compare with the experimental data of ([41]). The predictions for the mean velocities are good but the streamwise turbulence intensity shows some wiggles, which is an indication of insufficient resolution. However, the low-resolution spectral simulations obtain an overall better agreement with the experimental data than the dissipative LES predictions reported in ([35]).

The results presented so far were obtained with only $N = 2$ Fourier modes employed along the cylinder span. Of interest is to examine the influence of the number of Fourier modes N on the mean velocity profiles presented above while retaining the same resolution in the $x - y$ planes. We performed additional simulations with $N = 8$ and 32 and also a two-dimensional simulation. As we see in figure 1.29 there is essentially no difference in the predicted mean *streamwise*

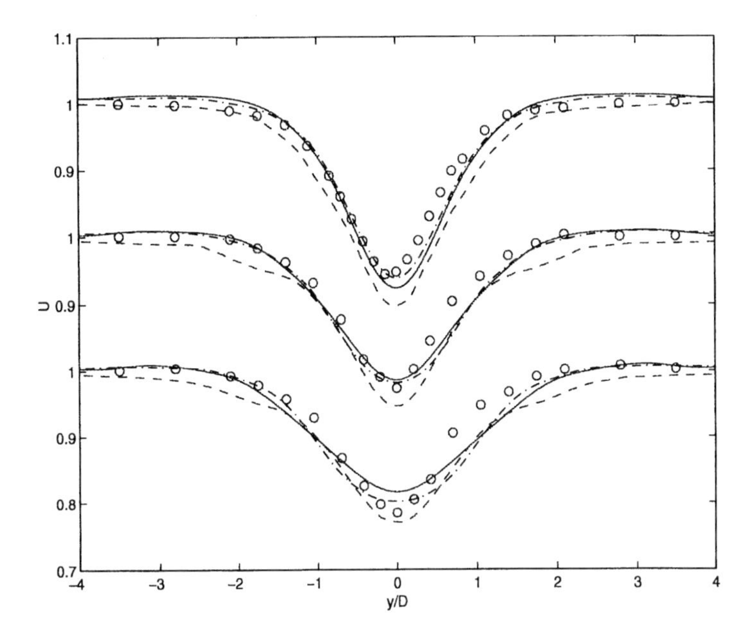

Figure 1.27. Mean velocity profiles at $x/D = 4, 7, 10$. Circles denote experimental data of Ong & Wallace, solid line DNS ($K = 412$; $P = 6$), dash-dot line DNS ($K = 902$; $P = 4$) and dash line LES of ([35]).

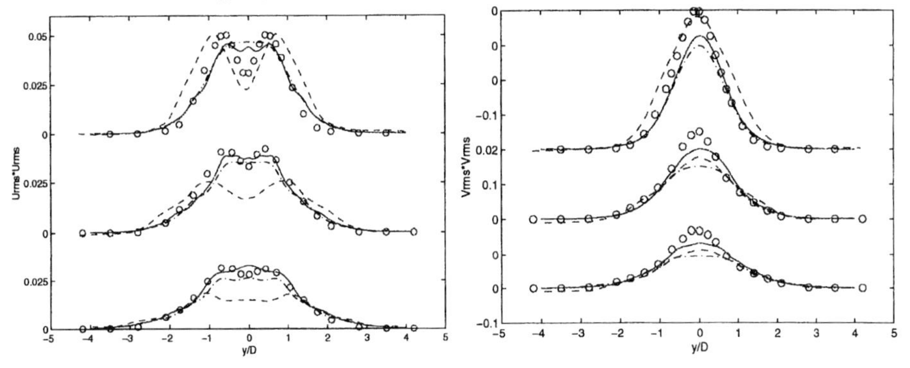

Figure 1.28. Left: Turbulent intensity of the streamwise velocity (u^2_{rms}) at $x/D = 4, 7, 10$. Right: Turbulent intensity of the cross-flow velocity (v^2_{rms}) at $x/D = 4, 7, 10$. Circles denote experimental data of Ong & Wallace, solid line DNS ($K = 412$; $P = 6$), dash-dot line DNS ($K = 902$; $P = 4$) and dash line LES of ([35]).

velocity profile from $N = 2$ to $N = 32$ but the two-dimensional prediction deviates substantially. The cases with $N = 8$ and $N = 32$ correspond to almost identical predictions suggesting convergence in the z-direction.

The results presented here indicate that the first Fourier mode carries most of the spanwise energy for the chosen span $L_z/D = \pi/2$, as it is evident by comparing with the two-dimensional results in figure 1.29. This has been inde-

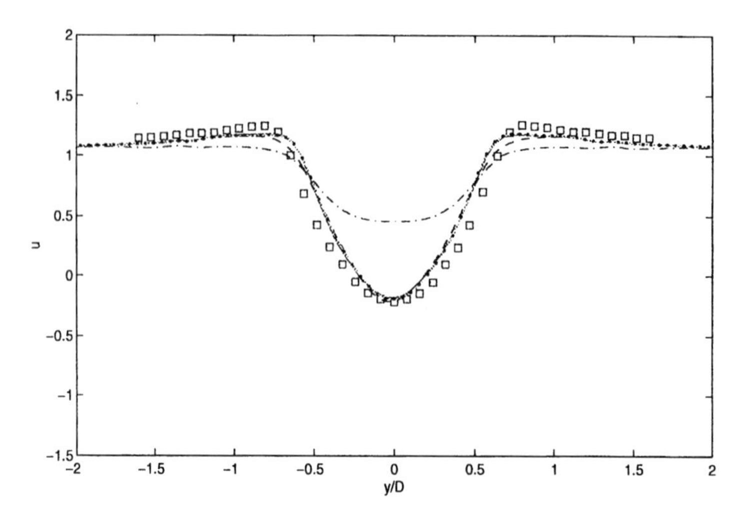

Figure 1.29. Streamwise mean velocity profile at $x/D = 1.06$ for different Fourier modes employed along the span. Squares denote experimental data of ([43]), dash-dot line 2D simulation, dash line $N = 2$, dot-solid line $N = 8$ and solid line $N = 32$ (coincides with $N = 8$).

pendently verified by computing the averaged plane-modal energy $E_{xy}(m) = \int_{xy}[u_m^2 + v_m^2 + w_m^2]dxdy$ and observe its decay with respect to the mode number.

Given the surprisingly good results with this low-resolution at $Re = 3,900$, we performed another set of simulations with the same low-resolution and with only $N = 2$ Fourier modes at $Re = 5,000$ for which we had available experimental data from the work of ([44]). In figure 1.30 we plot the mean velocity profile and the streamwise turbulent intensity $\overline{u'^2}$ at station $x/D = 10$. Again, we see that despite some wiggles in the numerical results the agreement with the experimental results is good.

5. Discussion

Spectral methods have been used with great success in simulating turbulent flows in periodic cubes and channel domains. The algorithmic developments of the last two decades have led to a new simulation capability for complex-geometry turbulent flows as well. Such capability, in conjunction with terascale computing at the PC cluster level, will undoubtly lead to significant advances in simulating turbulence in more realistic configurations and in realistic operating conditions. In this chapter, we have summarized some of these develop-

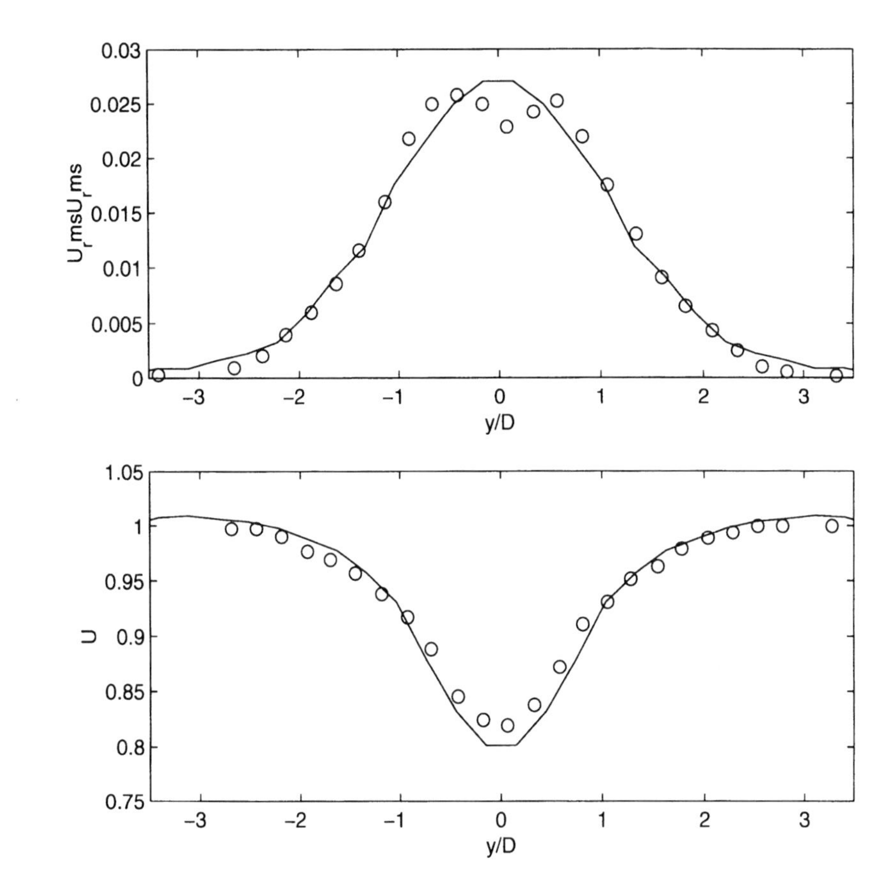

Figure 1.30. Streamwise mean velocity profile (bottom) and turbulent fluctuation (top) at $x/D = 10$ and $Re = 5,000$. The experimental data (circles) are from ([44]).

ments and have presented results for several transitional and turbulent flows in complex-geometry domains.

Great care has to be exercised however in interpreting results from DNS or LES in simple-geometry flows at high Reynolds number or in complex-geometry flows with new physics. Many of these simulations may be under-resolved at some level; for example, although the velocity mean and variance may be correctly predicted, the high-order statistics or the dissipation spectrum may be erroneous. It is important, therefore, to have a diverse set of diagnostic tools to characterize numerical uncertainty in these situations. In particular, understanding how numerical methods behave in DNS and LES for many prototype flows provides an insight into uncertainities and their origin in large-scale simulations. Validation and verification ([1]) of a turbulent simulation is a very difficult task that, unfortunately, cannot be handled solely by error estimators which are based primarily on extensions of linear concepts ([45, 46, 47]).

Under-resolved simulations are not useless if the numerical uncertainty is properly characterized ([48, 49, 50]), i.e. quantified with a properly constructed error bar ([51]). In many cases such under-resolved simulations may contain the answer that we seek, e.g. an averaged lift or drag coefficient and at the accuracy level that we expect. The type of flow that we simulate is important in that respect. We have presented here, for example, low-resolution simulations of the cylinder turbulent wake which lead to results that match the experiments at $Re = 5,000$. Flows with inherent low-dimensionality, such as the cylinder wake for which the vortex shedding process dominates the dynamics, may then be easier to simulate. Moreover, quantification of numerical uncertainty in hierarchical methods, such as the spectral/hp element method, is also easier to achieve and offer the possibility of obtaining multiple solutions with relatively simple P-refinement without the need for re-meshing which, typically, is a large overhead component in a turbulence simulation.

Acknowledgments

This work was supported by ONR, AFOSR, DOE, DARPA and NSF. It is based partially on findings in the Ph.D theses of S.J. Sherwin, I. Lomtev and X. Ma. The simulations were performed on the IBM SP2/3 at the San Diego Supercomputing Center (NPACI), at the Maui High Performance Computing Center, at the Brown University Center for Scientific Computing and Visualization, and on the SGI Origin 2000 at the NCSA, University of Illinois at Urbana Champaign.

References

[1] P.J. Roache: 1998. Verification and validation in computational space and engineering. *Hermosa.*

[2] R. Zores: 1989. Numerische untersuchungen met einem grobauflosenden simulationsmodell fur die turbulente kanalstromung. *Institut fur Theoretische Stromungsmechanik.* DLR-Gotting-Germany

[3] G.E. Karniadakis, S.A. Orszag: 1993. Nodes, modes and flow codes. *Physics Today* **46**, 34.

[4] S. Ghosal, P. Moin: 1995. The basic equations for the large-eddy simulation of turbulent flow in complex geometry. *J. Comp. Phys.* **118**, 24.

[5] G.E. Karniadakis, S.J. Sherwin: 1999. Spectral/hp element methods for CFD. *Oxford University Press.*

[6] A.T. Patera: 1984. A spectral method for fluid dynamics: laminar flow in a channel expansion. *J. Comp. Phys.* **54**, 468

[7] D.C. Chu, G.E. Karniadakis: 1993. A direct numerical simulation of laminar and turbulent flow over riblet-mounted surfaces. *J. Fluid Mech.* **250**, 1

[8] R.D. Henderson, G.E. Karniadakis: 1995. Unstructured spectral element methods for simulation of turbulent flows. *J. Comp. Phys.* **122**, 191

[9] S.J. Sherwin, G.E. Karniadakis: 1995. A triangular spectral element method; applications to the incompressible Navier-Stokes equations. *Comp. Meth. Appl. Mech. Eng.* **23**, 83

[10] S.J. Sherwin, G.E. Karniadakis: 1995. A new triangular and tetrahedral basis for high-order finite element methods *J. Num. Meth. Eng.* **38**, 3775

[11] D.J. Mavripilis, V. Venkatakrishnan: 1995 A unified multigrid solver for the Navier-Stokes equations on mixed element meshes. *AIAA-95-1666, San Diego, CA*

[12] V. Parthasarathy, Y. Kallinderis, K. Nakajima: 1995. Hybrid adaptation method and directional viscous multigrid with presmatic-tetrahedral meshes. *AIAA-95-0670, Reno, NV.*

[13] G.E. Karniadakis, M. Israeli, S.A. Orszag: 1991. High-order splitting methods for the incompressible Navier-Stokes equations. *J. Comp. Phys.* **97**, 414

[14] N.A. Petterson: 2001. Stability of pressure boundary conditions for Stokes and Navier-Stokes equations. *J. Comp. Phys.* **172**, 40

[15] B. Cockburn, G.E. Karniadakis, C.W. Shu: 2000. The development of discontinuous Galerkin methods. *Discontinuous Galerkin methods: Theory, Computation and Applications*, Springer-Verlag

[16] D. Xiu and G.E. Karniadakis, A Semi-Lagrangian High-Order Method for Navier-Stokes Equations, J. Comput. Phys. 172, 658-684 (2001)

[17] I. Lomtev, G. Quillen, G.E. Karniadakis: 1998. Spectral/hp methods for viscous compressible flows on unstructured 2D meshes. *J. Comp. Phys.* **144**, 325

[18] C. Johnson: 1994. Numerical solution of partial differential equations by the finite element method. *Cambridge University Press*

[19] G. Jiang, C.W. Shu: 1994. On a cell entropy inequality for discontinuous Galerkin methods. *Math. Comp.* **62**, 531

[20] R.M. Kirby: 2002. Dynamic spectral/hp refinement: algorithms and applications to flow-structure interactions. *PhD-thesis: Brown University: Applied Mathematics*

[21] R. Verstappen, A. Veldman: 2002. Preserving symmetry in convection-diffusion schemes. *This volume*

[22] W. Couzy: 1995. Spectral element discretization of the unsteady Navier-Stokes equations and its iterative solution in parallel computers. *PhD-thesis: Ecole Polytechnique Federale de Lausanne*

[23] E.M. Ronquist: 1996. Convection treatment using spectral elements of different order. *Int. J. Numer. Methd. Fluids* **22**, 241

[24] I. Lomtev, C. Quillen, G.E. Karniadakis: 1998. A discontinuous Galerkin ALE method for compressible viscous flow in moving domains. *J. Comp. Phys.* **155**, 128

[25] E. Tadmor: 1989. Convergence of spectral methods for nonlinear conservation laws. *SIAM J. Numer. Anal.* **26**, 30

[26] G.S. Karamanos, G.E. Karniadakis: 2000. A spectral vanishing viscosity method for large-eddy simulations. *J. Comp. Phys.* **162**, 22

[27] E. Tadmor: 1993. Total variation and error estimates for spectral viscosity approximations. *Math. Comp.* **60**, 245

[28] D.S. Henningson, A. Lundblach, A.V. Johansson: 1993. A mechanism for bypass transition from localized disturbances in wall-bounded shear flows. *J. Fluid Mech.* **250**, 169

[29] L. Kaiktsis, G.E. Karniadakis, S.A. Orszag: 1991. Onset of three-dimensionality, equilibria, and early transition in flow over a backward-facing step. *J. Fluid Mech.* **191**, 501

[30] L. Kaiktsis, G.E. Karniadakis, S.A. Orszag: 1996. Unsteadiness and convective instabilities in two-dimensional flow over a backward-facing step. *J. Fluid Mech.* **321**, 157

[31] J.A. Sethian, A.F. Ghoniem: 1988. Validation of vortex methods. *J. Comp. Phys.* **74**, 283

[32] Y.H. Qian and S. Succi and F. Massaioli and S.A. Orszag: 1993. A benchmark for lattice BGK model: flow over a backward-facing step. *Pattern formation and lattice-gas automata* A. Lawniczak, R. Kapral (Eds.), June 7-12, Waterloo.

[33] M.R. Visbal, J.S. Shang: 1989. Investigation of the flow structure around a rapidly pitching airfoil. *AIAA J.* **27**, 1044

[34] R. Beam, R. Warming: 1978. An implicit factored scheme for the compressible Navier-Stokes equations. *AIAA J.* **16**, 393

[35] P. Beaudan, P.Moin: 1994. Numerical experiments on the flow past a circular cylinder at sub-critical Reynolds number. *Report No. TF-62, Stanford University*

[36] R. Mittal, P. Moin: 1996. Large-eddy simulation of flow past a circular cylinder. *APS Bulletin: 49-th CFD Meeting* **41**

[37] A.G. Kravchenko, P. Moin: 1998. B-spline methods and zonal grids for numerical simulations of turbulent flows. *Stanford University: Report No. TF-73*

[38] J. Frohlich, W. Rodi, Ph. Kessler, S. Parpais, J.P. Bertoglio, D. Laurence: 1998. Large-eddy simulation of flow around circular cylinders on structured and unstructured grids. *Notes on numerical fluid mechanics* E.H. Hirschel (Ed.)

[39] R. Mittal: 1996. Progress on LES of flow past a circular cylinder. *CTR Annual research briefs, Stanford*, 233

[40] X. Ma, G.S. Karamanos, G.E. Karniadakis: 2000. Dynamics and low-dimensionality in the turbulent near-wake. *J. Fluid Mech.* **410**, 29

[41] L. Ong, J. Wallace: 1996. The velocity field of the turbulent very near wake of a circular cylinder. *Experiments in fluids* **40**, 441

[42] G.S. Karamanos: 1999. Large eddy simulation using unstructured spectral/hp elements. *PhD-thesis: Imperial College*

[43] L.M. Lourenco, C. Shih: Characteristics of the plane turbulent near wake of a circular cylinder. A particle image velocimetry study. (Unpublished results taken from Beaudan and Moin 1994.)

[44] Y. Zhou, R.A. Antonia: 1993. A study of turbulent vortices in the near wake of a cylinder. *J. Fluid Mech.* **253**, 643

[45] T.J. Oden, W. Wu, M. Ainsworth: 1994. An a porsteriori error estimate for finite element approximations of the Navier-Stokes equations. *Comp. Meth. Appl. Mech. Eng.* **111**, 185

[46] L. Machiels, J. Peraire, A.T. Patera: 2001. A posteriori finite element output bounds for the incompressible Navier-Stokes equations: applications to a natural convection problem. *J. Comp. Phys.*, to appear

[47] M. Ainsworth, T.J. Oden: 2000. A posteriori error estimation in finite element analysis. *John Wiley & Sons*

[48] A.J. Chorin, A.P. Kast, R. Kupferman: 1998. Optimal prediction of underresolved dynamics. *Proc. Natl. Acad. Sci. USA* **95**, 4094

[49] Y. Maday and A.T. Patera and J. Peraire: 1999. A general formulation for a posteriori bounds for output functionals of partial differential equations. *C.R. Acad. Sci. Paris, Series I*, **328**, 823

[50] J. Glimm, D.H. Sharp: 1999. Prediction and the quantification of uncertainty. *Physica D* **133**, 152

[51] G.E. Karniadakis: 1995. Towards an error bar in CFD. *J. Fluids Eng.* **117**

Chapter 2

HIGH RESOLUTION METHODS FOR COMPUTING TURBULENT FLOWS

William J. Rider

Computer and Computational Sciences Division, MS D413, Los Alamos National Laboratory, Los Alamos, NM 87545 USA.

wjr@lanl.gov

Dimitris Drikakis

Queen Mary, University of London, Department of Engineering, London E1 4NS, United Kingdom.

d.drikakis@qmul.ac.uk

Abstract Over the past decade there has been an increasing amount of evidence that high resolution numerical methods for hyperbolic partial differential equations have an embedded (or "implicit") turbulence model. The present chapter describes this general class of methods and outlines the basic structure of high resolution methods as an effective turbulence model in the context of large eddy simulation (LES). This discussion is an extension of the MILES concept introduced by Boris, where monotone numerical algorithms are used for LES (MILES is an acronym for monotone integrated LES). We show that the implicit modeling includes elements of nonlinear eddy viscosity, scale-similarity and an effective dynamic model. In addition, we give examples of both success and failures with currently available methods and examine the effects of the embedded modeling in contrast to widely used subgrid scale (SGS) models.

Keywords: Turbulence, Large Eddy Simulation, High Resolution Methods, Subgrid Models, MILES, Compressible Flows, Incompressible Flows.

1. Motivation

The development of highly accurate and efficient methods for the computation of turbulent flows is motivated by the broad spectrum of applications in science and engineering in which turbulence appears. High resolution or non-

D. Drikakis and B.J. Geurts (eds.), Turbulent Flow Computation, 43–74.

oscillatory methods are used to simulate a broad variety of physical processes including unstable flows that are dominated by vorticity leading to turbulence, and the mixing of materials (Boris 1989).

Following Harten's definition (Harten 1983), we classify as high resolution methods those with the following properties: i) provide at least second order of accuracy in smooth areas of the flow, ii) produce numerical solutions (relatively) free from spurious oscillations, and iii) in the case of discontinuities, the number of grid points in the transition zone containing the shock wave is smaller in comparison with that of first-order monotone methods. High-resolution can be realized by numerically reconstructing the variables at the cell faces of a computational volume and computing the fluxes via the solution of a Riemann problem. This can be achieved by using, for example, Godunov-type methods which offer an ingenious computational framework for developing schemes that adjust the amount of numerical dissipation locally, i.e., at the cell faces, in order to maintain monotonicity and conservation. Previous studies (Oran & Boris 2001; Margolin et. al. 1998; Margolin & Rider 2001; Drikakis 2001a) have shown that the properties of high resolution methods appear to infuse the simulations with many of the characteristics of turbulent flows.

2. Multimaterial Mixing

The true litmus test for a methodology is its ability to replicate phenomena from the physical world. An essential activity in using a numerical method is its validation against experimental data. We will present two examples from multimaterial mixing where the details of the experimentally observed mixing are compared in detail with calculations.

2.1 Rayleigh-Taylor Instability

Our first example is based on Rayleigh-Taylor experiments (Dimonte 1999). In this experiment often referred to as the linear electric motor (LEM) experiments, a box is accelerated using electromagnetic rails. A number of cameras are stationed along the the travel of the box containing the mixing fluids to image the evolution of the flow. The LEM allows the time dependent acceleration of the interface to be tailored in a variety of ways. We will consider comparison with the "constant" acceleration history where water and hexane are the mixing fluids. These fluids give an Atwood number, $A = (\rho_{\max} - \rho_{\min}) / (\rho_{\max} + \rho_{\min}) = 0.5$. Time and depth of mixing are measured by the expected self-similar profile expected for the growth of the bubble dome (the edge of the light fluid mixing into the heavy) from its initial interface position, $h(t) = \alpha g t^2$, where α is a proportionality constant that is one principle object of the experimental investigation (in LEM $\alpha = 0.054$).

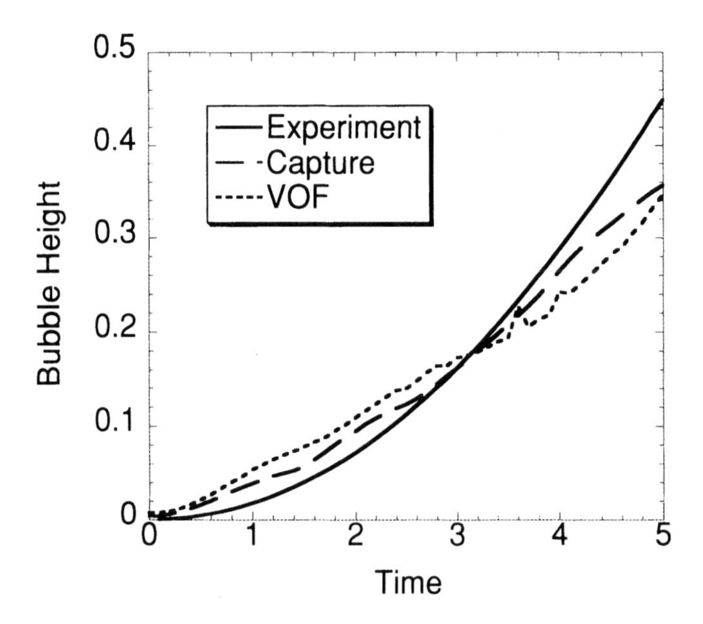

Figure 2.1. The comparison of the integral scale growth from the LEM experiment and comparable simulations. Early in time the compututational α for the bubble height is higher then experiment, but it becomes less than the experimentally measured value late in time. The 99 percent integral volume fraction was used to compute the bubble height in each case. The experimental value is $\alpha = 0.054$, the captured interface (labeled as "captured") yields $\alpha = 0.043$, and the volume tracking (laveled "VOF") gives $\alpha = 0.042$.

The Reynolds number of the experiment is strongly time-dependent (with a t^3 dependence) with a values of $\approx 10^5$ at the end of the acceleration.

The flow is essentially incompressible and best approximated using methods for incompressible flow with immissible interfaces. The flow solver is a high resolution Godunov solver using an approximate projection for the pressure-velocity coupling combined with interface tracking (Puckett et. al 1997), the actual methodology is detailed in (Rider 1994). In this work, we conduct the comparison with a genuinely unsplit method based on limited gradients computed on multidimensional stencils (as opposed to one-dimensional strips). The time integration method is a genuinely multidimensional "Hancock" method (van Leer 1984; 17). The volume of fluid (VOF) interface tracking method (Rider & Kothe 1998) is most appropriate for fluids that behave immisibly. We use an ensemble small amplitude ($\ll \Delta x$) multimode perturbations with random phase in the material interface to initialize the instability. It then evolves via constant acceleration through approximately five generations of bubble merger. Lattice Boltzmann results were provided by Tim Clark (LANL, private communication).

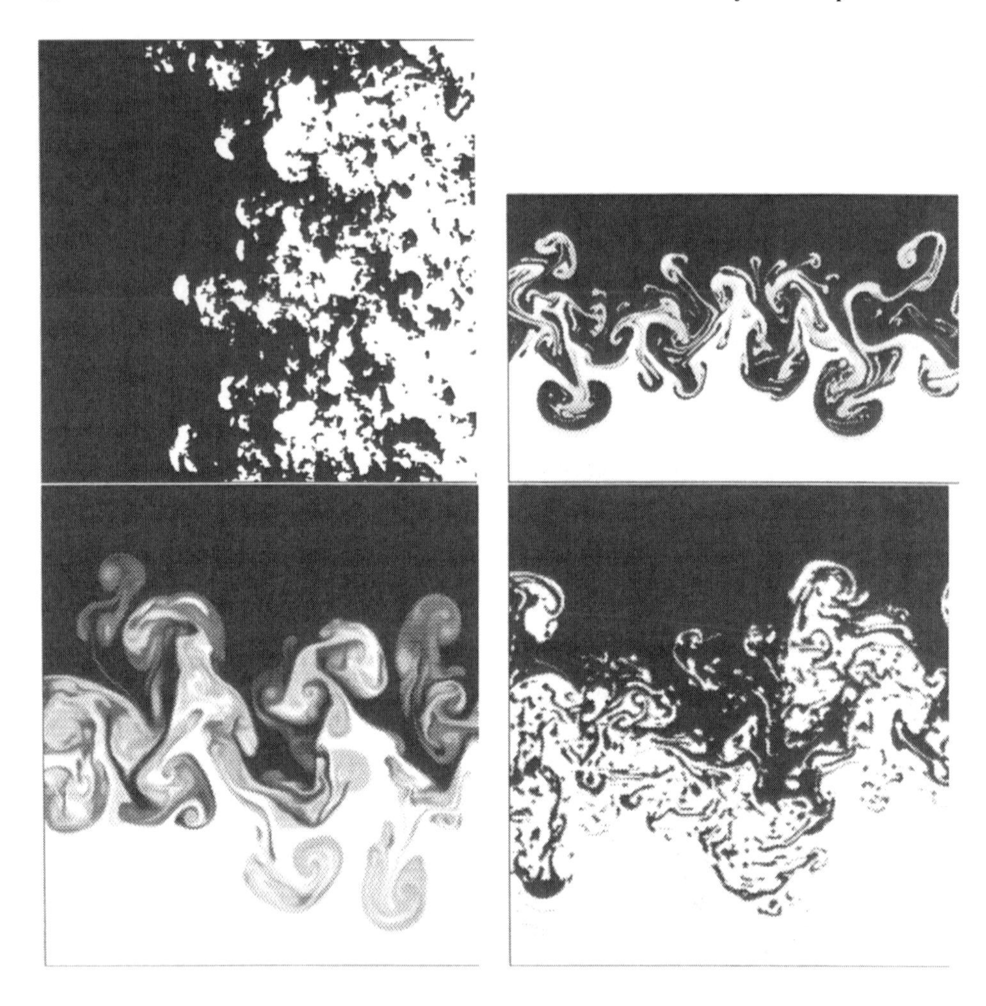

Figure 2.2. The comparison of the mixing calculations of Rayleigh-Taylor starting top-to-bottom, left-to-right, experimental image, lattice Boltzmann, interface capturing and interface tracking (volume-of-fluid, VOF).

First, we give the comparison with the integral scale growth rate for the bubble height that is measured experimentally. This is shown in Fig. (2.1). As most existing calculations (the "alpha" group Dimonte 2001), the α computed computationally is less than the experimentally measured value. Our results here are no different in this respect. With LEM we have detailed two-dimensional slices of experimental structure. This is shown in Fig. (2.2) along side the numerical simulations examined. Using the fractal dimension as a statistical measurement tool we directly compare the calculations. This is shown in Fig. (2.3). In this case the qualitative and quantitative results are best computed with the high res-

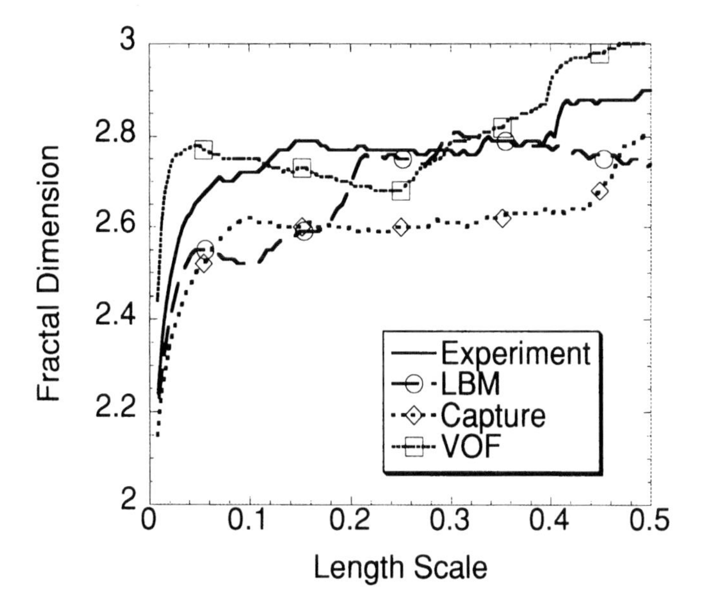

Figure 2.3. The comparison of the local fractal dimension of the experimental data and computations. LBM stands for the Lattice Boltzmann Method.

olution Godunov method with interface tracking (i.e., numerically immissible). Indeed, the results seem to show that the data can be considered to be fractal at large scales as indicated by the constant D_f at large scales (greater than 0.1 of the integral scale).

2.2 Richtmyer-Meshkov Instability

Our next results are based upon the gas curtain, Richtmyer-Meshkov experiments conducted by Rightley et al. (Rightley et al. 1999) at Los Alamos. The experimental apparatus is a 5.5 m shock tube with a 75 mm square test section. The driver section is pressurized before the shot, and the rupturing of a polypropylene diaphragm produces a Mach 1.2 planar shock. In the test section, a vertical curtain of SF_6 is injected through a nozzle in the top, and removed through an exhaust plenum at the bottom. Interchangeable nozzles containing different contours impose perturbations on the cross section of the curtain, which has a downward velocity of $\sim$ 10 cm/s. The evolving flow is imaged by a horizontal laser light sheet. A tracer material consisting of glycol fog (with a typical droplet dimension of 0.5μm) is added to the curtain to greatly improve the dynamic range of the images, which are captured by CCD camera.

As discussed later in detail limiters used in high resolution schemes produce effective differential forms. We will briefly describe the general idea here as it strongly impacts the results for simulating this experiment. The standard

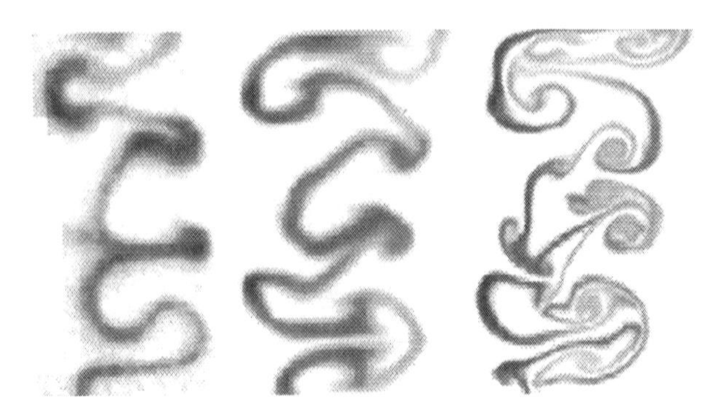

Figure 2.4. SF_6 volume fraction at 400 μs for the experimental image (left) the dynamic Smagorinsky Cuervo result (center), the second-order standard MUSCL-type scheme (right).

monotone limiter provides a modified equation form like $1 - \Delta x\, |u_{xx}/u_x|$, one can design a limiter that produces a Smagorinsky dissipation, $1 - \Delta x\, |u_x/u|$. In order to accomplish this end, the limiter should switch off when the flow is resolved. This is determined by the switching function,

$$\psi_{i,j} = 1 - (\max u_{i,j} - \min u_{i,j}) \,/\, |\min u_{i,j} + \max u_{i,j}|$$

The overall limiter is then chosen to give the dynamic feature,

$$\partial_x u = \min(1, 2\psi_{i,j})\, \partial_c u$$

. Thus, when the switch ψ is greater than $\frac{1}{2}$ in value, the centered gradient will be chosen to approximate, $\partial_c u$. The flow is considered to be "resolved" by the limiter when $\psi \leq \frac{1}{2}$.

We now examine the performance of this limiter when used to replace our standard monotone limiter in simulating the gas curtain. Our results are computed using a modified version of the Cuervo code (Rider 1999). Our results are shown in Fig. (2.4) and statistically in Fig. (2.5). The statistical comparison of the standard high resolution method is also shown in Fig. (2.5). In a statistical sense the results with the dynamic Smagorinsky limiter are better than the standard method. The standard method on the other hand shows characteristics that are intrinsically different than the experimental data.

In summary, we find that all standard higher order methods used to simulate the gas curtain compare poorly with the experimental data when quantified with these spatial statistics. Moreover, the comparisons degrade under mesh refinement. This occurs despite the fact that the integral scale comparison is acceptable and consistent with the expectations from this class of methods. The most surprising result is that a first-order Godunov method does produce a good comparison relative to the assumed to be higher-order methods. However,

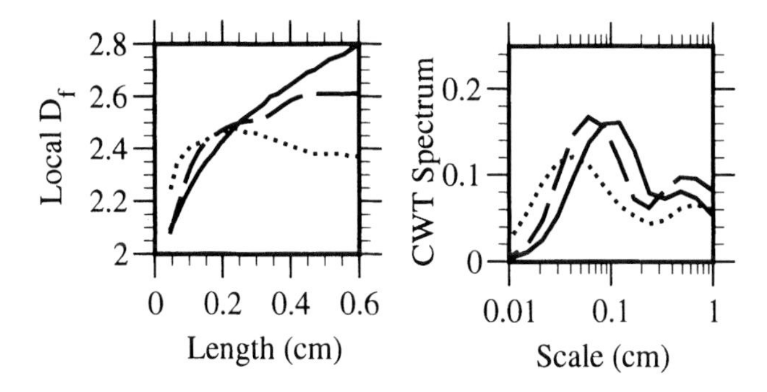

Figure 2.5. Local fractal dimension (left) and continuous wavelet spectrum (CWT) (right) for the images of Fig. (2.4) as functions of scale. The experimental result is a solid line, and the Smagorinsky-like limiter result is a dashed line. The dotted lines shows the result with a standard high resolution method.

we find that a simple modification mimicking turbulence modeling improves results greatly while retaining second-order accuracy.

These results are not meant to imply that high-order methods should not be used; rather, our viewpoint is that the high-order methods are not presently suitable for calculations of these experiments if statistical congruence is desired. They need to be modified to improve results. We speculate that some aspects of high-resolution methods for multidimensional compressible flows may be undermining the statistical scaling in the gas-curtain Richtmyer-Meshkov simulations.

Lastly, we observe that experiments with high fidelity diagnostics were critical to uncovering these issues. At present, our numerical results for shock-driven mixing are far out outside the experimental error and show qualitatively different details. Of course, we would be delighted to have flow diagnostics of even higher spatial resolution. Nonetheless, we argue that quantitative "apples to apples" comparisons of experimental data with numerical results provide the most meaningful and compelling measure of the capabilities of numerical simulation.

3. Numerical Methods as LES Models

Previous computations using high resolution methods for turbulent (or turbulent-like) flows have suggested that: i) these methods appear to achieve many of the properties of subgrid models (e.g., Oran & Boris 2001; Margolin & Rider 2001; Margolin et. al. 1998) used in LES (Eyink 1995; Pope 2000), and ii) the accuracy of simulations depends on the details of the numerical scheme employed (Bagabir & Drikakis 2001; Drikakis 2001b). Examination of the modified equations associated with high-resolution methods as applied to the gas dynamic equations has yielded the enticing hint that characteristics implicit in these methods may mimic certain aspects of turbulent flow modeling. The implicit turbulent modeling aspects of high-resolution methods may provide a fruitful avenue by which to gain understanding of the simulation of unstable flows of compressible fluids. Below, we describe this relation in some detail.

Some evidence concerning the first point was presented in the previous section. We demonstrate below the dependence of simulations on the method for hyperbolic equations employed using as an example the interaction of a planar shock wave with a cylindrical bubble. Experiments for this problem have been presented in (Haas and Sturtevant 1987) while there have also been various computational studies on the basis of these experiments (Picone & Boris 1988; Quirk and Karni 1996; Bagabir & Drikakis 2001). Figure (2.6) shows (instantaneous) isodensity contours as obtained by different Riemann solvers while Fig. (2.7) presents comparisons between simulations and experiment (Bagabir 2000).

Although one can argue that uncertainties arising from numerical diffusion make difficult a direct comparison between such simulations and real-world effects of molecular diffusion and turbulent mixing, the results of Figs. (2.6) and (2.7) suggest the following: i) the details of the simulated flow depend on the numerical scheme employed; ii) there are strong similarities between the predictions of certain methods, e.g. the solutions obtained by the CBM-FVS (Zółtak & Drikakis 1998), HLLC (Toro et al. 1994) and Roe (Roe & Pike 1984) schemes show stronger similarities than the solutions obtained by the (modified) SW-FVS (Zółtak & Drikakis 1998), VL-FVS (Thomas et al. 1985), HLL (Harten 1983) and Rusanov (Rusanov 1961) schemes; iii) even though the solutions obtained by the former group of methods are clearly less diffusive than those obtained by the second group, all schemes provide very similar results for the position of the upstream and downstream bubble interface and also in close agreement with the experiment Fig. (2.7). We note that for a grid containing 500×100 cells, the results obtained by different high resolution schemes differ in absolute value about 0.7% and are therefore indistinguishable on the plot of Fig. (2.7).

Perhaps more importantly, these methods are often necessary to ensure a stable computation under difficult physical circumstances. We take the view that physical phenomena, associated models and their numerical solution are intertwined and should be developed and solved together. Indeed, the modeling and their solution cannot be separated. This approach is not generally adopted instead models are developed independently from their numerical solution, under the assumption that it is error-free. We consider the case that may often exist in turbulence modeling and its solution where the errors are the same differential order as the model.

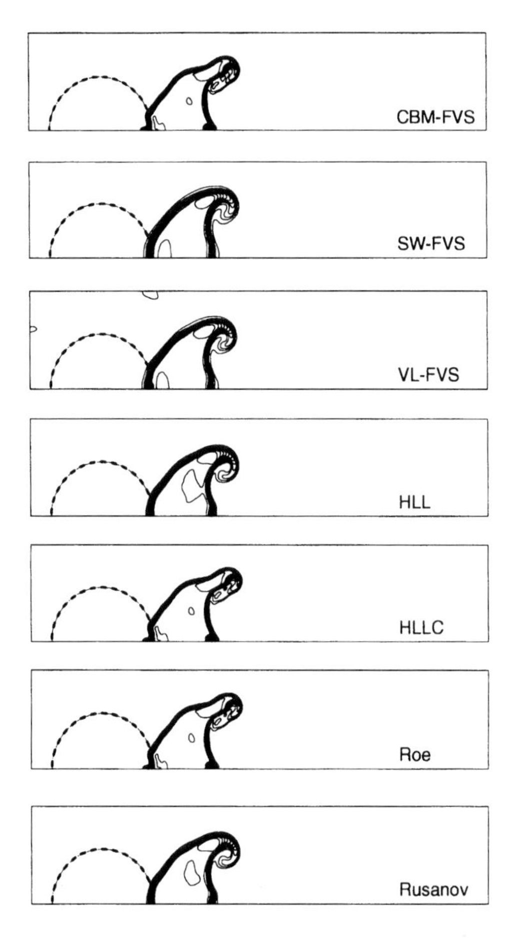

Figure 2.6. Isodensity contours for the interaction of a shock wave with a dense bubble (R22) as obtained by different high resolution methods (Bagabir & Drikakis 2001; Bagabir 2000).

Here, we focus on the numerical approximation of the hyperbolic part of the fluid dynamic equations (i.e., the nonlinear transport terms that are responsible for turbulence). The nonlinear interactions induced by these terms naturally cause scale-to-scale transfer of energy, resulting ultimately in entropy production by viscosity at small length scales. Shock waves are the prototypical example of this. At large scales in the inertial range, the flow behaves (nearly) independently of viscosity. In addition to the viscous dissipation, important energy transfers known as *backscatter* move energy from small scales toward larger scales.

For example, rarefactions produce transfer of energy from small to large scales, hyperbolically. This process is not represented in those turbulence models that are purely dissipative, but found in models exhibiting self-similarity. Conversely, this effect is naturally present in the hyperbolic terms which dis-

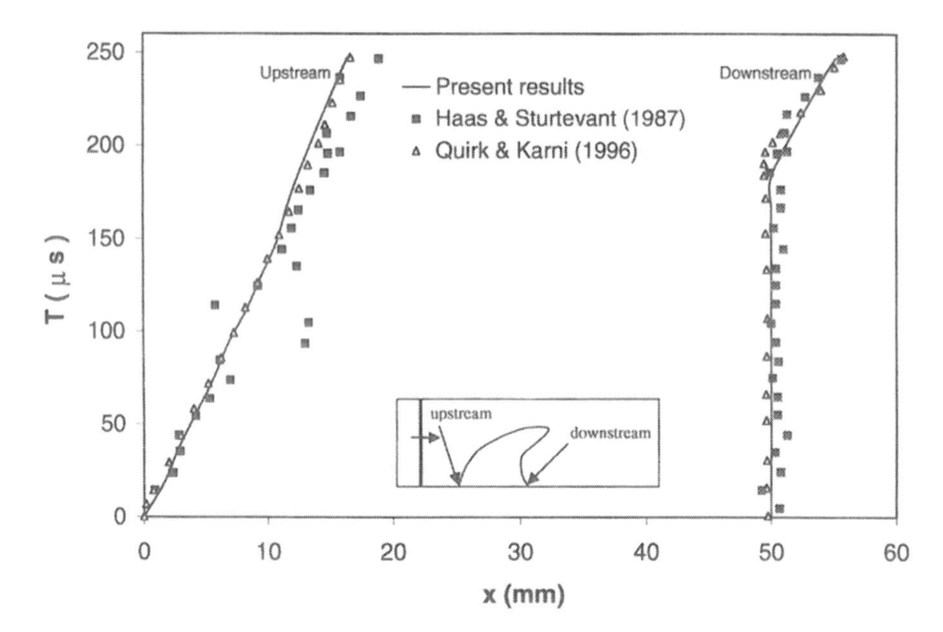

Figure 2.7. $x - t$ diagram for the interaction of a shock wave $M_s = 1.22$ with a dense bubble (Bagabir 2000; Bagabir & Drikakis 2002); comparison with the experimental results of (Haas and Sturtevant 1987) and adaptive-grid high fidelity simulations of (Quirk and Karni 1996).

play a natural scale invariance. Indeed self-similarity is embedded in many high resolution methods through the use of Riemann solvers (either exact or approximate in nature). The Riemann solution provides much of the dissipation used to stabilize the numerical method.

It is the dominance of the transport (hyperbolic) terms that leads to turbulence. As hyperbolic terms become more important, the problems become more sensitive to initial conditions in the presence of hydrodynamic instability. In compressible flows, the scale-changing phenomena cause wave steepening and shock waves. As an example of the nature of scale changing phenomena consider Burgers' equation,

$$\frac{\partial u}{\partial t} + u\frac{\partial u}{\partial x} = 0 \rightarrow \frac{\partial u}{\partial t} + \frac{\partial}{\partial x}\left(\frac{1}{2}u^2\right) = 0,$$

and derive an equation for the evolution of the gradient. Under the assumption of sufficient smoothness (indeed this equation is used to determine when smoothness breaks down),

$$\frac{\partial}{\partial t}\left(\frac{\partial u}{\partial x}\right) + u\frac{\partial^2 u}{\partial x^2} + \left(\frac{\partial u}{\partial x}\right)^2 = 0.$$

It is the final term that controls the breakdown of smoothness when $\partial u/\partial x < 0$, that it is compressive. Here, a shock will form. Otherwise the flow expands in a rarefaction. The shock formation pushes information to small scales where dissipation asymptotically operates to create entropy. In rarefactions the information moves to larger scales where smoothness is not threatened.

High resolution schemes combine the action of limiters to judiciously allow high accuracy while still relying upon a combination of conservation form (Lax & Wendroff 1960), and entropy production to produce a unique weak solution. As shown in (Margolin & Rider 2001) the conservation naturally produces the equations for the evolution of a control volume of fluid. It is the evolution equation for the finite volume of fluid that then is solved by the numerical method. Dissipation acts to regularize the flow, thereby allowing shock propagation to proceed physically even while it is unresolved on the computational mesh. The key assumption is that the desired goal is to propagate the discontinuity on a fixed number of discrete mesh cells.

Here, we use the assumptions of large eddy simulation (LES). For example, we assume the numerical solution resolves the energy-containing range of the flow, i.e., that the grid scale is in the inertial range. In the following text, we will develop a description of the effective model defined by high resolution methods for hyperbolic PDEs. This will make use of the modified equations (Warming & Hyett 1974) describing the action of limiters in terms of nonlinear differential terms. For example, the modified equation can be used to show that upwind differencing has a leading order error that is dissipative in nature, i.e., proportional to u_{xx}. This will result in a description of nonlinear eddy viscosity, solution adaptivity and scale-similarity arising naturally from a broad class of numerical methods.

At the outset of this exposition it is useful to contrast certain aspects of turbulence modeling with numerical methods for hyperbolic PDEs. One computes hyperbolic PDEs with two competing criteria in mind: a desire for high accuracy coupled with protections against catastrophic failure due to nonlinear wave steepening or unresolved features. Nonlinear mechanisms (limiters) guard the method from such catastrophic failures by triggering entropy producing mechanisms that safeguard the calculation when the need arises. The question is the criteria used to design the nonlinear mechanism that triggers the entropy production. Godunov's theorem (Godunov 1959) provides the theoretical reasoning for having to employing nonlinear limiters. The theorem can be restated as the requirement that a scheme must be nonlinear in order to be higher than first-order accurate and monotone.

Another key point to emphasize about the numerical methods for hyperbolic PDEs: the theory for both the numerical methods and more importantly for the physics of the flow is quite well developed albeit in one dimension. The details of this physical theory have been well developed and are described by (Menikoff & Plohr 1989). There the connection between a thermodynamically consistent equation of state and hyperbolic wave structure is elucidated. This follows the mathematical description due to Lax (Lax 1971; Lax 1972) leading to the current numerical theory and analysis (Leveque 1990). This combination has culminated with the availability of powerful numerical methods for several decades pervading many application areas in physics and engineering. Open questions still exist in two or three dimensions, for example: are some multi-dimensional well-posed, i.e., stable (Lax & Liu 1998). In the case where the solution involves a vortex sheet, the solution to the two-dimensional Riemann problem shows a progression of greater and greater complexity as the mesh is refined. Could these sorts of solutions be related to turbulence?

Two issues need to be considered for the advancement of this concept: empirical evidence from the successful use of the idea and some sort of theoretical structure to build upon. Empirical evidence for this idea has been building for more than a decade as summarized by (Oran & Boris 2001). There several issues are discussed: the character of an ideal subgrid model, and some fortunate circumstances for high resolution schemes arising from physics. High resolution schemes have a number of convenient properties that make them attractive as general tools for modeling turbulence. This of course assumes that there is some accuracy and fidelity in such a model. We will develop some structural evidence that strong connections exist with current practice in LES modeling. Next, we detail some theoretical scaling arguments linking conceptually the physics of shock waves (implicitly built into the numerical methods focused upon here) and turbulence (explicitly built into turbulence modeling).

4. Theoretical Basis

As motivation for considering the utility of shock-capturing methods for turbulence, we consider the following similarity among many theoretical models. During the early 1940s, similar forms of dissipation were derived on both sides of the Atlantic. Kolmogorov (Kolmogorov 1941a; Kolmogorov 1941b) defined a dissipation of kinetic energy that was independent of the coefficient of viscosity in the limit of infinite Reynolds number; this theory was refined in (Kolmogorov 1962). In this form, the average time-rate-of-change of dissipation of kinetic energy, K, is given as

$$\langle K_t \rangle \, \mathcal{L} = \frac{5}{4} \left\langle (\Delta u)^3 \right\rangle . \tag{2.1}$$

In homogeneous, isotropic turbulence, this term is proportional to the average velocity difference at a length scale, $\mathcal{L}$, cubed. Note that this theory is analytic and independent of viscosity (although subtle arguments about viscous corrections at large, but finite Reynolds number persist today). Moreover, this theory provides a basis for the functional form of nonlinear eddy viscosity, i.e., (Smagorinsky 1963); this is discussed in more detail later.

In 1942, Bethe (Bethe 1998) derived the dissipation rate due to the passage of a shock wave (for a modern perspective on this relation see Menikoff & Plohr 1989). This rate depends on the curvature of the isentrope, $\mathcal{G}$, and on the cube of the jump of dependent variables across the shock:

$$T\Delta S = \frac{\mathcal{G}\rho^3 c^2}{6} (\Delta V)^3 \ , \tag{2.2}$$

where ρ is the density and c is the sound speed. Bethe defined this jump in terms of specific volume, V, but this can be restated in terms of velocity by applying the Rankine-Hugoniot conditions, $s\Delta V = -\Delta u$, where s is the shock speed. Both of these results are analytic. In each of these cases, the flow experiences an intrinsic asymmetry since the dissipative forces arise predominantly where velocity gradients are negative (i.e., compressive).

For Burgers' equation a similar result may be obtained (Gurbatov et. al. 1997; Bec et al. 2000),

$$\langle K_t \rangle \ \mathcal{L} = \frac{1}{12} \left\langle (\Delta u)^3 \right\rangle \ . \tag{2.3}$$

Again, this is an analytic result through the application of integration by parts, and the shock jump conditions. Next, we display the congruence of high resolution numerical methods with this theory. In a sense (2.3) is an entropy condition for "Burgers' turbulence" describing the minimum integral amount of inviscid dissipation for a physically meaningful solution. This dissipation is produced at the shocks and is a consequence of and proportional to the jump in dependent variables.

We will briefly discuss the context of these theoretical structures in thinking about subgrid modeling and the design of numerical methods with implicit models. Eyink (Eyink 1995) studied a conjecture by Kraichnan that the dissipation of kinetic energy as defined by the Kolmogorov similarity is both local as well as integral in nature (by definition, the shock dissipation is local). These regularizations are the essence of the physical conditions that numerical methods must reproduce correctly. It is this idea, viz., the existence of a finite rate of dissipation independent of viscosity with an inherently local nature, that numerical methods can reproduce. Modern high-resolution methods have an effective subgrid model that is inherently local. In addition, the algebraic form of the high resolution methods has a great deal in common with scale-similarity forms of LES subgrid models coupled with a nonlinear eddy viscosity. This creates a coherent tie between the modern high resolution shock capturing methods and LES subgrid models.

Next, we will describe the similarities between modern numerical methods and LES models. One can show that control volume differencing can be viewed as a form of implicit spatial filtering. The consequence of control volume differencing is that the cell values are the cell average values for quantities, thus filtering the point values. The difference between point values and cell averages start a second-order,

$$\langle u\left(x\right)\rangle \approx u\left(x\right) + \frac{\partial^2 u}{\partial x^2} + \text{H.O.T.}$$

Furthermore, control volume differencing naturally produces terms that are analogous to scale-similarity subgrid models. This analogy is predicated upon the structure of the modified equations for this class of methods (Margolin & Rider 2001).

5. Modified Equations for Limiters

5.1 Background

Limiters are the general nonlinear mechanism that distinguishes modern methods from classical linear schemes. These are sometimes referred to as flux limiters or slope limiters, but their role is similar: to act as a nonlinear switch between more than one underlying linear methods thus adapting the choice of numerical method based upon the behavior of the local solution. The general practice is to base the analysis of the nonlinear method on the linear analysis of the available methods to be chosen by the limiter. Here, we depart from this view and include the limiter in the analysis providing a nonlinear truncation error analysis even when the equation being solved in linear. Moreover, this shadows Godunov's theorem which motivates the utilization of nonlinear methods for linear equation in order to achieve second-order accuracy simultaneously with monotonicity.

The truncation error of a method is an important indicator of its performance. Here, we advocate a more active point of view: that one should *design* the non-linear truncation error to better mimic turbulence modeling. The "modified equation" is the PDE that a numerical method effectively solves, including leading-order estimates of the truncation error. In this sense, the modified equation describes specific *continuum* properties of a scheme. This analysis substitutes the Taylor series for the finite difference approximation (see Warming & Hyett 1974). Usually the differential equation is used to convert all time derivatives to space derivatives.

While such properties are important, one must also remain mindful of important *discrete* properties to maintain such as *monotonicity*, *total Variation Diminishing (TVD)*, *conservation*, and *entropy condition*. A numerical scheme is considered as monotone if it does not lead to an oscillatory behavior of the numerical solution. The monotonicity condition is also strongly linked to the concept of bounded total variation which says that the total variation (TV)

$$\text{TV} = \int |\frac{\partial u}{\partial x}| dx$$

of any physically admissible solution u does not increase in time. Conservation implies that the mass, momentum and total (internal plus kinetic) energy integrated over the computational space do not change due to the algorithm.

Finally, the *entropy condition* is related to the fact that certain numerical schemes provide solutions not physically acceptable, since they are associated with a decrease in entropy which is not, however, allowed by the second law of thermodynamics. The importance of designing methods with exact local conservation with proper entropy production is founded upon the Lax-Wendroff theorem (Lax & Wendroff 1960). This theorem that if one has a method in discrete conservation form and it converges, it converges to a weak solution. The entropy condition is necessary to choose the physically meaningful weak solution among the infinite set of possible weak solutions. One possible way for designing schemes which satisfy the above properties and at the same time provide higher order of accuracy is the flux limiter approach (Sweby 1984). Note that monotone methods are at most first order accurate (Godunov 1959). The concept of flux limiters for designing TVD schemes is linked to artificial viscosity method, invented by von Neumann and Richtmyer (von Neumann & Richtmyer 1950). In both cases the objective is the regularization of the numerical solution to that selected via vanishing viscosity.

There is a close connection of the differential forms used in the von Neumann-Richtmyer artificial viscosity and the Smagorinsky eddy viscosity often used in LES. It is often noted that Smagorinsky viscosity simplifies to von Neumann-Richtmyer in one dimension. Indeed, this connection is more explicit than is commonly appreciated, as the original motivation for Smagorinsky's viscosity was to use a nonlinear von Neumann-Richtmyer viscosity to stabilize calculations (Smagorinsky 1983). These two forms of dissipation differ chiefly in the detailed form of their nonlinear terms in multiple dimensions.

Since its conception the von Neumann-Richtmyer viscosity has become significantly more sophisticated, for a good recent example see (Caramanna et al. 1998). For example, approximately coincidentally with the time when it was suggested by Charney that Smagorinsky employ an artificial viscosity in weather simulations, Rosenbluth suggested that artificial viscosity be turned off in rarefactions (usually on expansion, $\nabla \vec{u} > 0$) (Richtmyer & Morton 1967). This acknowledges that rarefactions are inherently adiabatic and dissipation is not physically motivated in an inviscid fluid (except in the sense of vanishing viscosity consistent with the second law of thermodynamics). Further extensions are noted in (Dukowicz 1986) where Kurapatenko's analysis of artificial viscosity with a shock Hugoniot suggests analytic coefficients for the viscosity. Other more recent developments are detailed in (Benson 1992) including limiters that turn a viscosity off should the flowfield be considered resolved by the numerical derivatives.

One last significant factor is the nature of the dissipation. The multidimensional extension of von Neumann viscosity was done isotropically by Smagorinsky. In general, the nonlinear viscosity resulting from artificial viscosity or high resolution methods is anisotropic. Isotropic models for artificial viscosity do however exist (Benson 1992). Typically, this anisotropic viscosity is proportional to the gradient of the normal velocity in the direction chosen. The general nature of this formalism itself distinguishes this nonlinear viscosity from the typically applied modeling approach. Other types of viscosity are patterned after physical Navier-Stokes viscosity having an isotropic multidimensional form or are rotated into the frame of a shock much like rotated Riemann solvers.

Further similarities exist with the form of the third-order terms found in the Camassa-Holm equations (Camassa & Holm 1993). Because the Camassa-Holm equations imply a dissipation that results from time-averaging determined by dynamical theory, there is a strong connection between the entropy production and the proper nonlinear dissipative form. Such observations suggest that these numerical methods as well as turbulence models all share common dynamical mechanisms for producing entropy. This in turn leads to a similar scaling of solutions and behavior.

5.2 Modified Equation Analysis for Limiters

Here, we will show how numerical flux limiters can act like a dynamic, self-adjusting model, modifying the numerical viscosity to produce a nonlinear eddy viscosity. The limiters, ϕ, can be cast in a differential form resulting from their modified equations. In the case of a sign-preserving limiter, the form is $1 - C\Delta x \left|u_x/u\right|$ and for a monotone (minmod) limiter it $1 - C\Delta x \left|u_{xx}/u_x\right|$. The sign-preserving limiter produces a form for the viscosity that is similar to Smagorinsky's nonlinear viscosity. With a small amount of reintepretation, a broad class of modern numerical methods can be viewed as dynamic mixed LES models.

Consider the construction of a monotone (minmod) limiter from three neighboring points u_{j-1}, u_j and u_{j+1}. The limiter has the form,

$$\frac{\partial u}{\partial x} \approx \max\left[0, \frac{S_j}{\Delta x}\min\left(\left|u_{j+1}-u_j\right|, \left|u_j - u_{j-1}\right|\right)\right], \tag{2.4}$$

where $S_j = \text{sign}\left(u_{j+1} - u_{j-1}\right)$. Replacing the discrete data with Taylor series and retaining the leading order of the expansion gives

$$\frac{\partial u}{\partial x} \approx S_j \max\left[0, \min\left(\left|\frac{\partial u}{\partial x} + \frac{\Delta x}{2}\frac{\partial^2 u}{\partial x^2}\right|, \left|\frac{\partial u}{\partial x} - \frac{\Delta x}{2}\frac{\partial^2 u}{\partial x^2}\right|\right)\right], \tag{2.5}$$

where $S_j = \text{sign}\left(\frac{\partial u}{\partial x}\right)$. Away from extrema and factoring out the derivative gives

$$\frac{\partial u}{\partial x} \approx \phi \frac{\partial u}{\partial x}, \tag{2.6}$$

where

$$\phi = 1 - \frac{\Delta x}{2} \left| \frac{\frac{\partial^2 u}{\partial x^2}}{\frac{\partial u}{\partial x}} \right| .$$

Of course at extrema a monotone limiter will choose $\partial u/\partial x \approx 0$. In the case of simplest second-order ENO scheme (Harten et al. 1987), the above limiter is used everywhere in the flow because the limiter effectively chooses the smallest derivative in absolute value.

Other limiters can be treated similarly. For example, the van Leer limiter,

$$\frac{\partial u}{\partial x} \approx \frac{2\left(u_{j+1} - u_j\right)\left(u_j - u_{j-1}\right)}{\Delta x \left[\left(u_j - u_{j-1}\right) + \left(u_{j+1} - u_j\right)\right]},$$

is essentially the harmonic mean of the gradients yields a modified equation for the limiter of

$$\phi = 1 - \left(\frac{u_{xx}}{2u_x \Delta x}\right)^2 . \tag{2.7}$$

The van Albada limiter can be similarly treated (it is the weighted average of local gradients) and gives a modified equation of

$$\phi = 1 - \frac{1}{2}\left(\frac{u_{xx}}{u_x \Delta x}\right)^2 . \tag{2.8}$$

We can show that these forms play an essential role in the effective dissipation of a scheme as well as acting as a trigger for a dynamic dissipative scheme. Also, we can see the implicit correspondence between some limiter forms to dynamic LES models in which the viscosity is adjusted locally based on whether the flow exhibits a similar structure at adjacent length scales. The limiters provide additional utility by comparing several local estimates of a derivative. If these estimates are close enough in magnitude, the flow is treated as being resolved, allowing the method to detect smooth (laminar) flow. This limiter can be most as

$$\frac{\partial u}{\partial x} \approx \frac{S_j}{\Delta x} \max\left[0, \min\left(2\left|u_{j+1} - u_j\right|, 2\left|u_j - u_{j-1}\right|, \frac{1}{2}\left|u_{j+1} - u_{j-1}\right|\right)\right], \tag{2.9}$$

following (2.4). Again substituting in the Taylor series expanding then factoring out the derivative gives

$$\phi = \min\left(1, 2 - \frac{\Delta x}{2}\left|\frac{\frac{\partial^2 u}{\partial x^2}}{\frac{\partial u}{\partial x}}\right|\right) \tag{2.10}$$

away from extrema.

One can also produce a limiter of lower differential order that preserves the sign of the data rather than monotonicity. This limiter has the general form,

$$\phi = 1 - \left| \frac{\Delta x \frac{\partial u}{\partial x}}{u} \right|, \tag{2.11}$$

In actuality this function is implemented using $\phi := \max(0, \phi)$. The same general form can be derived for the multidimensional positive definite advection transport algorithm (MPDATA) scheme using the modified equations analysis (Smolarkiewicz& Margolin 1998; Margolin & Rider 2001). This limiter also arises from considering a sign-preserving extension of a monotone limiter (Rider & Margolin 2001). In one dimension the sign-preserving limiter is written

$$\frac{\partial u}{\partial x} \approx \frac{S_j}{\Delta x} \min\left(\frac{1}{2} \left| u_{j+1} - u_{j-1} \right|, 2u_j \right) \tag{2.12}$$

assuming that u_j is positive. This can be converted to a limiter as before,

$$\phi = \min\left(1, \left| \Delta x \frac{2u}{\frac{\partial u}{\partial x}} \right| \right). \tag{2.13}$$

So when this limiter is active it will degrade accuracy although away from zero the limiter will preserve the design accuracy of the linear scheme. One can show that this scheme forms an envelope for the earlier sign-preserving schemes. The following expression intersects the bounding scheme,

$$\phi = 1 - \left| \frac{\Delta x \frac{\partial u}{\partial x}}{8u} \right|. \tag{2.14}$$

The major difference in the bounding scheme and the initial scheme is that the bounding scheme creates a constant derivative estimate equal to the cell value while the original scheme gradually degrades to first-order as the gradients become steeper as normalized by the cell values (and more completely preserve second-order accuracy).

One can analyze these schemes in a similar manner to Sweby's analysis of TVD schemes (Sweby 1984). There the limiter was defined in terms of the ratio of gradients, $r = (u_{j+1} - u_j) / (u_j - u_{j-1})$. We will use the difference used for the gradient normalized by the cell value, $q = 2(u_j) / (u_{j+1} - u_j - 1)$, in the place of r. The bounding sign-preserving scheme is then $\phi = \min(1, 2/|q|)$.

Recently, flux limiters for third-order upwind TVD schemes have been proposed (Drikakis 2001b). The analytic derivation of limiters for nonlinear equations such as the Navier-Stokes equations seems a rather impossible task, but may be in reach with symbolic algebra programs; however, following Sweby's

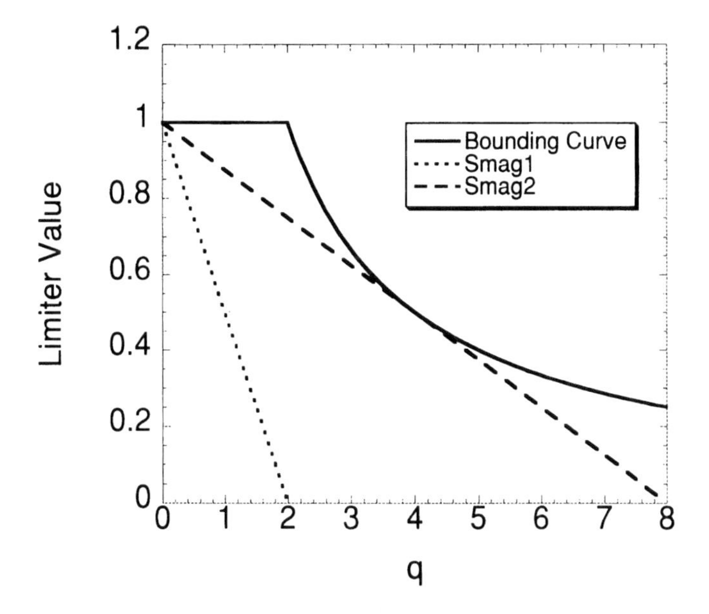

Figure 2.8. The plot of the sign preserving limiter against the bounding function, (2.13). The labels "Smag1" and "Smag2" stand for limiters that produce the Smagorinsky-like nonlinear diffusion. These limiters are shown as Equations (2.12) and (2.14) respectively in the text.

analysis (Sweby 1984) limiters can be derived on the basis of linear equations such as the linear advection equation: $u_t + f_x = 0$, where $f = au$ and a is a constant.

For an upwind Godunov scheme the flux f can be discretized by

$$f^{hi} = \frac{1}{2}\Big([1 + sign(a)]f_L + [1 - sign(a)]f_R\Big), \tag{2.15}$$

where f_L and f_R can be second- or third-order interpolated values of the flux f, e.g.

$$f_L = \frac{1}{6}(5f_i - f_{i-1} + 2f_{i+1}), \qquad f_R = \frac{1}{6}(5f_{i+1} - f_{i+2} + 2f_i)\,, \tag{2.16}$$

The TVD formulation of f will then be written as

$$f^{TVD}_{i+1/2} = f^{lo}_{i+1/2} + \psi_{i+1/2}(f^{hi}_{i+1/2} - f^{lo}_{i+1/2}), \tag{2.17}$$

where $f^{hi}_{i+1/2}$ is the high-order flux (2.15) and $f^{lo}_{i+1/2}$ is the flux of a first-order monotone scheme, e.g. Lax-Friedrichs flux (Lax 1954), and $\psi_{i+1/2}$ is a flux limiter function. Further analysis (Drikakis 2001b) leads to the derivation of a *characteristic-based limiter*

$$\psi_{i+1/2} = \begin{cases} \dfrac{3(1-|C|)}{3-|C|} & \text{if} \quad r \leq 0 \\ \dfrac{3(1-|C|)}{3-|C|}(r+1) & \text{if} \quad 0 \leq r \leq \dfrac{|C|(7-3|C|)}{(3-2|C|)(1-|C|)} \\ \dfrac{3(1+|C|)}{3-2|C|} & \text{if} \quad r > \dfrac{|C|(7-3|C|)}{(3-2|C|)(1-|C|)} \end{cases} \tag{2.18}$$

where

$$r = \frac{\Delta_{upw}}{\Delta_{loc}} = \begin{cases} \dfrac{u_i^n - u_{i-1}^n}{u_{i+1}^n - u_i^n}, & a > 0\,, \\ \dfrac{u_{i+2}^n - u_{i+1}^n}{u_{i+1}^n - u_i^n}, & a < 0\,. \end{cases} \tag{2.19}$$

and $C = a\Delta t/\Delta x$ is the Courant-Friedrichs-Lewy (CFL) number. Results using (2.18) are presented in Section 7.

6. Towards an Implicit Turbulence Model

It is the common (mis) perception that the implicit numerical viscosity is all that high resolution methods have to offer. We will show that methods can have much more in common with LES models than simple nonlinear viscosity. Indeed various elements of other types of subgrid models are found when looking at these method through the lens of nonlinear truncation error.

Below, we expand on this idea and show how high resolution methods contain an implicit subgrid model that can be viewed as a dynamic mixed self similarity model of sorts. The dynamic aspect is associated with limiters whose effects vanish in resolved flows. The limiters are essential in modifying second-order dissipative terms into high-order nonlinear viscosity so commonly associated with this class of methods. Self-similarity comes from the slope limited interpolation (or similar concepts) which dynamically changes the nature of the high-order flux used locally.

To fully assess the similarities of high-resolution methods with LES models, we will focus on the numerical interfacial flux, F, constructed in the fashion of a Godunov method, in these methods as

$$F_{LR} = \frac{1}{2}(F_L + F_R) - \frac{1}{2}|A|(U_R - U_L)\,, \tag{2.20}$$

where L and R denote the left and right states in the Riemann solution, U is the array of flow variables, and A is the flux Jacobian. The absolute value of

A can be found via an eigen-decomposition, $A = \mathcal{R}\,|\lambda|\,\mathcal{L}$, where are the right eigenvectors, eigenvalues and left eigenvectors, respectively, of A. This is a fairly generic and standard manner to introduce upwinding into a numerical method (17). The states can be accessed via interpolation from cell centered to the edges this is the reconstruction step of a high resolution Godunov method. The left state at a cell edge is

$$U_{j+\frac{1}{2},L} = U_j + \frac{h}{2}\frac{\partial U_j}{\partial x},$$

and the right state is

$$U_{j+\frac{1}{2},R} = U_{j+1} - \frac{h}{2}\frac{\partial U_{j+1}}{\partial x},$$

Thus, the states are extrapolated to a common cell edge from adjacent cell centers. Note, that the interpolation will usually be limited so that

$$\frac{\partial U_j}{\partial x} := \phi_j \frac{\partial U_j}{\partial x},$$

with limiters that have the above stated properties. The flux can be generally decomposed into terms that are hyperbolic and that are dissipative in nature. The portion that is the sum of the local contributions (the mean flux) is hyperbolic, while those proportional to the difference in the variables is dissipative with a magnitude proportional to the coefficient of numerical viscosity. We can use this effective decomposition to identify what the physical effect of various algorithmic components are.

It is important to note where the above stated mis-perception arose with respect to these methods only providing nonlinear dissipation. Perhaps the most commonly known category of these methods are TVD methods often written in the following form (in semi-discrete form),

$$F_{j+\frac{1}{2}} = \frac{1}{2}\left(F_j + F_{j+1}\right) - \frac{1}{2}\left|A\right|\left(1-\phi\right)\left(U_{j+1} - U_j\right). \tag{2.21}$$

Basically, this is viewed in the following way: the flux is a second-order centered flux with a dissipative term that yields first-order upwinding that is triggered by a limiter, ϕ. Thus, the action of the high resolution scheme is entirely dependent upon the limiter acting on the numerical viscosity. These certainly are valid and useful schemes, but there is more variety to the effective subgrid models as we now elaborate.

First, consider the nonlinear truncation error arising from the discretization of the following equation

$$\frac{\partial U}{\partial t} + \frac{\partial F\left(U\right)}{\partial x} = 0 \rightarrow \frac{\partial U}{\partial t} + \frac{\partial F\left(U\right)}{\partial U}\frac{\partial U}{\partial x} = 0. \tag{2.22}$$

This equation using a first-order upwind scheme including the leading order (spatial) truncation error is

$$\frac{\partial U}{\partial t} + \frac{\partial F(U)}{\partial U}\frac{\partial U}{\partial x} = \frac{\Delta x}{2}\left[\left|\frac{\partial F(U)}{\partial U}\right|\frac{\partial^2 U}{\partial x^2} + \text{sign}\left(\frac{\partial F(U)}{\partial U}\right)\frac{\partial^2 F(U)}{\partial U^2}\left(\frac{\partial U}{\partial x}\right)^2\right] \tag{2.23}$$

The first term on the right hand side is the second-order dissipation most commonly associated with this method whereas the second term is primarily dispersive in character and produces oscillations near discontinuities. Perhaps most notable about this form for the truncation error and those that follow is that the nonlinear terms have no counterparts for the linear truncation error. Since many high order methods were designed with linear analysis, their behavior on general nonlinear problems may be unpredictable or even unstable.

Next, we show the leading order spatial truncation error for Lax-Wendroff (second-order in space) differencing,

$$\frac{\partial U}{\partial t} + \frac{\partial F(U)}{\partial U}\frac{\partial U}{\partial x} = \Delta x^2\left[-\frac{1}{6}\frac{\partial F(U)}{\partial U}\frac{\partial^3 U}{\partial x^3} - \frac{1}{2}\frac{\partial^2 F(U)}{\partial U^2}\frac{\partial U}{\partial x}\frac{\partial^2 U}{\partial x^2} - \frac{1}{6}\frac{\partial^3 F(U)}{\partial U^3}\left(\frac{\partial U}{\partial x}\right)^3\right]. \tag{2.24}$$

The terms on the right hand side have mixed effect although it is dominated by dispersive effects. The last linear scheme we describe is Fromm's scheme which gives a spatial truncation error of,

$$\frac{\partial U}{\partial t} + \frac{\partial F(U)}{\partial U}\frac{\partial U}{\partial x} = \Delta x^2\left[-\frac{1}{12}\frac{\partial F(U)}{\partial U}\frac{\partial^3 U}{\partial x^3} + \frac{1}{24}\frac{\partial^3 F(U)}{\partial U^3}\left(\frac{\partial U}{\partial x}\right)^3\right]. \tag{2.25}$$

Again, the errors are dominated by primarily dispersive terms although they are smaller in magnitude than Lax-Wendroff.

By using a limiter to hybridize the upwind and Lax-Wendroff method we can accentuate the dissipative effects seen in the leading order spatial truncation error. The MPDATA (Smolarkiewicz& Margolin 1998) or sign-preserving limiter has the form $(1-\phi)$ where $\phi \approx \left|\Delta x \frac{\partial U}{\partial x}/U\right|$ and produces a nonlinear truncation error similar to that derived in (Margolin & Rider 2001),

$$\frac{\partial U}{\partial t} + \frac{\partial F(U)}{\partial U}\frac{\partial U}{\partial x} = \tag{2.26}$$

$$\Delta x^2\left[-\frac{1}{6}\frac{\partial F(U)}{\partial U}\frac{\partial^3 U}{\partial x^3} - \frac{1}{2}\frac{\partial^2 F(U)}{\partial U^2}\frac{\partial U}{\partial x}\frac{\partial^2 U}{\partial x^2} - \frac{1}{6}\frac{\partial^3 F(U)}{\partial U^3}\left(\frac{\partial U}{\partial x}\right)^3\right.$$

$$+\frac{1}{2}\left|\frac{\partial F(U)}{\partial U}\frac{\partial U}{\partial x}\right|\frac{\partial^2 U}{\partial x^2} + \frac{1}{2}\text{sign}\left(\frac{\partial F(U)}{\partial U}\right)\frac{\partial^2 F(U)}{\partial U^2}\left(\frac{\partial U}{\partial x}\right)^3 \Bigg]$$

Two new terms are added to the error. One is chiefly dissipative and the other is more dispersive. The nonlinear viscous term augments the $U_x U_{xx}$ term when the gradient is negative and counteracts the anti-dissipation when the gradient is positive. Note that the present *ad hoc* form will be mismatched if $\frac{\partial^2 F(U)}{\partial U^2} \neq 1$.

The reason for the somewhat more complicated form for the dissipative term will become clear as we introduce limiters to the analysis. There limiters will be seen to introduce standard sorts of nonlinear eddy viscosities through the numerical flux. If the limiters are not active, the dissipative term will vanish and in the case of second-order methods a fourth-order dissipation will result. Next recognize that this term will be paired with another flux, $F_{j-\frac{1}{2}}$ to update p_j and divided by h. This will give an approximation to the differential term (providing consistency), and some additional terms that have potential physical significance. Let us now examine each of these terms for their commonality with LES models.

We can easily see that the terms in the hyperbolic portion of the flux are like those used in the self-similar gradient model (Pope 2000). This can be seen by writing the $U_x U_{xx}$ term in conservation form, $\left[(U_x)^2\right]_x$. This term may have the effect of dissipation or anti-dissipation dependent upon the local gradients. These terms are present in Fromm's scheme, but cancel when the divergence of fluxes is taken. In the same vein, a limiter can provide an eddy viscosity term, like $[|U_x|\, U_x]_x$. For a stable integration these terms must be controlled by the dissipation found in the second term in the flux. This decomposition of states is analogous to a derivation of Reynolds' stresses.

One key difference is the use of nonlinear limiters in the decomposition adding a dynamic nature to the approximations. It is also important to recognize that these terms combine both a dissipative and dispersive character. To see this consider the evolution equation for $S = U_x$,

$$\frac{\partial S}{\partial t} + \frac{\partial F(U)}{\partial U}\frac{\partial S}{\partial x} + \frac{\partial^2 F(U)}{\partial U^2} S^2 = 0. \tag{2.27}$$

It is the term $\frac{\partial^2 F(U)}{\partial U^2} S^2$ that leads to a finite time singularity via a process like a Ricatti ODE depending on the sign of S and the curvature of the flux function. Now regularize the original PDE, (2.22), using $\left[(U_x)^2\right]_x$, and convert this to a modification of (2.27). The regularized equation for the evolution of S now reads

$$\frac{\partial S}{\partial t} + \frac{\partial F(U)}{\partial U}\frac{\partial S}{\partial x} + \frac{\partial^2 F(U)}{\partial U^2} S^2 = |S|\,\frac{\partial^2 S}{\partial x^2} + \text{sign}(S)\,(S_x)^2\,. \tag{2.28}$$

The right hand side of (2.28) now contains two terms one that is clearly dissipative, and a second term that is dispersive. It is structured much like the leading order error terms in upwinding and the evolution of the gradient will behave similarly. A similar analysis can be done on the term,

$$\frac{\partial F(U)}{\partial U}\frac{\partial^3 U}{\partial x^3}.$$

This yields the following form in terms of the evolution of S,

$$\frac{\partial F(U)}{\partial U}\frac{\partial^3 S}{\partial x^3} + \frac{\partial^2 F(U)}{\partial U^2} S \frac{\partial^2 S}{\partial x^2}$$

showing that it has both the dispersive behavior (the first term) along with (anti) dissipative behavior depending on the curvature of $F(U)$ and the sign of S.

The limiter is a function of local data. The expansions using the limiters here are carried out differently in each computational cell, thereby making the effective modeling inherently local. In modern methods, these limiters are nonlinear "switches" that are used to preserve monotonicity; see (Leveque 1990) for a review). These methods can also be viewed as a mechanism to change computational stencils advantageously throughout a calculation on the basic of local solution behavior.

This provides a local subgrid model although it will provide a global nature if combined with an implicit solution. This is most acutely true if one is solving incompressible flow with such a method. The effect of the subgrid model is globalized through the pressure solution.

7. Numerical Experiments: Burgers' Equation

Our first example uses an initial condition that evolves into a single shock, $u(x,0) = \sin(2\pi x), x \in [0,1]$. This example will produce a shock, for the purposes of this study we examine the congruence of the numerical results with (2.3). The flow is evolved to a time of one with the shock forming at $t = 1/2\pi$. The results are shown in space and statistically in Fig. (2.9). All of the high resolution methods show a good degree of agreement with the theory. One can also view this result as a sort of entropy condition where the law defined by (2.3) is the minimum dissipation that is physical. As the mesh is refined, the minimum for the function approaches one, in agreement with the inviscid theory, and at larger scales the values are converging toward the DNS result (the solid black line). This demonstrates that the methods produce physically meaningful solutions in the appropriate limits.

Next, we show Fig. (2.10) the results for a multimode initial condition, defined by random Gaussian numbers. The flow is evolved to a time of one (shocks form virtually instantaneously). The solution and congruence with

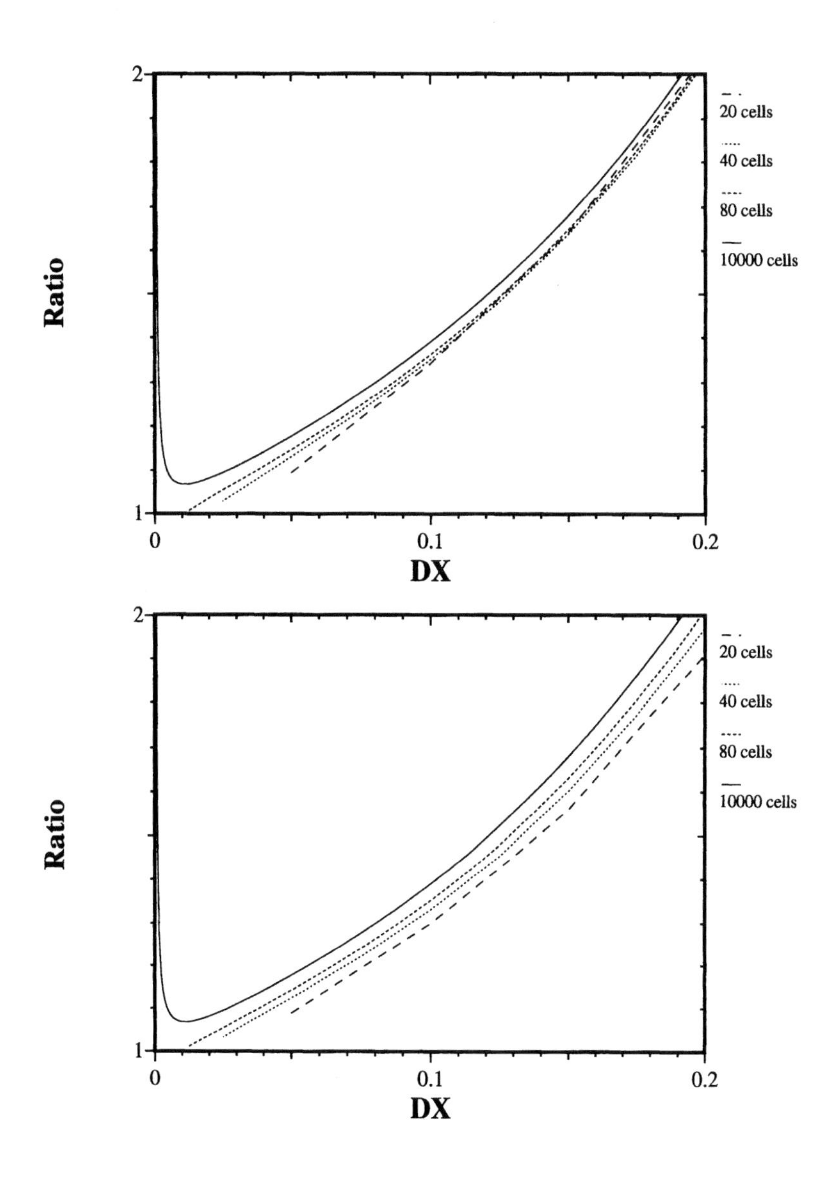

Figure 2.9. The single mode solution and agreement with (2.3) for several high resolution schemes. One the top a minmod TVD scheme is shown and on the bottom is MPDATA.

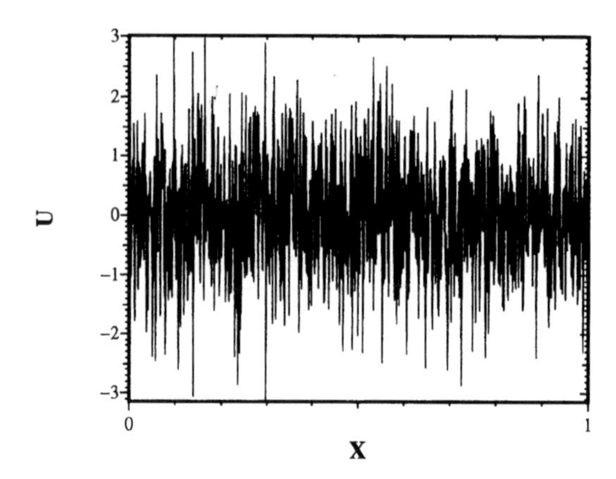

Figure 2.10. The random multimode initial condition where the data is normalized to give a magnitude of unity.

(2.3) is shown in Fig. (2.11). Again, the high resolution methods all display agreement with the analytical result.

The behavior of high resolution methods and SGS models in turbulent flow simulations can be examined on the basis of Burgers' turbulence (Kraichnan 1968). We have considered a random initial condition for the velocity Fig. (2.12) which exhibits maximum value of the wave spectrum at $log(k) = 1.283$. The velocity has become dimensionless by defining a characteristic length scale $L_o = 1/log^{-1}(1.283)L = 0.052L$ (where L is an arbitrary unit of length; here $L = 1$), and a characteristic velocity u_o as the root mean square of the initial condition. The viscosity ν can then be defined by $\nu = (L_o u_o)/Re$, where Re is the Reynolds number. We have conducted simulations for $Re = 6,000$ in a domain of length $l = 12L = 12$, using a very fine grid (9,000 grid points) and a very small time step ($\Delta t = 0.0001$). The obtained solution (henceforth labeled DNS or "Direct Numerical Solution") is grid and time-step independent and can thereby be considered as the exact solution.

We have carried out coarsely resolved simulations on a 700×100 space-time grid using different numerical schemes with and without different SGS models. Specifically, we have used: i) the characteristic-based (Godunov-CB) scheme of (Drikakis 2001a) without a SGS model; ii) the TVD-CB scheme of (Drikakis 2001b) without a SGS model; iii) the CB scheme in conjunction with the modified version of the dynamic SGS model (Lilly 1992) – the solution is labeled "D-Model"; iv) the CB scheme in conjunction with the structure-function SGS model (Métais & Lesieur 1992) – the solution is labeled "SF-Model". The results for the kurtosis distribution Fig. (2.12) reveal that: i) modeling the unresolved scales through a SGS model does not always improve the results; for example, compare the Godunov-CB solutions with and without the dynamic model; ii) high resolution schemes designed to satisfy the total variation dimin-

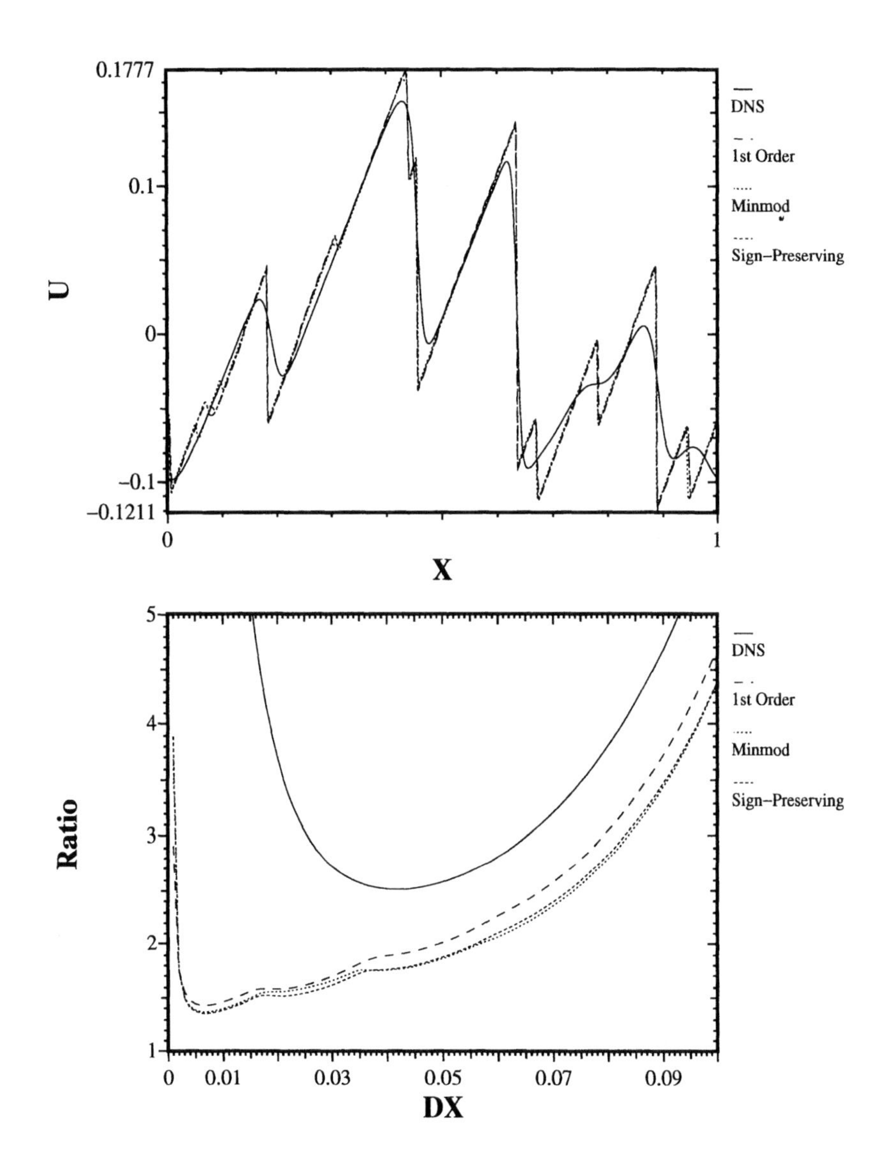

Figure 2.11. The random multimode solution and agreement with (2.3) for several high resolution schemes.

ishing (TVD) condition can significantly improve the predictions without even using a SGS. Further, in Table 1 we compare the average kurtosis value for the various simulations including the case where the TVD-CB scheme is used in conjunction with the structure-function SGS model. The simulation based upon the TVD-CB scheme without a SGS model gives the closest agreement with the DNS solution.

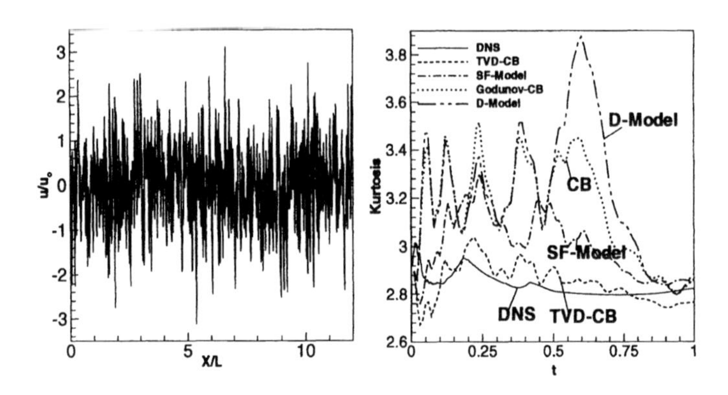

Figure 2.12. Simulation of Burgers' turbulence (Drikakis 2001b). The left plot shows the initial condition and the right plot the kurtosis distributions using various combinations of numerical schemes and SGS models (see text for details).

Table 2.1. Average kurtosis for different Godunov-type schemes and SGS models.

Method	DNS	TVD-CB+SF	TVD-CB	CB	CB+SF	CB+D
Average Kurtosis	2.8371	2.8131	2.8487	3.1515	3.	3.2192

We have investigated the impact the effective differential form for the numerical methods that we introduced above in (Margolin & Rider 2001). original equations. As demonstrated in (Margolin & Rider 2001), one can simulate the modified equations directly. In this manner we can investigate the performance of the subgrid model in direct manner. The equation that is simulated is

$$\begin{aligned}\frac{\partial u}{\partial t} + \frac{1}{2}\frac{\partial}{\partial x}\left(u^2\right) - \nu\frac{\partial^2 u}{\partial x^2} &= \frac{\partial}{\partial x}\left(\alpha\left(\frac{\partial u}{\partial x}\right)^2\right) \\ + \frac{\partial}{\partial x}\left(\beta\left|\frac{\partial u}{\partial x}\right|\frac{\partial u}{\partial x}\right) &+ \frac{\partial}{\partial x}\left(\left|\frac{u u_{xx}}{u_x}\right|\frac{\partial u}{\partial x}\right) + \frac{\partial}{\partial x}\left(\delta u\frac{\partial^2 u}{\partial x^2}\right) \\ &+ \frac{\partial}{\partial x}\left(\xi\left|u\right|\frac{\partial^3 u}{\partial x^3}\right),\end{aligned} \tag{2.29}$$

where the left-hand side is the viscous Burgers' equation and the right hand side are the modeling/truncation error terms. For more details on the investigation of this equation see (Margolin & Rider 2001).

8. Closing Remarks

We have described both the current state and the challenge of using high resolution methods for turblent flows. Experimental validation shows both success and failure in the challenging regime of multimaterial mixing. Theoretical development of effective models embedded in numerical methods has been described. Future work must focus on endowing these effective models with greater physical relevance and insight. Moreover, numerical challenges arising from the presence of wall boundaries must be investigated. The end target for these methods are flows where the theory of turbulent flows is lacking due to material interfaces, compressibility and other complications. Ultimately the utility of these methods will be measured in these regimes and in more ideal circumstances where our theoretical understanding is more complete.

Acknowledgments

This work is available as Los Alamos National Laboratory report LA–UR-01–5196, and WJR's work was performed at Los Alamos National Laboratory, which is operated by the University of California for the United States Department of Energy under contract W-7405-ENG-36. WJR would like to thank James Kamm, Len Margolin, Darryl Holm, Robert Lowrie, and Beth Wingate for useful comments, and fruitful collaborations on related research. WJR especially thanks Tim Clark for the LBM results and Guy Dimonte for the experimental images from the LEM. DD would like to thank Ahmed Bagabir for fruitful collaboration.

References

A. Bagabir (2000), On the accuracy and efficiency of Godunov -type methods in various compressible flows, *PhD Dissertation*, Queen Mary, University of London.

A. Bagabir and D. Drikakis (2001), Mach number effects on shock-bubble interaction, *Shock Waves J.*,**11**:209-218.

A. Bagabir and D. Drikakis (2002), Performance of Godunov-type methods in time-dependent compressible flows, in preparation.

J. Bec, U. Frisch, K. Khanin (2000), Kicked Burgers' turbulence, *J. Fluid Mech.*, **416**:239–267.

D. J. Benson (1992), Computational Methods in Lagrangian and Eulerian Hydrocodes, *Computer Methods in Applied Mechanics and Engineering*, **99**: 235–394.

H. A. Bethe (1998), On the Theory of Shock Waves for An Arbitrary Equation of State, in *Classic Papers in Shock Compression Science*, J.N. Johnson & R. Cheret (eds.), Springer-Verlag, Berlin.

J. P. Boris (1989), in *Whither Turbulence? Turbulence at the Crossroads*, J. L. Lumley (ed.), Springer-Verlag, Berlin.

R. Camassa and D.D. Holm (1993), An integrable shallow-water equation with peaked solitons, *Phys. Rev. Lett.*, **71**:1661–1664.

E. J. Caramana, M. J. Shashkov and P. P. Whalen, Formulations of Artificial Viscosity for Multi-Dimensional Shock Wave Computations, *J. Comput. Phys.*, **144**:70–97.

G. Dimonte (1999), Nonlinear Evolution of Rayleigh-Taylor and Richtmyer-Morton Instabilities, *Phys. Plasmas*, **6**: 209–215.

G. Dimonte (2001), Personal Communication.

D. Drikakis (2001a), Uniformly high-order methods for unsteady incompressible flows, *Godunov Methods: Theory and Applications*, Kluwer Academic Publishers (ed. E.F. Toro), 263-283.

D. Drikakis (2001b), Numerical issues in very large eddy simulation, CD Rom Proceedings of the *ECCOMAS CFD 2001 Conference*, submitted to the International Journal for Numerical Methods in Fluids.

J. K. Dukowicz (1986), A General Non-Iterative Riemann Solver for Godunov's Method, *J. Comput. Phys.*, **61**:119–137.

G. Eyink (1995), Local energy flux and the refined similarity hypothesis, *J. Stat. Phys.*, **78**:335–351.

S. K. Godunov (1959), Finite Difference Method for Numerical Computation of Discontinuous Solutions of the Equations of Fluid Dynamics, *Matematicheski Sbornik*, **47**:271–306.

S. N. Gurbatov, S. I. Simdyankin, E. Aurell, U. Frisch, G. Toth (1997), On the decay of Burgers' turbulence, *J. Fluid Mech.*, **344**:339–374.

J.F. Haas and B. Sturtevant (1987), Interaction of weak shock waves with cylindrical and spherical gas inhomogeneities, *J. Fluid Mech.*, **181**: 41-76.

A. Harten, P. D. Lax and B. van Leer (1983), On Upstream Differencing and Godunov-Type Schemes for Hyperbolic Conservation Laws, *SIAM Review*, **25**: 35–61.

A. Harten (1983), High resolution schemes for hyperbolic conservation laws, *J. Comput. Phys.*, **49**:357–393.

A. Harten, B. Engquist, S. Osher, S. Chakravarthy (1987), Uniformly High Order Accurate Essentially Non-oscillatory Schemes III *J. Comp. Phys.*, **71**: 231–303.

A. N. Kolmogorov (1941), The local structure of turbulence in incompressible viscous fluid at very high Reynolds number, *Dokl. Akad. Nauk SSSR*, **30**:538-541 (Reprinted Kolmogorov Anniversary Edition, 1991 *Proc. Roy. Soc. Lond.*, **A434**).

A. N. Kolmogorov (1941), Energy decay in locally isotropic turbulence, *Dokl. Akad. Nauk SSSR*, **31**:16-18 (Reprinted Kolmogorov Anniversary Edition, 1991 *Proc. Roy. Soc. Lond.*, **A434**).

A. N. Kolmogorov (1962), A refinement of previous hypotheses concerning the local structure of turbulence in a viscous incompressible fluid at high Reynolds number, *J. Fluid Mech.*, **13**:82–85.

R.H. Kraichnan (1968), Lagrangian-history statistical theory for Burgers' equation, *Phys. Fluids*, **11**: 265–277.

P.D. Lax (1954) Weak solutions of nonlinear hyperbolic equations and their numerical computation *Comm. Pure Appl. Math.*, **VII**: 159-193.

P. D. Lax (1971), Shock Waves and Entropy, *Contributions to Nonlinear Functional Analysis*, Academic Press, E. H. Zarantonello, Ed. 603–634.

P. D. Lax (1972), *Hyperbolic Systems of Conservation Laws and the Mathematical Theory of Shock Waves*, SIAM, Philadelphia, USA.

P. D. Lax & X.-D. Liu (1998), Solution of two-dimensional Riemann problems of gas dynamics by positive schemes, *SIAM J. Sci. Comput.* **19**(2):319–340.

P. D. Lax and B. Wendroff (1960), Systems of Conservation Laws, Communications in Pure and Applied Mathematics, **13**:217–237.

R. J. Leveque (1990), *Numerical Methods for Conservation Laws*, Birkhauser-Verlag, Basel.

D.K. Lilly (1992), A proposed modification of the Germano subgrid-scale closure method, *Phys. Fluids*, **4**: 633-635.

L. G. Margolin, P. K. Smolarkiewicz & Z. Sorbjan (1998), Large eddy simulations of convective boundary layers using nonoscillatory differencing, *Physica D*, **133**:390–397.

L. G. Margolin & W. J. Rider (2001), *A Rationale for Implicit Turbulence Modeling*, ECCOMAS September 2001, Submitted to the International Journal for Numerical Methods in Fluids, also see LA-UR-01-793.

O. Métais and M. Lesieur (1992), Spectral large-eddy simulations of isotropic and stably-stratified turbulence, *J. Fluid Mech.*, **239**:157-194.

R. Menikoff and B. J. Plohr (1989), The Riemann Problem for Fluid Flow of Real Materials, *Reviews in Modern Physics*, **61**:75–129.

E. S. Oran and J. P. Boris (2001), *Numerical Simulation of Reactive Flow*, Elsiever.

J.M. Picone and J.P. Boris (1988), Vorticity Generation by Shock Propagation through Bubbles in a Gas. *J. Fluid Mech.*, **189**: 23-51.

S. B. Pope (2000), *Turbulent Flows*, Cambridge University Press, Cambridge.

E. G. Puckett, A. S. Almgren, J. B. Bell, D. L. Marcus, W. J. Rider (1997), A Second-Order Projection Method for Tracking Fluid Interfaces in Variable Density Incompressible Flows, *J. Comp. Phys.*, **130**: 269–282.

J.J. Quirk and S. Karni (1996), On the dynamics of a shock-bubble interaction. *J. Fluid Mech.*, **318**: 129-163.

R. D. Richtmyer and K. W. Morton (1967), Difference Methods for Initial Value Problems, Wiley-Interscience.

W. J. Rider (1994), Approximate Projection Methods for Incompressible Flow: Implementation, Variants and Robustness, Los Alamos National Laboratory Report, LA–UR–94–2000.

W. J. Rider and D. B. Kothe (1998), Reconstructing Volume Tracking, *J. Comp. Phys.*, **141**: 112-152.

W. J. Rider (1999), An Adaptive Riemann Solver Using a Two-Shock Approximation, *Computer & Fluids*, **28**, pp. 741-777.

P. M. Rightley, P. Vorobieff, R. Martin & R. F. Benjamin (1999), Experimental-observations of the mixing transition in a shock-accelerated gas curtain, *Phys. Fluids*, **11**:186–200.

P.L. Roe, J. Pike (1984), Efficient construction and utilisation of approximate Riemann solutions, In *Computing Methods in Applied Science and Engineering*, North-Holland.

W. J. Rider and L. G. Margolin (2001), Simple Extensions of Monotone Limiters, to be published in J. Comput. Phys.

V.V. Rusanov (1961), Calculation of interaction of non-steady shock waves with obstacles, *J. Comput. Math. Phys. USSR*, **1**: 267-279.

J. Smagorinsky (1983), The Beginnings of Numerical Weather Prediction and General Circulation Modeling: Early Recollections, *Advances in Geophysics*, **25**: 3–37.

J. Smagorinsky (1963), General circulation experiments with the primitive equations. I. the basic experiment, *Mon. Wea. Rev.*, **101**:99–164.

P.K. Smolarkiewicz and L.G. Margolin (1998), MPDATA: a finite difference solver for geophysical flows, *J. Comput. Phys.* **140**:459–480.

P. K. Sweby (1984), High Resolution Schemes using Flux Limiters for Hyperbolic Conservation Laws, *SIAM J. Num. Anal.*, **21**:995–1011.

J.L. Thomas, B. van Leer, R.W. Walters (1985), Implicit flux split scheme for the Euler equations, *AIAA-Paper 85-1680.*

E.F. Toro, M. Spruce, W. Speares (1994), Restoration of the contact surface in the HLL-Riemann solver, *Shock Waves J.*, **4**: 25-34

E. F. Toro (1997), *Riemann Solvers and Numerical Methods for Fluid Dynamics: A Practical Introduction*, Springer-Verlag.

B. van Leer (1984), On the Relation Between the Upwind-Differencing Schemes of Godunov, Enquist-Osher and Roe, *SIAM J. Sci. Comp.*, **5**: 1-20.

J. von Neumann and R. D. Richtmyer (1950), A method for the numerical calculation of hydrodynamic shocks, *J. Appl. Phys.*, **21**:232-237.

R. F. Warming and B. J. Hyett (1974), The modified equation approach to the stability and accuracy analysis of finite-difference methods, *J. Comput. Phys.*, **14**:159-179.

J. Zółtak and D. Drikakis (1998), Hybrid upwind methods for the simulation of unsteady shock-wave diffraction over a cylinder, *Comput. Meth. in Appl. Mech. & Engrg.*, **162**: 165-185.

Chapter 3

PRESERVING SYMMETRY IN CONVECTION-DIFFUSION SCHEMES

R.W.C.P. Verstappen
verstappen@math.rug.nl

A.E.P. Veldman
Research Institute for Mathematics and Computing Science, University of Groningen
P.O.Box 800, 9700 AV Groningen, The Netherlands.
veldman@math.rug.nl

Abstract We propose to perform turbulent flow simulations in such manner that the difference operators do have the same symmetry properties as the corresponding differential operators. That is, the convective operator is represented by a skew-symmetric difference operator and the diffusive operator is approximated by a symmetric, positive-definite matrix. Mimicing crucial properties of differential operators forms in itself a motivation for discretizing them in a certain manner. We give it a concrete form by noting that a symmetry-preserving discretization of the Navier-Stokes equations is conservative, *i.e.* it conserves the (total) mass, momentum and kinetic energy (when the physical dissipation is turned off); a symmetry-preserving discretization of the Navier-Stokes equations is stable on any grid. Because the numerical scheme is stable on any grid, the choice of the grid spacing can be based on the required accuracy. We investigate the accuracy of a fourth-order, symmetry-preserving discretization for the turbulent flow in a channel. The Reynolds number (based on the channel width and the mean bulk velocity) is equal to $5,600$. It is shown that with the fourth-order, symmetry-preserving method a $64 \times 64 \times 32$ grid suffices to perform an accurate simulation.

Keywords: Direct Numerical Simulation, Turbulence, Conservation properties and stability, Channel flow.

1. Introduction

In the first half of the nineteenth century, Claude Navier (1822) and George Stokes (1845) derived the equation that governs turbulent flow. 'Their' equation states that the velocity $\boldsymbol{u}$ and pressure $\boldsymbol{p}$ (in an incompressible fluid) are given

D. Drikakis and B.J. Geurts (eds.), Turbulent Flow Computation, 75–100.

by

$$\partial_t \boldsymbol{u} + (\boldsymbol{u} \cdot \nabla)\, \boldsymbol{u} - \tfrac{1}{\mathrm{Re}} \nabla \cdot \nabla \boldsymbol{u} + \nabla p = \boldsymbol{0}, \qquad \nabla \cdot \boldsymbol{u} = 0, \tag{3.1}$$

where the parameter Re denotes the Reynolds number.

Turbulence is created by the non-linear, convective term in this equation. To illustrate this, we consider a velocity field with a x-component given by

$$u = \mathrm{e}^{\mathrm{i}\omega x},$$

and all other components equal zero, for simplicity. This wave (eddy) transports momentum. Its portion is governed by the x-component of the convective term in the Navier-Stokes equations:

$$u \partial_x u = \mathrm{i}\omega \mathrm{e}^{2\mathrm{i}\omega x}$$

Strikingly, the wave-length of this contribution is half that of the velocity u. Via the time-derivative in the Navier-Stokes equations this shorter wave-length becomes part of the velocity itself, and thus a smaller scale of motion is created. This process continues, and smaller and smaller scales of motion originate. The cascade ends when the diffusive forces become sufficiently strong to damp the small scales of motion. In our example, the diffusive term in the Navier-Stokes equations reads

$$\tfrac{1}{\mathrm{Re}} \partial_{xx} u = -\tfrac{1}{\mathrm{Re}} \omega^2 \mathrm{e}^{\mathrm{i}\omega x}.$$

As this contribution grows quadratically in terms of ω, it can overtake the convective term, which depends 'only' linearly on ω. The wave-length at which this happens is the smallest wave-length in the flow. In 1922, the meteorologist Lewis Fry Richardson described this process as follows

> *Big whorls have little whorls,*
> *Which feed on their velocity,*
> *And little whorls have lesser whorls,*
> *And so on to viscosity.*

So far, our arguments are heuristic, and not entirely correct, since we have left the time scales out of consideration. This leads to the wrong suggestion that the smallest length scale behaves like Re^{-1}. In 1941, Kolmogorov has considered both time and length scales. He argued that the diffusive term at a somewhat larger length scale, proportional to $\mathrm{Re}^{-3/4}$, is sufficiently strong to end the cascade to smaller scales.

To capture the essence of turbulence in a direct numerical simulation (DNS), the convective term in the Navier-Stokes equations need be discretized with care. The subtle balance between convective transport and diffusive dissipation may be disturbed if the discretization of the convective derivative is stabilized by means of numerical (artificial) diffusion. With this in mind, we consider the discretization of the convective term in the Navier-Stokes equations. As convection is described by a first-order differential operator, this leads to the apparently simple question how the discretize a first-order derivative.

In mathematical terms: given three values of a smooth function u, say $u_{i-1} = u(x_{i-1})$, $u_i = u(x_i)$ and $u_{i+1} = u(x_{i+1})$ with $x_{i-1} < x_i < x_{i+1}$, find an approximation of the (spatial) derivative of u at x_i. Almost any textbook on numerical analysis answers this question by combining Taylor-series expansions of u around $x = x_i$ in such a manner that as many as possible low-order terms cancel. After some algebraic work this results into the following approximation

$$\partial_x u(x_i) \approx \frac{\delta x_i^2 u_{i+1} + (\delta x_{i+1}^2 - \delta x_i^2) u_i - \delta x_{i+1}^2 u_{i-1}}{\delta x_{i+1} \delta x_i (\delta x_{i+1} + \delta x_i)}, \tag{3.2}$$

where the local spacing of the mesh is denoted by $\delta x_i = x_i - x_{i-1}$. This expression may also be derived by constructing a parabola through the three given data points and differentiating that parabola at $x = x_i$. Expression (3.2) is motivated by the fact that it minimizes the local truncation error at the grid point x_i. But, is this criterion based on sound physical principles? Recalling that the convective term in the Navier-Stokes equations transports energy without dissipating any, and that this transport ends at the scale where diffusion is powerful enough to counterbalance any further transport to smaller scales, we would like that convection conserves the total energy in the discrete form too. This minicing of crucial properties, however, forms a different criterion for discretizing the differential operators in the Navier-Stokes equations, see [1].

Rather than concentrating on reducing local truncation error, we propose to discretize in such a manner that the symmetry of the underlying differential operators is preserved. That is, the convective operator is replaced by a skew-symmetric difference-operator and the diffusive operator is approximated by a symmetric, positive-definite operator. We will show that such a symmetry-preserving discretization of the Navier-Stokes equations is stable on any grid, and conserves the total mass, momentum and kinetic energy (if the physical dissipation is turned off).

Conservation properties of numerical schemes for the (incompressible) Navier-Stokes equations are currently also pursued at other research institutes, in particular in Stanford [2]-[3], at Cerfacs [4], and at Delft University where a variant of our symmetry-preserving discretization for collocated grids has been developed [5]-[6]. Another approach that considers properties such as symmetry, conservation, stability and the relationships between the gradient, divergence and curl operator can be found in [7].

The next section concerns the incompressible Navier-Stokes equations. In this introductory section, we will sketch the main lines of symmetry-preserving discretization by means of the following, one-dimensional, linear, convection-diffusion equation

$$\partial_t u + \bar{u}\partial_x u - \tfrac{1}{\mathrm{Re}}\partial_{xx} u = 0, \tag{3.3}$$

where the convective transport velocity $\bar{u}$ is taken constant. The time-evolution of the semi-discrete velocity $u_i(t)$ at the grid point x_i reads

$$\boldsymbol{\Omega}_0 \frac{\mathrm{d}\boldsymbol{u}_h}{\mathrm{d}t} + \boldsymbol{C}_0(\bar{u})\boldsymbol{u}_h + \boldsymbol{D}_0\boldsymbol{u}_h = \mathbf{0}, \tag{3.4}$$

where the discrete velocities u_i form the vector $\boldsymbol{u}_h$. The diagonal matrix $\boldsymbol{\Omega}_0$ contains the local spacings of the mesh, that is $(\boldsymbol{\Omega}_0)_{i,i} = (x_{i+1} - x_{i-1})/2$. The coefficient matrix $\boldsymbol{C}_0(\bar{u})$ represents the convective operator. When the discretization of the derivative is taken as in (3.2), $\boldsymbol{C}_0(\bar{u})$ becomes a tri-diagonal matrix with entries

$$\boldsymbol{C}_0(\bar{u})_{i,i-1} = -\frac{\bar{u}\delta x_{i+1}}{2\delta x_i}, \qquad \boldsymbol{C}_0(\bar{u})_{i,i+1} = \frac{\bar{u}\delta x_i}{2\delta x_{i+1}}$$

and

$$\boldsymbol{C}_0(\bar{u})_{i,i} = \frac{\bar{u}\delta x_{i+1}}{2\delta x_i} - \frac{\bar{u}\delta x_i}{2\delta x_{i+1}}. \tag{3.5}$$

In the absence of diffusion, that is for $\boldsymbol{D}_0 = \mathbf{0}$, the kinetic energy $||\boldsymbol{u}_h||^2 = \boldsymbol{u}_h^* \boldsymbol{\Omega}_0 \boldsymbol{u}_h$ of any solution $\boldsymbol{u}_h$ of the dynamical system (3.4) evolves in time according to

$$\frac{\mathrm{d}}{\mathrm{d}t}||\boldsymbol{u}_h||^2 = -\boldsymbol{u}_h^* \left(\boldsymbol{C}_0(\bar{u}) + \boldsymbol{C}_0^*(\bar{u})\right)\boldsymbol{u}_h.$$

The right hand-side of this expression equals zero for all discrete velocities $\boldsymbol{u}_h$, *i.e.* the energy is conserved unconditionally, if and only if the coefficient matrix $\boldsymbol{C}_0(\bar{u})$ is skew-symmetric:

$$\boldsymbol{C}_0(\bar{u}) + \boldsymbol{C}_0^*(\bar{u}) = \mathbf{0}, \tag{3.6}$$

To avoid possible confusion, it may be noted that we use the adjective 'skew-symmetric' to describe a property of the coefficient matrix $\boldsymbol{C}_0(\bar{u})$ of the discrete convective operator. In the literature, the adjective 'skew-symmetric' is also related to a differential formulation of the convective term in the Navier-Stokes equations. The convective term may be written in four different ways (provided that the continuity equation is satisfied). These differential forms are referred to as divergence, advective, skew-symmetric and rotational form. We do not use the adjective 'skew-symmetric' in this context. Note that in our linear example, with a constant convective transport velocity, all differential forms coincide.

To conserve the energy during the convective cascade the coefficient matrix $\boldsymbol{C}_0(\bar{u})$ of the discrete, convective operator has to be a skew-symmetric matrix.

We see immediately that the traditional discretization scheme (3.2) leads to a coefficient matrix that is not skew-symmetric on a non-uniform grid. Indeed, the diagonal entry given by (3.5) is non-zero (unless the grid is uniform). Thus, if the discretization scheme is constructed to minimize the local truncation error, the skew-symmetry of the convective term is lost on non-uniform grids, and quantities that are conserved in the continuous formulation, like the kinetic energy, are not conserved in the discrete formulation.

In general, the symmetric part of $\boldsymbol{C}_0(\bar{u})$ will have both positive and negative eigenvalues. If the discrete velocity $\boldsymbol{u}_h$ is given by a linear combination of eigenvectors corresponding to negative eigenvalues of $\boldsymbol{C}_0(\bar{u}) + \boldsymbol{C}_0^*(\bar{u})$, the kinetic energy increases exponentially in time. Thus, an unconditionally stable solution of the discrete set of equations can not be obtained, unless a damping mechanism is added. Such a mechanism may interfere with the subtle balance between the production of turbulence and its dissipation at the smallest length scales. For that reason, we consider a symmetry-preserving discretization.

To obtain a skew-symmetric, discrete representation of the convective operator, we approximate the convective derivative by

$$\bar{u}\,\partial_x u(x_i) \approx \bar{u}\,\frac{u_{i+1} - u_{i-1}}{x_{i+1} - x_{i-1}} = \left(\boldsymbol{\Omega}_0^{-1}\boldsymbol{C}_0(\bar{u})\boldsymbol{u}_h\right)_i. \tag{3.7}$$

The resulting coefficient matrix $\boldsymbol{C}_0(\bar{u})$ is skew-symmetric on any grid:

$$\boldsymbol{C}_0(\bar{u})_{i,i-1} = -\frac{1}{2}\bar{u}, \qquad \boldsymbol{C}_0(\bar{u})_{i,i} = 0, \qquad \boldsymbol{C}_0(\bar{u})_{i,i+1} = \frac{1}{2}\bar{u}. \tag{3.8}$$

The two ways of discretization, given by (3.2) and (3.7), are illustrated in Figure 3.1. In the symmetry-preserving discretization (3.7) the derivative

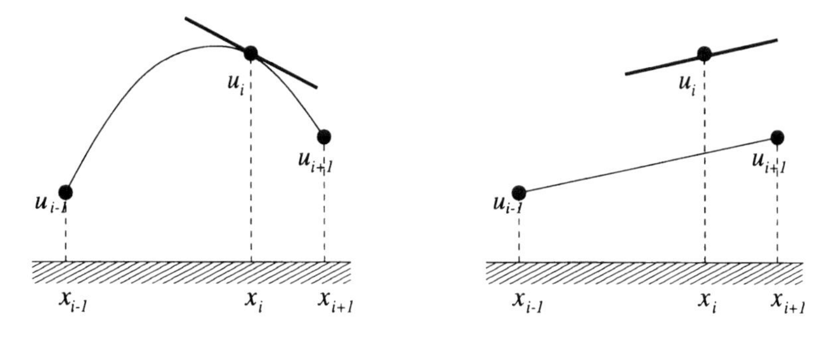

Figure 3.1. Two ways of approximating $\partial_x u$. In the left-hand figure the derivative is approximated by means of a Lagrangian interpolation, that is by Eq. (3.2). In the right-hand figure the symmetry-preserving discretization (3.7) is applied.

$\partial_x u(x_i)$ is simply approximated by drawing a straight line from (x_{i-1}, u_{i-1})

to (x_{i+1}, u_{i+1}). This gives the proper symmetry, but one may get the feeling that this approach is not very accurate. The local truncation error in the approximation of the derivative in (3.7), that is

$$\tau_h(x_i) = \frac{1}{2}(\delta x_{i+1} - \delta x_i)\partial_{xx}u(x_i), + \mathcal{O}(\delta x^2_{\max}) \tag{3.9}$$

is only first-order (unless the grid is almost uniform). Given stability, a sufficient condition for second-order accuracy of the discrete solution u_i is that the local truncation error be second order. Yet, this is not a necessary condition, as is emphasized by Manteufel and White [8]. They have proven that the approximation (3.7) yields second-order accurate solutions on uniform as well as on non-uniform meshes, even though its local truncation error τ_h is formally only first-order on non-uniform meshes. The standard proof (which uses stability and consistency to imply convergence) is inadequate to handle non-uniform meshes. Instead, Manteufel and White [8] argue that the error ϵ_i in the approximation of $u(x_i)$ in (3.7) satisfies $\boldsymbol{C}_0(\bar{u})\boldsymbol{\epsilon}_h = \boldsymbol{\Omega}_0\boldsymbol{\tau}_h$, or written out per element

$$\tfrac{\bar{u}}{2}\epsilon_{i+1} - \tfrac{\bar{u}}{2}\epsilon_{i-1} \stackrel{(3.7)+(3.9)}{=} \frac{1}{2}(\delta x^2_{i+1} - \delta x^2_i)\partial_{xx}u(x_i) + (\delta x_{i+1} + \delta x_i)\,\mathcal{O}(\delta x^2_{\max}).$$

The left hand-side of this expression may be written as $\epsilon_{i+1/2} - \epsilon_{i-1/2}$, where $\epsilon_{i+1/2} = (\epsilon_{i+1} + \epsilon_i)\bar{u}/2$. Recurring this error-equation back to an error at a boundary, say $\epsilon_{1/2}$, we have

$$\begin{aligned}
\epsilon_{i+1/2} - \epsilon_{1/2} &= \frac{1}{2}\sum_{k=1}^{i}\left(\delta x^2_{k+1} - \delta x^2_k\right)\partial_{xx}u(x_k) + \mathcal{O}(\delta x^2_{\max}) \\
&= \frac{1}{2}\sum_{k=1}^{i-1}\delta x^2_k\left(\partial_{xx}u(x_{k-1}) - \partial_{xx}u(x_k)\right) + \mathcal{O}(\delta x^2_{\max}).
\end{aligned}$$

The final sum is itself $\mathcal{O}(\delta x^2_{\max})$. Thus, the error $\epsilon_{i+1/2}$ is second-order in spite of the first-order truncation error. This implies that ϵ_i itself is second-order.

Here, it may be noted that the skew-symmetric coefficient matrix $C_0(\bar{u})$ in (3.7) may also be derived from a Galerkin finite element method. In that approach the velocity is written as

$$u(x,t) = \sum_i u_i(t)\psi_i(x)$$

where the basis functions $\psi_i(x)$ are piecewise linear functions with $\psi_i(x_i) = 1$ and $\psi_i(x_j) = 0$ for $i \neq j$. For the linear problem (3.3), the coefficients

$$\boldsymbol{C}_0(\bar{u})_{i,j} = \bar{u}\int_0^1 \psi_i(x)\partial_x\psi_j(x)dx$$

are identical to those given by (3.8), and satisfy the symmetry property (3.6) by construction. The difference with the 'finite volume/difference' method (3.4) is that the mass matrix $\tilde{\mathbf{\Omega}}_0 = \int_0^1 \psi_i(x)\psi_j(x)dx$ of the finite element method is not a diagonal, but a tri-diagonal matrix. Both methods are identical, when the off-diagonal entries of $\tilde{\mathbf{\Omega}}_0$ are lumped to the diagonal.

To construct the coefficient matrix $\boldsymbol{D}_0$ of the diffusive term in (3.4), we rewrite the second-order differential equation (3.4) as a system of two first-order differential equations

$$\partial_t u + \bar{u}\partial_x u - \tfrac{1}{\mathrm{Re}}\partial_x \phi = 0 \qquad \phi = \partial_x u. \tag{3.10}$$

The diffusive flux ϕ is discretized in a standard way:

$$\phi_{i+1/2} = \left(\mathbf{\Lambda}_0^{-1}\mathbf{\Delta}_0 \boldsymbol{u}_h\right)_i,$$

where the difference matrix $\mathbf{\Delta}_0$ is defined by $(\mathbf{\Delta}_0 \boldsymbol{u}_h)_i = u_i - u_{i-1}$, and the non-zero entries of the diagonal matrix $\mathbf{\Lambda}_0$ read $(\mathbf{\Lambda}_0)_{i,i} = x_i - x_{i-1}$. The derivative of ϕ is approximated according to

$$\partial_x \phi(x_i) \approx \left(\mathbf{\Omega}_0^{-1}\mathbf{\Delta}_0^* \boldsymbol{\phi}_h\right)_i,$$

where the vector $\boldsymbol{\phi}_h$ consists of the discrete values of ϕ at the mid-points $x_{i-1/2}$. Eliminating these auxiliary unknowns from the expressions above gives

$$\boldsymbol{D}_0 = \tfrac{1}{\mathrm{Re}}\,\mathbf{\Delta}_0^* \mathbf{\Lambda}_0^{-1}\mathbf{\Delta}_0. \tag{3.11}$$

The quadratic form $\boldsymbol{u}_h^* \boldsymbol{D}_0 \boldsymbol{u}_h = \frac{1}{\mathrm{Re}}\left(\mathbf{\Delta}_0 \boldsymbol{u}_h\right)^* \mathbf{\Lambda}_0^{-1}\left(\mathbf{\Delta}_0 \boldsymbol{u}_h\right)$ is strictly positive for all $\mathbf{\Delta}_0 \boldsymbol{u}_h \neq \mathbf{0}$ (*i.e.* for all $\boldsymbol{u}_h \neq \alpha\mathbf{1}$, where α is an arbitrary constant), since the entries of $\mathbf{\Lambda}_0$ are positive. The quadratic form $\boldsymbol{u}_h^* \boldsymbol{D}_0 \boldsymbol{u}_h$ is equal to zero if $\boldsymbol{u}_h = \alpha\mathbf{1}$. Thus, the matrix $\boldsymbol{D}_0$ is positive-definite, like the underlying differential operator $-\partial_{xx}$.

The symmetric part of $\boldsymbol{C}_0(\bar{u}) + \boldsymbol{D}_0$ is only determined by diffusion, and hence is positive-definite. Under this condition, the evolution of the kinetic energy $||\boldsymbol{u}_h||_h^2 = \boldsymbol{u}_h^* \mathbf{\Omega}_0 \boldsymbol{u}_h$ of any discrete solution $\boldsymbol{u}_h$ of (3.4) is governed by

$$\begin{aligned}
\frac{\mathrm{d}}{\mathrm{dt}}(\boldsymbol{u}_h^* \mathbf{\Omega}_0 \boldsymbol{u}_h) &\overset{(3.4)}{=} -\boldsymbol{u}_h^*(\boldsymbol{C}_0(\bar{u}) + \boldsymbol{C}_0^*(\bar{u}))\boldsymbol{u}_h - \boldsymbol{u}_h^*(\boldsymbol{D}_0 + \boldsymbol{D}_0^*)\boldsymbol{u}_h \\
&\overset{(3.6)}{=} -\boldsymbol{u}_h^*(\boldsymbol{D}_0 + \boldsymbol{D}_0^*)\boldsymbol{u}_h \leq 0,
\end{aligned}$$

where the right-hand side is zero if and only if $\boldsymbol{u}_h$ lies in the null space of $\boldsymbol{D}_0 + \boldsymbol{D}_0^*$. Consequently, a stable solution can be obtained on any grid.

As the eigenvalues of $\boldsymbol{C}_0(\bar{u}) + \boldsymbol{D}_0$ lie in the stable half-plane, this matrix is regular, which is important for the relationship between the global and local truncation error. To illustrate this, we consider the stationary equivalent of Eq. (3.3): $\bar{u}\partial_x u - \frac{1}{\mathrm{Re}}\partial_{xx}u = 0$. As before, this equation is approximated by $(\boldsymbol{C}_0(\bar{u}) + \boldsymbol{D}_0)\,\boldsymbol{u}_h = \boldsymbol{0}$. To define the global truncation error, we restrict the exact solution to the grid points, first. The vector of these values is denoted by $\boldsymbol{u}$. The global truncation error, defined as $\boldsymbol{u} - \boldsymbol{u}_h$, is equal to the product of the inverse of the discrete operator $\boldsymbol{C}_0(\bar{u}) + \boldsymbol{D}_0$ and the local truncation error. Therefore, a (nearly) singular discrete operator can destroy favourable properties of the local truncation error. Examples of this (for non-symmetry-preserving discretizations!) can be found in [9].

2. Symmetry-preserving discretization

In the preceding section, we saw that the conservation properties and the stability of the spatial discretization of a simple, one-dimensional, convection-diffusion equation (3.3) may be improved when less emphasis is laid upon the local truncation error, so that the symmetry of the underlying differential operators can be respected. In this section, we will extend the symmetry-preserving discretization to the incompressible, Navier-Stokes equations (in two spatial directions only, as the extension to 3D is straightforward).

On a uniform grid the traditional aim, minimize the local truncation error, need not break the symmetry. The well-known, second-order scheme of Harlow and Welsh [10] forms an example of this. In Section 2.1, we will generalize Harlow and Welsh's scheme to non-uniform meshes in such a manner that the symmetries of the convective and diffusive operator are not broken. The conservation properties and stability of the resulting, second-order scheme are discussed in Section 2.2. After that (Section 2.3), we will improve the order of the basic scheme by means of a Richardson extrapolation, just like in [11]. This results into a fourth-order, symmetry-preserving discretization. The last section (Sec. 2.4) concerns the treatment of the boundary conditions.

2.1 Basic, second-order method

In this section we will apply symmetry-preserving discretization to the incompressible Navier-Stokes equations (3.1) in two spatial dimensions. For that, we will use a staggered grid and adopt the notations of Harlow & Welsh [10]. Figure 3.2 illustrates the definition of the discrete velocities $(u_{i,j}, v_{i,j})$.

For an incompressible fluid the mass of any control volume $\Omega_{i,j} = [x_{i-1}, x_i] \times [y_{j-1}, y_j]$ is conserved:

$$\bar{u}_{i,j} + \bar{v}_{i,j} - \bar{u}_{i-1,j} - \bar{v}_{i,j-1} = 0, \tag{3.12}$$

where $\bar{u}_{i,j}$ denotes the mass flux through the face $y = y_j$ of the grid cell $\Omega_{i,j}$ and $\bar{v}_{i,j}$ stands for the mass flux through the grid face $x = x_i$:

$$\bar{u}_{i,j} = \int_{y_{j-1}}^{y_j} u(x_i, y, t)dy \quad \text{and} \quad \bar{v}_{i,j} = \int_{x_{i-1}}^{x_i} v(x, y_j, t)dx. \tag{3.13}$$

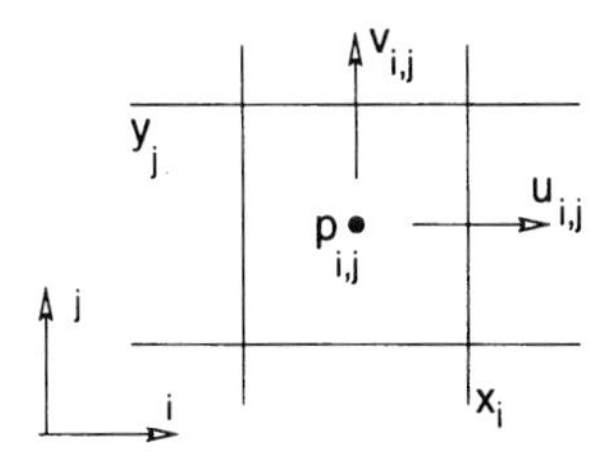

Figure 3.2. The location of the discrete velocities.

The combination (3.12)+(3.13) does not contain a discretization error, since the integrals in (3.13) have not yet been discretized. We postpone their discretization till later in this section. Till then we view the velocities $(u_{i,j}, v_{i,j})$ as the unknowns and the mass fluxes $(\bar{u}_{i,j}, \bar{v}_{i,j})$ as being given such that (3.12) holds.

As mass and momentum are transported at equal velocity, the mass flux is used to discretize the transport velocity of momentum. The (spatial) discretization of the transport of momentum of a region $\Omega_{i+1/2,j} = [x_{i-1/2}, x_{i+1/2}] \times [y_{j-1}, y_j]$ becomes

$$\begin{aligned} |\Omega_{i+1/2,j}|\frac{du_{i,j}}{dt} &+ \bar{u}_{i+1/2,j}u_{i+1/2,j} + \bar{v}_{i+1/2,j}u_{i,j+1/2} \\ &- \bar{u}_{i-1/2,j}u_{i-1/2,j} - \bar{v}_{i+1/2,j-1}u_{i,j-1/2}. \end{aligned} \tag{3.14}$$

The non-integer indices in (3.14) refer to the faces of $\Omega_{i+1/2,j}$. For example, $u_{i-1/2,j}$ stands for the u-velocity at the interface of $\Omega_{i-1/2,j}$ and $\Omega_{i+1/2,j}$. The velocity at a control face is approximated by the average of the velocity at both sides of it:

$$u_{i+1/2,j} = \frac{1}{2}(u_{i+1,j} + u_{i,j}) \quad \text{and} \quad u_{i,j+1/2} = \frac{1}{2}(u_{i,j+1} + u_{i,j}). \tag{3.15}$$

In addition to the set of equations for the u-component of the velocity (3.14)-(3.15), there is an analogous set for the v-component:

$$\begin{aligned} |\Omega_{i,j+1/2}|\frac{dv_{i,j}}{dt} &+ \bar{v}_{i,j+1/2}v_{i,j+1/2} + \bar{u}_{i,j+1/2}v_{i+1/2,j} \\ &- \bar{v}_{i,j-1/2}v_{i,j-1/2} - \bar{u}_{i-1,j+1/2}v_{i-1/2,j}, \end{aligned} \tag{3.16}$$

with

$$v_{i+1/2,j} = \frac{1}{2}(v_{i+1,j} + v_{i,j}) \quad \text{and} \quad v_{i,j+1/2} = \frac{1}{2}(v_{i,j+1} + v_{i,j}). \tag{3.17}$$

We conceive Eqs. (3.14)-(3.17) as expressions for the velocities, where the mass fluxes $\bar{u}$ and $\bar{v}$ form the coefficients. Thus, we can write the (semi-

)discretization in matrix-vector notation as

$$\boldsymbol{\Omega}_1 \frac{\mathrm{d}\boldsymbol{u}_h}{\mathrm{d}t} + \boldsymbol{C}_1(\bar{\boldsymbol{u}})\boldsymbol{u}_h,$$

where $\boldsymbol{u}_h$ denotes the discrete velocity-vector (which consists of both the $u_{i,j}$'s and $v_{i,j}$'s), $\boldsymbol{\Omega}_1$ is a (positive-definite) diagonal matrix representing the sizes of the control volumes $|\Omega_{i+1/2,j}|$ and $|\Omega_{i,j+1/2}|$, whereas $\boldsymbol{C}_1(\bar{\boldsymbol{u}})$ is built from the flux contributions through the control faces, *i.e.* $\boldsymbol{C}_1$ depends on the mass fluxes $\bar{u}$ and $\bar{v}$ at the control faces.

With no (in- or external) force, the discrete transport equation

$$\boldsymbol{\Omega}_1 \frac{\mathrm{d}\boldsymbol{u}_h}{\mathrm{d}t} + \boldsymbol{C}_1(\bar{\boldsymbol{u}})\boldsymbol{u}_h = \boldsymbol{0} \tag{3.18}$$

conserves the discrete energy $\boldsymbol{u}_h^*\boldsymbol{\Omega}_1\boldsymbol{u}_h$ (of any discrete velocity field $\boldsymbol{u}_h$), that is

$$\frac{\mathrm{d}}{\mathrm{d}t}\left(\boldsymbol{u}_h^*\boldsymbol{\Omega}_1\boldsymbol{u}_h\right) \overset{(3.18)}{=} -\boldsymbol{u}_h^*\left(\boldsymbol{C}_1(\bar{\boldsymbol{u}}) + \boldsymbol{C}_1^*(\bar{\boldsymbol{u}})\right)\boldsymbol{u}_h = 0, \tag{3.19}$$

if and only if the coefficient matrix $\boldsymbol{C}_1(\bar{\boldsymbol{u}})$ is skew-symmetric:

$$\boldsymbol{C}_1(\bar{\boldsymbol{u}}) + \boldsymbol{C}_1^*(\bar{\boldsymbol{u}}) = \boldsymbol{0}. \tag{3.20}$$

This condition is verified in two steps. To start, we consider the off-diagonal elements. The matrix $\boldsymbol{C}_1(\bar{\boldsymbol{u}}) - \mathrm{diag}(\boldsymbol{C}_1(\bar{\boldsymbol{u}}))$ is skew-symmetric if and only if the weights in the interpolations (3.15) and (3.17) of the discrete velocities are taken constant. On a non-uniform grid one would be tempted to tune the weights $\frac{1}{2}$ in Eqs. (3.15) and (3.17) to the local mesh sizes to minimize the local truncation error. Yet, this breaks the skew-symmetry. Indeed, suppose we would follow the Lagrangian approach by taking

$$u_{i+1/2,j} = (1-\omega_{i,j})u_{i+1,j} + \omega_{i,j}u_{i,j}$$

instead of (3.15), where the coefficient $\omega_{i,j}$ depends on the local mesh sizes. Then, by substituting this mesh-dependent interpolation rule into Eq. (3.14) we see that the coefficient of $u_{i+1,j}$ becomes $(1-\omega_{i,j})\bar{u}_{i+1/2,j}$, while the term $u_{i-1/2,j}\bar{u}_{i-1/2,j}$ in (3.14) with i replaced by $i+1$ yields the coefficient $-\omega_{i,j}\bar{u}_{i+1/2,j}$ for $u_{i,j}$. For skew-symmetry, these two coefficients should be of opposite sign. That is, we should have

$$(1-\omega_{i,j})\bar{u}_{i+1/2,j} = \omega_{i,j}\bar{u}_{i+1/2,j},$$

for all mass fluxes $\bar{u}_{i+1/2,j}$. This can only be achieved when the weight $\omega_{i,j}$ is taken equal to the uniform weight $\omega_{i,j} = 1/2$, hence independent of the grid location. Therefore we take constant weights in Eqs. (3.15) and (3.17), also on non-uniform grids. Here, it may be noted that it is either one or the other: either the discretization is selected on basis of its formal, local truncation error (that is, the interpolation is adapted to the local grid spacings) or the skew-symmetry is preserved. For skew-symmetry, the convective flux through the common interface between two neighbouring control volumes has to be computed independent of the control volume in which it is considered.

Next, we consider the diagonal of $\boldsymbol{C}_1$. In the notation above, we have suppressed the argument $\bar{\boldsymbol{u}}$ of $\boldsymbol{C}_1$, because $\boldsymbol{C}_1 - \mathrm{diag}(\boldsymbol{C}_1)$ is skew-symmetric for all $\bar{\boldsymbol{u}}$. The interpolation rule for the mass fluxes $\bar{u}$ and $\bar{v}$ through the faces of the control volumes is determined by the requirement that the diagonal of $\boldsymbol{C}_1$ has to be zero. Then, we have (3.20). By substituting (3.15) into (3.14) we obtain the diagonal element

$$\frac{1}{2}\left(\bar{u}_{i+1/2,j} + \bar{v}_{i+1/2,j} - \bar{u}_{i-1/2,j} - \bar{v}_{i+1/2,j-1}\right). \tag{3.21}$$

This expression is equal to a linear combination of left-hand sides of Eq. (3.12) if the mass fluxes in (3.14) are interpolated to the faces of a u-cell according to

$$\bar{u}_{i+1/2,j} = \frac{1}{2}(\bar{u}_{i+1,j} + \bar{u}_{i,j}) \quad \text{and} \quad \bar{v}_{i+1/2,j} = \frac{1}{2}(\bar{v}_{i+1,j} + \bar{v}_{i,j}). \tag{3.22}$$

It goes without saying that this interpolation rule is also applied in the j-direction to approximate the mass flux through the faces of v-cells. Thus, the coefficient matrix $\boldsymbol{C}_1$ is skew-symmetric if Eq. (3.12) holds, and if the discrete velocities $\boldsymbol{u}_h$ and fluxes $\bar{\boldsymbol{u}}$ are interpolated to the surfaces of control cells with weights $\frac{1}{2}$, as in Eqs. (3.15) and (3.22).

The matrix $\boldsymbol{C}_1(\bar{\boldsymbol{u}})$ is skew-symmetric for any relation between $\bar{\boldsymbol{u}}$ and $\boldsymbol{u}_h$. Obviously, the mass flux $\bar{\boldsymbol{u}}$ has to be expressed in terms of the discrete velocity vector $\boldsymbol{u}_h$ in order to close the system of equations (3.18). The coefficient matrix $\boldsymbol{C}_1(\bar{\boldsymbol{u}})$ becomes a function of the discrete velocity $\boldsymbol{u}_h$ then. We will make liberal use of its name, and denote the resulting coefficient matrix by $\boldsymbol{C}_1(\boldsymbol{u}_h)$. The mass fluxes $\bar{u}_{i,j}$ and $\bar{v}_{i,j}$ are approximated by means of the mid-point rule:

$$\bar{u}_{i,j} = (y_j - y_{j-1})u_{i,j} \quad \text{and} \quad \bar{v}_{i,j} = (x_i - x_{i-1})v_{i,j}. \tag{3.23}$$

The continuity equation (3.12) may then be written in terms of the discrete velocity vector $\boldsymbol{u}_h$. We will denote the coefficient matrix by $\boldsymbol{M}_1$. Hence, the discretization of the continuity equation reads $\boldsymbol{M}_1\boldsymbol{u}_h =$ given, where the right-hand side depends upon the boundary conditions. It is formed by those parts of (3.12) that correspond to mass fluxes through the boundary of the computational domain. To keep the expressions simple, we take the right-hand side equal to zero, *i.e.* we consider no-slip or periodical boundary conditions. Other boundary conditions can be treated likewise (at the expense of some additional terms in the expressions to follow).

The surface integrals that result after that the pressure gradient in the Navier-Stokes equations is integrated over the control volumes for the discrete velocities are discretized by the same rule that is applied to discretize the mass fluxes $\bar{u}_{i,j}$ and $\bar{v}_{i,j}$. Then, the coefficient matrix of the discrete pressure gradient becomes $-(\boldsymbol{\Omega}_1)^{-1}\boldsymbol{M}_1^*$. That is, apart from a diagonal scaling (by $-(\boldsymbol{\Omega}_1)^{-1}$), the coefficient matrix of the discrete gradient operator is given by the transpose of the discrete divergence.

In the continuous case diffusion corresponds to a symmetric, positive-definite operator. In our approach we want this property to hold also for the discrete diffusive operator. To that end, we view the underlying, second-order differential operator as the product of two first-order differential operators, a divergence and a gradient. We discretize the divergence operator. The discrete gradient is constructed from that by taking the transpose of the discrete divergence and multiplying that by a diagonal scaling. This leads to a symmetric, positive-definite, approximation of the diffusive fluxes. We will work this out for the diffusive flux through the faces of the control volume $\Omega_{i+1/2,j}$ for the discrete velocity $u_{i,j}$. To start, we introduce the fluxes

$$\bar{\phi}_{i+1/2,j} = \int_{y_{j-1}}^{y_j} \phi\left(x_{i+1/2}, y\right) dy \quad \text{and} \quad \bar{\psi}_{i,j} = \int_{x_{i-1/2}}^{x_{i+1/2}} \psi\left(x, y_j\right) dx,$$

where $\phi = \partial_x u$ and $\psi = \partial_y u$. In terms of these surface integrals the diffusive flux through the faces of the control volume $\Omega_{i+1/2,j}$ of $u_{i,j}$ reads

$$\tfrac{1}{\mathrm{Re}}\left(\bar{\phi}_{i+1/2,j} - \bar{\phi}_{i-1/2,j} + \bar{\psi}_{i,j} - \bar{\psi}_{i,j-1}\right).$$

The surface integrals in this expression are approximated according to

$$\bar{\phi}_{i+1/2,j} = (y_j - y_{j-1})\phi_{i+1/2,j} \quad \text{and} \quad \bar{\psi}_{i,j} = (x_{i+1/2} - x_{i-1/2})\psi_{i,j}.$$

In matrix-vector notation, the diffusive flux through the faces of u-cells is given by $\boldsymbol{M}_1^u\boldsymbol{\phi}_h$, where the vector $\boldsymbol{\phi}_h$ consists of the $\phi_{i+1/2,j}$'s and $\psi_{i,j}$'s. The coefficient matrix $\boldsymbol{M}_1^u$ may be constructed out of $\boldsymbol{M}_1$, by lowering $\boldsymbol{M}_1$'s dimension in the x-direction by one, and replacing $x_i - x_{i-1}$ by $x_{i+1/2} - x_{i-1/2}$. The difference between $\boldsymbol{M}_1^u$ and $\boldsymbol{M}_1$ is due to the staggering of the grid: the discrete divergence operator $\boldsymbol{M}_1^u$ works on the control cells $\Omega_{i+1/2,j}$ for u-momentum, whereas $\boldsymbol{M}_1$ operates on the grid cells $\Omega_{i,j}$. The gradient operator relating ϕ and ψ to the velocity component u is discretized by $-(\boldsymbol{\Omega}_1^u)^{-1}(\boldsymbol{M}_1^u)^*$, where the entries of the diagonal matrix $\boldsymbol{\Omega}_1^u$ are given by $|\Omega_{i,j}|$ and $|\Omega_{i,j+1/2}|$. We need to introduce this diagonal matrix, because the staggering of the grid yields different control volumes for the transport of mass and momentum: $\boldsymbol{\Omega}_1^u$ may be constructed out of $\boldsymbol{\Omega}_1$ by replacing the entries $|\Omega_{i+1/2,j}|$ with $|\Omega_{i,j}|$. The diffusive flux through v-cells is approximated similarly. It's coefficient matrix reads $\frac{1}{\mathrm{Re}}\boldsymbol{M}_1^v\left(\boldsymbol{\Omega}_1^v\right)^{-1}\left(\boldsymbol{M}_1^v\right)^*$, where $\boldsymbol{M}_1^v$ may be obtained from $\boldsymbol{M}_1$ by lowering $\boldsymbol{M}_1$'s y-dimension by one, and replacing $y_j - y_{j-1}$ by $y_{j+1/2} - y_{j-1/2}$; The diagonal matrix $\boldsymbol{\Omega}_1^v$ represents the sizes $|\Omega_{i+1/2,j}|$ and $|\Omega_{i,j}|$. So, the symmetric, positive-definite differential operator $-\frac{1}{\mathrm{Re}}\nabla\cdot\nabla u$ in the Navier-Stokes equations (3.1) is discretized by $\boldsymbol{\Omega}_1^{-1}\boldsymbol{D}_1\boldsymbol{u}_h$, where the coefficient matrix $\boldsymbol{D}_1$ is given by

$$\boldsymbol{D}_1 = \tfrac{1}{\mathrm{Re}}\boldsymbol{\Delta}_1^*\boldsymbol{\Lambda}_1^{-1}\boldsymbol{\Delta}_1 \quad \text{with} \quad \boldsymbol{\Delta}_1^* = \begin{pmatrix} \boldsymbol{M}_1^u & \boldsymbol{0} \\ \boldsymbol{0} & \boldsymbol{M}_1^v \end{pmatrix} \tag{3.24}$$

and $\boldsymbol{\Lambda}_1 = \mathrm{diag}\left(\boldsymbol{\Omega}_1^u, \boldsymbol{\Omega}_1^v\right)$. The matrix $\boldsymbol{D}_1$ is symmetric, (weakly) diagonal dominant, has positive entries at its diagonal, and negative off-diagonal elements. Hence, $\boldsymbol{D}_1$ is an M-matrix.

By adding viscous and pressure forces to the discrete transport equation (3.18), we obtain the following semi-discrete representation of the incompressible Navier-Stokes equations

$$\boldsymbol{\Omega}_1\frac{\mathrm{d}\boldsymbol{u}_h}{\mathrm{d}t} + \boldsymbol{C}_1(\boldsymbol{u}_h)\boldsymbol{u}_h + \boldsymbol{D}_1\boldsymbol{u}_h - \boldsymbol{M}_1^*\boldsymbol{p}_h = \boldsymbol{0}, \qquad \boldsymbol{M}_1\boldsymbol{u}_h = 0, \tag{3.25}$$

where the vector $\boldsymbol{p}_h$ represents the discrete pressure.

2.2 Conservation properties and stability

The total mass and momentum of a flow are conserved analytically. Without diffusion, the kinetic energy is conserved too. With diffusion, the kinetic energy decreases in time. The coefficient matrices in the semi-discretization (3.25) are constructed such that these conservation and stability properties hold also for the discrete solution, as will be shown in this section.

The total mass of the semi-discrete flow is trivially conserved. Its total amount of momentum evolves in time according to

$$\frac{\mathrm{d}}{\mathrm{d}t}\left(\boldsymbol{1}^*\boldsymbol{\Omega}_1\boldsymbol{u}_h\right) \overset{(3.25)}{=} -\boldsymbol{1}^*\left(\boldsymbol{C}_1(\boldsymbol{u}_h) + \boldsymbol{D}_1\right)\boldsymbol{u}_h + \boldsymbol{1}^*\boldsymbol{M}_1^*\boldsymbol{p}_h,$$

where the vector $\boldsymbol{1}$ has as many entries as there are control volumes for the discrete velocity components $u_{i,j}$ and $v_{i,j}$. Hence, momentum is conserved for any discrete velocity $\boldsymbol{u}_h$ and discrete pressure $\boldsymbol{p}_h$, if the coefficient matrices $\boldsymbol{C}_1(\boldsymbol{u}_h)$, $\boldsymbol{D}_1$ and $\boldsymbol{M}_1$ satisfy

$$\boldsymbol{C}_1^*(\boldsymbol{u}_h)\boldsymbol{1} = \boldsymbol{0} \qquad \boldsymbol{D}_1^*\boldsymbol{1} = \boldsymbol{0} \qquad \text{and} \qquad \boldsymbol{M}_1\boldsymbol{1} = \boldsymbol{0}. \tag{3.26}$$

The latter of these three conditions expresses that a constant (discrete) velocity field has to satisfy the law of conservation of mass. Obviously, this condition is satisfied. The first two conditions in Eq. (3.26) can be viewed as consistency conditions too. Indeed, we may leave the transposition in these conditions away, since $\boldsymbol{C}_1(\boldsymbol{u}_h)$ is skew-symmetric and $\boldsymbol{D}_1$ is symmetric. So it suffices to verify that the row-sums of $\boldsymbol{C}_1(\boldsymbol{u}_h)$ and $\boldsymbol{D}_1$ are zero. Those of $\boldsymbol{D}_1$ are zero by definition. The row-sums of $\boldsymbol{C}_1(\boldsymbol{u}_h)$ can be worked out from (3.14)+(3.15). Each row-sum is equal to two times the corresponding diagonal element, and thus zero, since $\boldsymbol{C}_1(\boldsymbol{u}_h)$ is skew-symmetric.

Without diffusion ($\boldsymbol{D}_1 = \boldsymbol{0}$), the kinetic energy $\boldsymbol{u}_h^* \boldsymbol{\Omega}_1 \boldsymbol{u}_h$ of any solution of (3.25) is conserved as the coefficient matrix $\boldsymbol{C}_1(\boldsymbol{u}_h)$ is skew-symmetric:

$$\begin{aligned}\frac{\mathrm{d}}{\mathrm{d}t}\left(\boldsymbol{u}_h^* \boldsymbol{\Omega}_1 \boldsymbol{u}_h\right) \overset{(3.25)}{=} & -\boldsymbol{u}_h^* \left(\boldsymbol{C}_1(\boldsymbol{u}_h) + \boldsymbol{C}_1^*(\boldsymbol{u}_h)\right) \boldsymbol{u}_h \\ & + \left(\boldsymbol{M}_1 \boldsymbol{u}_h\right)^* \boldsymbol{p}_h + \boldsymbol{p}_h^* \left(\boldsymbol{M}_1 \boldsymbol{u}_h\right) = 0.\end{aligned}$$

The two conditions (3.20) and (3.26) imposed on $\boldsymbol{C}_1(\boldsymbol{u}_h)$ reflect that it represents a discrete gradient: its null space consists of the vectors $\alpha\boldsymbol{1}$, with α constant, and $\boldsymbol{C}_1(\boldsymbol{u}_h)$ is skew-symmetric, like a first-order differential operator.

Furthermore, it may be remarked that the pressure does not effect the evolution of the kinetic energy, because the discrete pressure gradient is represented by the transpose of the coefficient matrix $\boldsymbol{M}_1$ of the law of conservation of mass. Formally, the contribution of the pressure to the evolution of the energy reads

$$\boldsymbol{u}_h^* \left(\boldsymbol{M}_1^* \boldsymbol{p}_h\right) + \left(\boldsymbol{M}_1^* \boldsymbol{p}_h\right)^* \boldsymbol{u}_h = \left(\boldsymbol{M}_1 \boldsymbol{u}_h\right)^* \boldsymbol{p}_h + \boldsymbol{p}_h^* \left(\boldsymbol{M}_1 \boldsymbol{u}_h\right).$$

As this expression equals zero (on condition that $\boldsymbol{M}_1 \boldsymbol{u}_h = \boldsymbol{0}$), the pressure can not unstabilize the spatial discretization.

The coefficient matrix $\boldsymbol{D}_1$ of the discrete diffusive operator inherits its symmetry and definiteness from the underlying Laplacian differential operator. Consequently, with diffusion (that is for $\boldsymbol{D}_1 \neq \boldsymbol{0}$) the energy $\boldsymbol{u}_h^* \boldsymbol{\Omega}_1 \boldsymbol{u}_h$ of any solution $\boldsymbol{u}_h$ of the semi-discrete system (3.25) decreases in time unconditionally:

$$\frac{\mathrm{d}}{\mathrm{d}t}(\boldsymbol{u}_h^* \boldsymbol{\Omega}_1 \boldsymbol{u}_h) = -\boldsymbol{u}_h^* (\boldsymbol{D}_1 + \boldsymbol{D}_1^*) \boldsymbol{u}_h < 0,$$

where the right-hand side is negative for all $\boldsymbol{u}_h$'s (except those that lie in the null space of $\boldsymbol{D}_1 + \boldsymbol{D}_1^*$), because the matrix $\boldsymbol{D}_1 + \boldsymbol{D}_1^*$ is positive-definite. This implies that the semi-discrete system (3.25) is stable. Since a solution can be obtained on any grid, we need not add an artificial dissipation mechanism. The grid may be chosen on basis of the required accuracy. But, how accurate is (3.25)? This question will be addressed in Section 3. First, we will further enhance its accuracy.

2.3 Higher-order, symmetry-preserving approximation

To turn Eq. (3.14) into a higher-order approximation, we write down the transport of momentum of a region $\Omega^{(3)}_{i+1/2,j} = [x_{i-3/2}, x_{i+3/2}] \times [y_{j-2}, y_{j+1}]$. Here, it may be noted that we can not blow up the 'original' volumes $\Omega_{i+1/2,j}$ by a factor of two (in all directions) since our grid is not collocated. On a staggered grid, three times larger volumes are the smallest ones possible for which the same discretization rule can be applied as for the 'original' volumes. This yields

$$
\begin{aligned}
|\Omega^{(3)}_{i+1/2,j}| \frac{du_{i,j}}{dt} &+ \bar{\bar{u}}_{i+3/2,j} u_{i+3/2,j} + \bar{\bar{v}}_{i+1/2,j+1} u_{i,j+3/2} \\
&- \bar{\bar{u}}_{i-3/2,j} u_{i-3/2,j} - \bar{\bar{v}}_{i+1/2,j-2} u_{i,j-3/2}, \qquad (3.27)
\end{aligned}
$$

where

$$\bar{\bar{u}}_{i,j} = \int_{y_{j-2}}^{y_{j+1}} u(x_i, y, t) dy \quad \text{and} \quad \bar{\bar{v}}_{i,j} = \int_{x_{i-2}}^{x_{i+1}} v(x, y_j, t) dy.$$

The velocities at the control faces of the large volumes are interpolated to the control faces in a way similar to that given by (3.15):

$$u_{i+3/2,j} = \frac{1}{2}(u_{i+3,j} + u_{i,j}) \quad \text{and} \quad u_{i,j+3/2} = \frac{1}{2}(u_{i,j+3} + u_{i,j}). \qquad (3.28)$$

We conceive Eq. (3.26) as an expression for the velocities, where the mass fluxes $\bar{\bar{u}}$ and $\bar{\bar{v}}$ form the coefficients. Considering it like that, we can recapitulate the equations above (together with the analogous set for the v-component) by

$$\boldsymbol{\Omega}_3 \frac{d\boldsymbol{u}_h}{dt} + \boldsymbol{C}_3(\bar{\bar{\boldsymbol{u}}}) \boldsymbol{u}_h, \qquad (3.29)$$

where the diagonal matrix $\boldsymbol{\Omega}_3$ represents the sizes of the large control volumes and $\boldsymbol{C}_3$ consists of flux contributions ($\bar{\bar{u}}$ and $\bar{\bar{v}}$) through the faces of these volumes.

On a uniform grid the local truncation errors in (3.18) and (3.29) are of the order $2 + d$, where $d = 2$ in two spatial dimensions and $d = 3$ in 3D. The leading term in the discretization error may be removed through a Richardson extrapolation (just like in [11]). This leads to the fourth-order approximation

$$\boldsymbol{\Omega} \frac{d\boldsymbol{u}_h}{dt} + \left(3^{2+d} \boldsymbol{C}_1(\bar{\boldsymbol{u}}) - \boldsymbol{C}_3(\bar{\bar{\boldsymbol{u}}})\right) \boldsymbol{u}_h,$$

where $\boldsymbol{\Omega} = 3^{2+d}\boldsymbol{\Omega}_1 - \boldsymbol{\Omega}_3$. The coefficient matrix of the convective operator depends on both $\bar{\boldsymbol{u}}$ and $\bar{\bar{\boldsymbol{u}}}$, since it is constructed out of $\boldsymbol{C}_1$ and $\boldsymbol{C}_3$. The diffusive term of the Navier-Stokes equations undergoes a similar treatment. This leads to a fourth-order coefficient matrix

$$D = \frac{1}{\mathrm{Re}}\left(3^{2+d}\boldsymbol{\Delta}_1 - \boldsymbol{\Delta}_3\right)^*\left(3^{2+d}\boldsymbol{\Lambda}_1 - \boldsymbol{\Lambda}_3\right)^{-1}\left(3^{2+d}\boldsymbol{\Delta}_1 - \boldsymbol{\Delta}_3\right)$$

where the difference matrix $\boldsymbol{\Delta}_3$ and the diagonal matrix $\boldsymbol{\Lambda}_3$ are the relatives of $\boldsymbol{\Delta}_1$ and $\boldsymbol{\Lambda}_1$ respectively, with the difference that they are defined on 3^d-times larger control volumes. In terms of the abbreviations $\boldsymbol{\Delta} = 3^{2+d}\boldsymbol{\Delta}_1 - \boldsymbol{\Delta}_3$ and $\boldsymbol{\Lambda} = 3^{2+d}\boldsymbol{\Lambda}_1 - \boldsymbol{\Lambda}_3$ we have $\boldsymbol{D} = \frac{1}{\mathrm{Re}}\boldsymbol{\Delta}^*\boldsymbol{\Lambda}^{-1}\boldsymbol{\Delta}$. The quadratic form

$$\boldsymbol{u}_h^*\boldsymbol{D}\boldsymbol{u}_h = \frac{1}{\mathrm{Re}}\left(\boldsymbol{\Delta}\boldsymbol{u}_h\right)^*\boldsymbol{\Lambda}^{-1}\left(\boldsymbol{\Delta}\boldsymbol{u}_h\right)$$

is non-negative provided that the entries of the diagonal matrix $\boldsymbol{\Lambda}$ are non-negative. Here, we assume that the grid is chosen such that this condition is satisfied. Note that $\boldsymbol{\Lambda}_{ii} < 0$ for some i implies that the grid is so irregular that is does not make sense to apply a fourth-order method; in that case the second-order method (3.25) should be applied. For $\boldsymbol{\Lambda} > \boldsymbol{0}$, the quadratic form $\boldsymbol{u}_h^*\boldsymbol{D}\boldsymbol{u}_h$ equals zero if and only if $\boldsymbol{\Delta}\boldsymbol{u}_h = \boldsymbol{0}$, that is if and only if the discrete gradient of the velocity equals zero. This is precisely the condition that need be satisfied in the continuous case. Indeed, there we have

$$-\int \boldsymbol{u}\nabla\cdot\nabla\boldsymbol{u}dV = \int|\nabla\boldsymbol{u}|^2dV = 0$$

if and only if $\nabla\boldsymbol{u} = \boldsymbol{0}$.

To eliminate the leading term of the discretization error in the continuity equation, we apply the law of conservation of mass to $\Omega_{i,j}^{(3)} = [x_{i-2}, x_{i+1}] \times [y_{j-2}, y_{j+1}]$:

$$\bar{\bar{u}}_{i+1,j} + \bar{\bar{v}}_{i,j+1} - \bar{\bar{u}}_{i-2,j} - \bar{\bar{v}}_{i,j-2} = 0. \tag{3.30}$$

As noted before, the matrix $\boldsymbol{C}_1 - \mathrm{diag}(\boldsymbol{C}_1)$ is skew-symmetric, because the velocities at the control faces are interpolated with constant coefficients. The same holds for $\boldsymbol{C}_3$. The matrix $\boldsymbol{C}_3 - \mathrm{diag}(\boldsymbol{C}_3)$ is skew-symmetric for all interpolations of $\bar{\bar{u}}$ and $\bar{\bar{v}}$ to the control faces, since the velocities at the control faces are interpolated with constant coefficients, see (3.28). Hence, without its diagonal the coefficient matrix $3^{2+d}\boldsymbol{C}_1(\bar{\boldsymbol{u}}) - \boldsymbol{C}_3(\bar{\bar{\boldsymbol{u}}})$ is skew-symmetric. By substituting the interpolation (3.28) into the semi-discretization (3.26), we obtain the diagonal element

$$\begin{aligned} 3^{d+2} \quad & \frac{1}{2}(\bar{u}_{i+1/2,j} + \bar{v}_{i+1/2,j} - \bar{u}_{i-1/2,j} - \bar{v}_{i+1/2,j-1}) \\ - \quad & \frac{1}{2}(\bar{\bar{u}}_{i+3/2,j} + \bar{\bar{v}}_{i+1/2,j+1} - \bar{\bar{u}}_{i-3/2,j} - \bar{\bar{v}}_{i+1/2,j-2}). \end{aligned} \tag{3.31}$$

For skew-symmetry the interpolation of the $\bar{u}$'s, $\bar{v}$'s, $\bar{\bar{u}}$'s and $\bar{\bar{v}}$'s to the control faces has to be performed in such a way that the diagonal entries of $3^{2+d}\boldsymbol{C}_1(\bar{\boldsymbol{u}}) - \boldsymbol{C}_3(\bar{\bar{\boldsymbol{u}}})$ become equal to zero, that is equal to linear combinations of (3.12) and (3.30). To achieve this, we interpolate $\bar{u}_{i+1/2,j}$ in the following manner

$$\bar{u}_{i+1/2,j} = \frac{1}{2}\alpha(\bar{u}_{i+1,j} + \bar{u}_{i,j}) + \frac{1}{2}(1-\alpha)(\bar{u}_{i+2,j} + \bar{u}_{i-1,j}) \tag{3.32}$$

where α is a constant, and interpolate $\bar{v}_{i+1/2,j}$, $\bar{\bar{u}}_{i+1/2,j}$ and $\bar{\bar{v}}_{i+1/2,j}$ likewise. We take $\alpha = 9/8$ because all interpolations are fourth-order accurate then (on a uniform grid). Note that we can not take $\alpha = 1$ here (as in Eq. (3.22)) since a Richardson extrapolation does not eliminate the leading term in the truncation error of $\bar{\bar{u}}_{i+1/2,j}$ and $\bar{\bar{v}}_{i+1/2,j}$. The interpolation rule (3.32) is also applied in the j-direction to approximate the flux through the faces of v-cells.

The fluxes $\bar{\bar{u}}_{i,j}$ and $\bar{\bar{v}}_{i,j}$ are approximated, so that they can be expressed in terms of the discrete velocities $u_{i,j}$ and $v_{i,j}$, respectively:

$$\bar{\bar{u}}_{i,j} = (y_{j+1} - y_{j-2})u_{i,j} \quad \text{and} \quad \bar{\bar{v}}_{i,j} = (x_{i+1} - x_{i-2})v_{i,j}. \tag{3.33}$$

Hence, on a uniform grid, the fluxes $\bar{\bar{u}}_{i,j}$ and $\bar{\bar{v}}_{i,j}$ are approximated by means of the mid-point rule. In matrix-vector notation, we may summarize the discretization of the law of conservation of mass applied to the volumes $\Omega^{(3)}_{i,j}$ by an expression of the form $\boldsymbol{M}_3\boldsymbol{u}_h = \boldsymbol{0}$. The fourth-order approximation of the law of conservation of mass becomes

$$\boldsymbol{M}\boldsymbol{u}_h = (3^{2+d}\boldsymbol{M}_1 - \boldsymbol{M}_3)\boldsymbol{u}_h = \boldsymbol{0}. \tag{3.34}$$

The weights 3^{2+d} and -1 are to be used on non-uniform grids too, since otherwise the symmetry of the underlying differential operator is lost.

After that the interpolation rule (3.32) is applied, and the flux is expressed in terms of the discrete velocity like in (3.23) and (3.33), the coefficient matrix $3^{2+d}\boldsymbol{C}_1(\bar{\boldsymbol{u}}) - \boldsymbol{C}_3(\bar{\bar{\boldsymbol{u}}})$ becomes a function of the discrete velocity vector $\boldsymbol{u}_h$ only. We will denote that function by $\boldsymbol{C}(\boldsymbol{u}_h)$. Then, the symmetry-preserving discretization of the Navier-Stokes equations (3.1) reads

$$\boldsymbol{\Omega}\frac{d\boldsymbol{u}_h}{dt} + \boldsymbol{C}(\boldsymbol{u}_h)\boldsymbol{u}_h + \boldsymbol{D}\boldsymbol{u}_h - \boldsymbol{M}^*\boldsymbol{p}_h = \boldsymbol{0}, \qquad \boldsymbol{M}\boldsymbol{u}_h = 0, \tag{3.35}$$

where the coefficient matrices $\boldsymbol{C}(\boldsymbol{u}_h)$, $\boldsymbol{D}$ and $\boldsymbol{M}$ are constructed such that the consistency

$$\boldsymbol{C}^*(\boldsymbol{u}_h)\boldsymbol{1} = \boldsymbol{0}, \quad \boldsymbol{D}^*\boldsymbol{1} = \boldsymbol{0} \quad \text{and} \quad \boldsymbol{M}\boldsymbol{1} = \boldsymbol{0} \tag{3.36}$$

and symmetry

$$\boldsymbol{C}(\boldsymbol{u}_h) + \boldsymbol{C}^*(\boldsymbol{u}_h) = \boldsymbol{0}, \qquad \boldsymbol{D} + \boldsymbol{D}^* \text{ positive-definite.} \tag{3.37}$$

conditions are fulfiled. These conditions guarantee that the discretization is fully conservative and stable.

2.4 Boundary conditions

So far, we have left the boundary conditions out of consideration. Their numerical treatment has to maintain the symmetry properties. In case of periodic conditions, the discretization can be extended up to the boundaries in a natural way. This does not break the symmetries of the coefficient matrices $\boldsymbol{C}$ and $\boldsymbol{D}$ nor does it conflict with the consistency conditions given in (3.36). Thus for periodic boundary conditions conservation properties are maintained.

For non-periodic boundary conditions, the requirement $\boldsymbol{M1} = \boldsymbol{0}$ can be met by defining the velocities that form part of the stencil (3.30) and fall outside the flow domain in such a way that (3.30) holds for a constant velocity. At a no-slip wall this can be achieved by mirroring both the grid and velocity normal to the wall. For example, at a wall $y = 0$ the missing, out-of-domain velocity is defined by $u(x, -y) = u(x, y)$. Implicitly, this also defines the out-of-domain pressures. Indeed, by defining $\boldsymbol{M}\boldsymbol{u}_h$ near a boundary we define $\boldsymbol{M}^*\boldsymbol{p}_h$ too.

The discretization of the convective fluxes near the boundaries has to be done such that (a) the skew-symmetry of $\boldsymbol{C}$ is preserved and (b) the row-sums of $\boldsymbol{C}$ are zero (provided that $\boldsymbol{M}\boldsymbol{u}_h = \boldsymbol{0}$). To satisfy these two conditions at a no-slip boundary, we mirror the velocity in the no-slip wall (as before). The mirroring of the velocity does not alter the row-sums of the coefficient matrix $\boldsymbol{C}$. Consequently, the row-sums remain equal to two times the corresponding diagonal entry, and thus it is sufficient to have a zero at the diagonal. We define the value of an out-of-domain convective flux such that the corresponding diagonal entry of $\boldsymbol{C}$ is zero. For example, near the wall $y = 0$ the out-of-domain mass flux $\bar{\bar{u}}_{-1/2,j}$ follows from the requirement that the diagonal entry (3.31) equals zero. That is, for $i = 1$:

$$\begin{aligned} \bar{\bar{u}}_{i-3/2,j} := \quad & -3^{d+2} \quad (\bar{u}_{i+1/2,j} + \bar{v}_{i+1/2,j} - \bar{u}_{i-1/2,j} - \bar{v}_{i+1/2,j-1}) \\ & + \quad (\bar{\bar{u}}_{i+3/2,j} + \bar{\bar{v}}_{i+1/2,j+1} - \bar{\bar{v}}_{i+1/2,j-2}). \end{aligned}$$

In this way the boundary conditions are built into the coefficient matrices $\boldsymbol{M}$ and $\boldsymbol{C}$ without violating (3.26) and (3.20). Thus also for non-periodic conditions, the mass, momentum and kinetic energy are conserved if $\boldsymbol{D} = \boldsymbol{0}$.

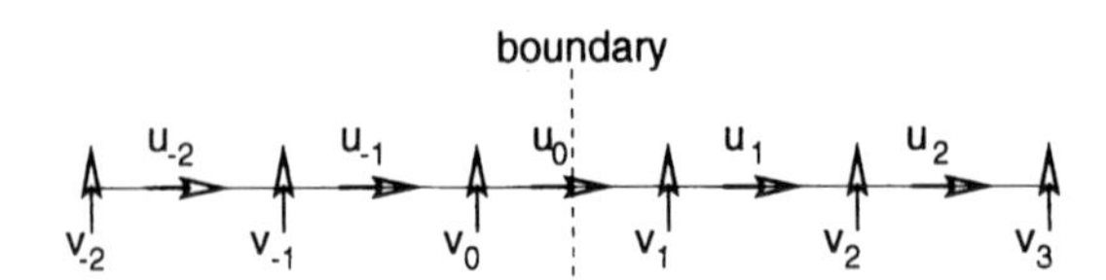

Figure 3.3. The location of the ghost velocities. Here, the velocity normal to the wall is denoted by u; v represents the tangential velocity. The discrete velocity u_0 lies at the wall. u_{-1}, u_{-2}, v_0 and v_{-1} are ghost velocities. The other discrete velocities lie in the fluid.

The diffusive fluxes through near-wall control faces are discretized such that the resulting coefficient matrix $\boldsymbol{D}$ is symmetric. The symmetry of $\boldsymbol{D}$ is preserved if the velocity-gradient is mirrored in a no-slip wall. We implement this condition by means of ghost velocities.

Figure 3.3 illustrates the positioning of the ghost velocities near a Dirichlet boundary $y = 0$. The velocity at the wall is given by (u_Γ, v_Γ). The grid is also mirrored in $y = 0$. The symmetry of the coefficient matrix $\boldsymbol{D}$ is unbroken if the near-boundary diffusive fluxes are computed with the help of

$$\begin{aligned} u_0 &= u_\Gamma & \qquad v_\Gamma - v_0 &= v_1 - v_\Gamma \\ u_\Gamma - u_{-1} &= u_1 - u_\Gamma & \qquad v_\Gamma - v_{-1} &= v_2 - v_\Gamma \\ u_\Gamma - u_{-2} &= u_2 - u_\Gamma & \qquad v_\Gamma - v_{-2} &= v_3 - v_\Gamma \end{aligned}$$

3. A test-case: turbulent channel flow

In this section, the symmetry-preserving discretization is tested for turbulent channel flow. The Reynolds number is set equal to Re = 5,600 (based on the channel width and the bulk velocity), a Reynolds number at which direct numerical simulations have been performed by several research groups; see [12]-[14]. In addition we can compare the numerical results to experimental data from Kreplin and Eckelmann [15].

As usual, the flow is assumed to be periodic in the stream- and span-wise direction. Consequently, the computational domain may be confined to a channel unit of dimension $2\pi \times 1 \times \pi$, where the width of the channel is normalized. All computations presented in this section have been performed with 64 (uniformly distributed) stream-wise grid points and 32 (uniformly distributed) span-wise points. In the lower-half of the channel, the wall-normal grid points are computed according to

$$y_j = \frac{\sinh(\gamma j / N_y)}{2 \sinh(\gamma/2)} \qquad \text{with} \quad j = 0, 1, ..., N_y/2,$$

where N_y denotes the number of grid points in the wall-normal direction. The stretching parameter γ is taken equal to 6.5. The grid points in the upper-half are computed by means of symmetry.

The temporal integration of (3.1) is performed with the help of a one-leg method that is tuned to improve its convective stability [16]. The non-dimensional time step is set equal to $\delta t = 1.25\ 10^{-3}$. Mean values of computational results are obtained by averaging the results over the directions of periodicity, the two symmetrical halves of the channel, and over time. The averaging over time starts after a start-up period. The start-up period as well as the time-span over which the results are averaged, 1500 non-dimensional time-units, are identical for all the results shown is this section. Figure 3.4 shows

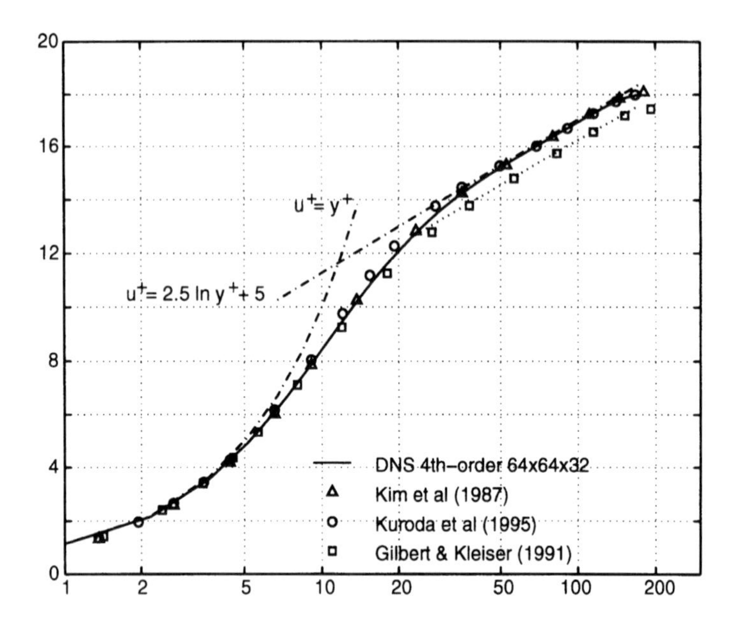

Figure 3.4. The mean stream-wise velocity u^+ versus y^+. The dashed lines represent the law of the wall and the log law. The markers represent DNS-results that are taken from the ERCOFTAC Database.

a comparison of the mean velocity profile as obtained from our fourth-order symmetry-preserving simulation ($N_y = 64$) with those of other direct numerical simulations. Here it may be stressed that the grids used by the DNS's that we compare with have typically about 128^3 grid points, that is 16 times more grid points than our grid has. Nevertheless, the agreement is excellent.

To investigate the convergence of the fourth-order method upon grid refinement, we have monitored the skin friction coefficient C_f as obtained from simulations on four different grids. We will denote these grids by A, B, C and D. Their spacings differ only in the direction normal to the wall. They have $N_y = 96$ (grid A), $N_y = 64$ (B), $N_y = 56$ (C) and $N_y = 48$ (D) points in the wall-normal direction, respectively. The first (counted from the wall) grid line used for the convergence study is located at $y_1^+ \approx 0.95$ (grid A), $y_1^+ \approx 1.4$ (B), $y_1^+ \approx 1.6$ (C), and $y_1^+ \approx 1.9$ (D), respectively. Figure 3.5 displays the skin friction coefficient C_f as function of the fourth power of y_1^+. The convergence study shows that the discretization scheme is indeed fourth-order accurate (on a non-uniform mesh). This indicates that the underlying physics is resolved when 48 or more grid points are used in the wall normal direction. In terms of the local grid spacing (measured by y_1^+), the skin friction coefficient is approximately given by $C_f = 0.00836 - 0.000004(y_1^+)^4$. The extrapolated value at $y_1^+ = 0$ lies in between the C_f reported by Kim *et al.* [12] and Dean's correlation of $C_f = 0.073\ Re^{-1/4} = 0.00844$ [17].

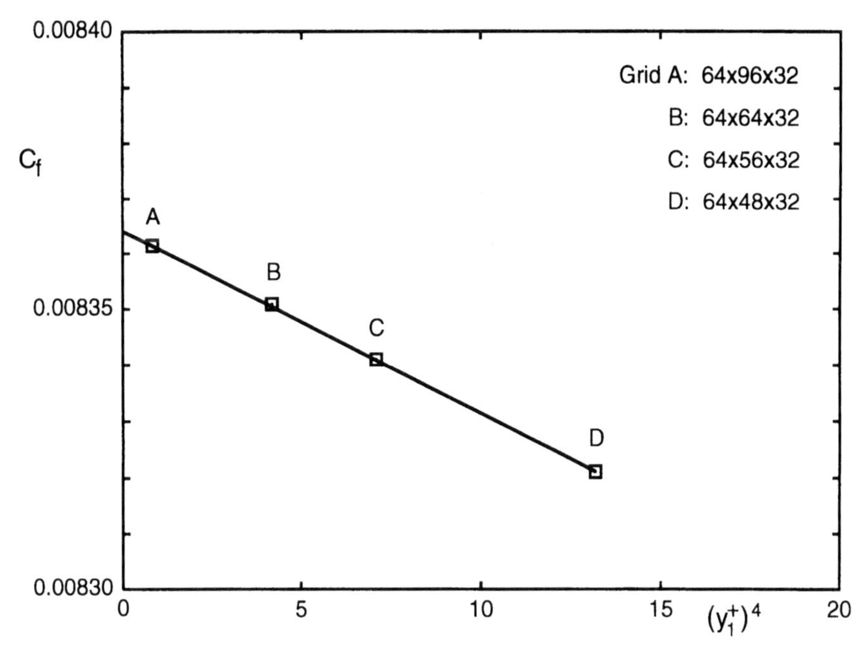

Figure 3.5. Convergence of the skin friction coefficient C_f upon grid refinement. The figure displays C_f versus the fourth power of the first grid point y_1^+.

The convergence of the fluctuating stream-wise velocity near the wall ($0 < y^+ < 20$) is presented in Figure 3.6. Here, we have added results obtained on three still coarser grids (with $N_y = 32$, $N_y = 24$ and $N_y = 16$ points in the wall-normal direction, respectively), since the results on the grids A, B, C and D fall almost on top of each other. The coarsest grid, with only $N_y = 16$ points to cover the channel width, is coarser than most of the grids used to perform a large-eddy simulation (LES) of this turbulent flow. Nevertheless, the $64 \times 16 \times 32$ solution is not that far off the solution on finer grids, in the near wall region. Further away from the wall, the turbulent fluctuations predicted on the coarse grids ($N_y \leq 32$) become too high compared to the fine grid solutions, as is shown in Figure 3.7.

The solution on the $64 \times 24 \times 32$, for example, forms an excellent starting point for a large-eddy simulation. The root-mean-square of the fluctuating stream-wise velocity is not far of the fine grid solution, and viewed through physical glasses, the energy of the resolved scales of motion, the coarse grid ($N_y = 24$) solution, is convected in a stable manner, because it is conserved by the discrete convective operator. Therefore, we think that the symmetry-preserving discretization forms a solid basis for testing sub-grid scale models. The discrete convective operator transports energy from a resolved scale of motion to other resolved scales without dissipating any energy, as it should do from a physical point of view. The test for a sub-grid scale model then

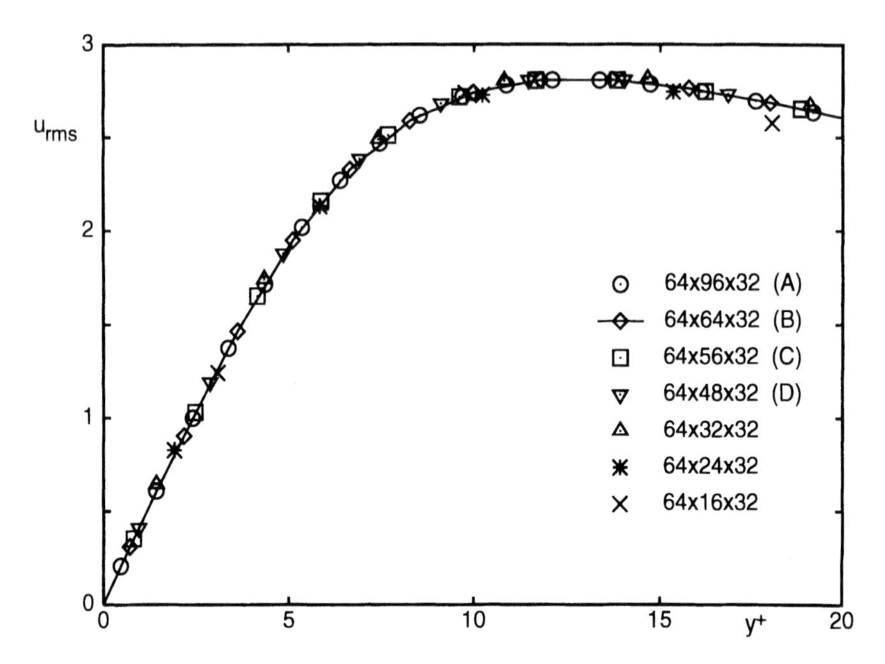

Figure 3.6. The root-mean-square velocity fluctuations normalized by the wall shear velocity as function of the wall coordinate y^+ on various grids for $y^+ \leq 20$. The markers correspond to the results obtained in the grid points. The solution on grid B is also represented by a continuous line.

reads: does the addition of the dissipative sub-grid model to the conservative convection of the resolved scales reduce the error in the computation of u_{rms}.

The results for the fluctuating stream-wise velocity u_{rms} are compared to the experimental data of Kreplin and Eckelmann [15] and to the numerical data of Kim *et al.* [12] in Fig. 3.8. This comparison confirms that the fourth-order, symmetry-preserving method is more accurate than the second-order method. With 48 or more grid points in the wall normal direction, the root-mean-square of the fluctuating velocity obtained by the fourth-order method is in close agreement with that computed by Kim *et al.* [12] for $y^+ > 20$ (Figure 3.8 shows this only for y^+ up to 40; yet, the agreement is also excellent for $y^+ > 40$). In the vicinity of the wall ($y^+ < 20$), the velocity fluctuations of the fourth-order simulation method fit the experiment data nicely, even up to very coarse grids with only 24 grid points in the wall-normal direction. However, the turbulence intensity in the sub-layer ($0 < y^+ < 5$) predicted by the simulations is higher than that in the experiment. According to the fourth-order simulation the root-mean-square approaches the wall like $u_{\text{rms}} \approx 0.38y^+$ ($N_y = 64$). The exact value of this slope is hard to pin-point experimentally. Hanratty *et al.* [18] have fitted experimental data of several investigators, and thus came to 0.3. Most direct numerical simulations yield higher values. Kim *et al.* [12] and

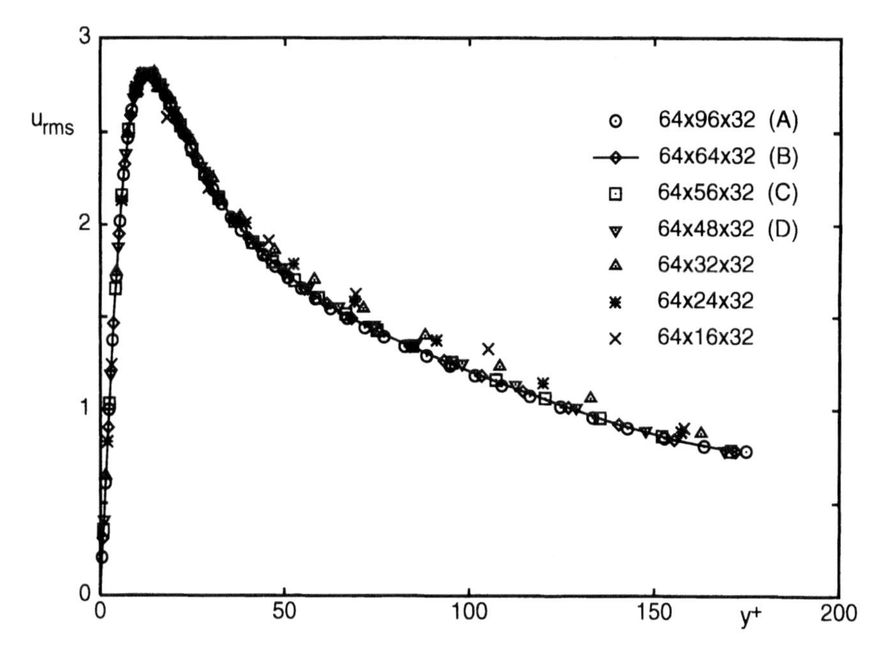

Figure 3.7. The root-mean-square velocity fluctuations normalized by the wall shear velocity for $y^+ \leq 200$ on various grids.

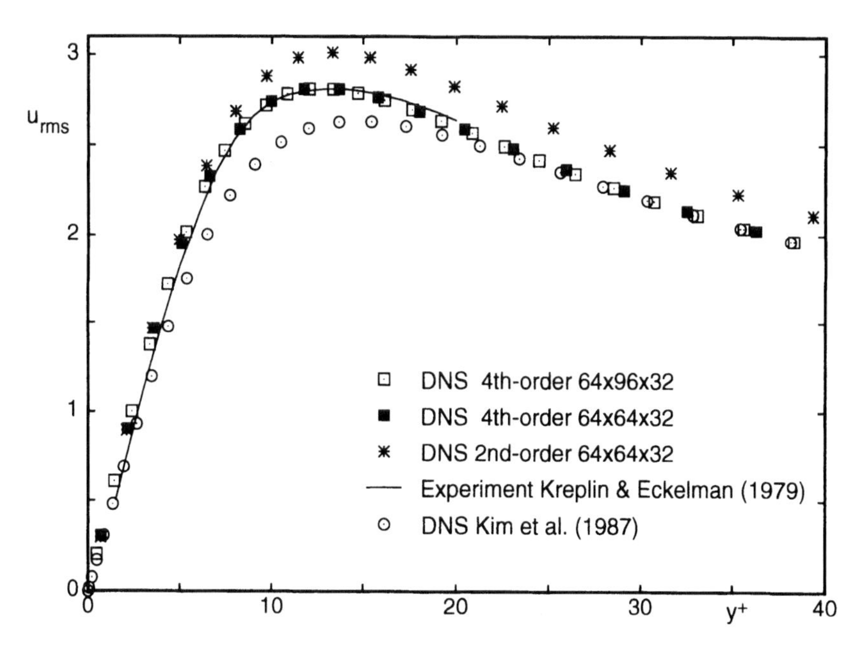

Figure 3.8. Comparison of the mean-square of the stream-wise fluctuating velocity as function of y^+.

Gilbert and Kleiser [13] have found slopes of 0.3637 and 0.3824 respectively, which is in close agreement with the present findings.

So, in conclusion, the results of the fourth-order symmetry-preserving discretization agree better with the available reference data than those of its second-order counterpart, and with the fourth-order method a $64 \times 64 \times 32$ grid suffices to perform an accurate DNS of a turbulent channel flow at Re=5,600.

4. Conclusions

The smallest scales of motion in a turbulent flow result from a subtle balance between convective transport and diffusive dissipation. In mathematical terms, the balance is an interplay between two differential operators differing in symmetry: the convective derivative is skew-symmetric, whereas diffusion is governed by a symmetric, positive-definite operator. With this in mind, we have developed a spatial discretization method which preserves the symmetries of the balancing differential operators. That is, convection is approximated by a skew-symmetric discrete operator, and diffusion is discretized by a symmetric, positive-definite operator. Second-order and fourth-order versions have been developed thus far, applicable to structured non-uniform grids. The resulting semi-discrete representation conserves mass, momentum and energy (in the absence of physical dissipation). As the coefficient matrices are stable and non-singular, a solution can be obtained on any grid, and we need not add an artificial damping mechanism that will inevitably interfere with the subtle balance between convection and diffusion at the smallest length scales. This forms our motivation to investigate symmetry-preserving discretizations for direct numerical simulation (DNS) of turbulent flow. Because stability is not an issue, the main question becomes how accurate is a symmetry-preserving discretization, or stated otherwise, how coarse may the grid be for a DNS? This question has been addressed for a turbulent channel flow. The outcomes show that with the fourth-order method a $64 \times 64 \times 32$ grid suffices to perform an accurate DNS of a turbulent channel flow at Re=5,600.

References

[1] R.W.C.P. Verstappen and A.E.P. Veldman, Spectro-consistent discretization of Navier-Stokes: a challenge to RANS and LES, *J. Engng. Math.* **34**, 163 (1998).

[2] Y. Morinishi, T.S. Lund, O.V. Vasilyev and P. Moin, Fully conservative higher order finite difference schemes for incompressible flow, *J. Comp. Phys.* **143**, 90 (1998).

[3] O.V. Vasilyev, High order finite difference schemes on non-uniform meshes with good conservation properties. *J. Comp. Phys.* **157**, 746

(2000).

[4] F. Ducros, F. Laporte, T. Soulères, V. Guinot, P. Moinat and B. Caruelle, High-order fluxes for conservative skew-symmetric-like schemes in structured meshes: application to compressible flows. *J. Comp. Phys.* **161**, 114 (2000).

[5] A. Twerda, A.E.P. Veldman and S.W. de Leeuw, High order schemes for colocated grids: Preliminary results. In: *Proc. 5th annual conference of the Advanced School for Computing and Imaging* 286 (1999).

[6] A. Twerda, Advanced computational methods for complex flow simulation. Delft University of Technology, PhD thesis (2000).

[7] J.M. Hyman, R.J. Knapp and J.C. Scovel, High order finite volume approximations of differential operators on nonuniform grids, *Physica D* **60**, 112 (1992).

[8] T.A. Manteufel and A.B. White, Jr., The numerical solution of second-order boundary value problems on nonuniform meshes, *Math. of Comp.* **47**, 511 (1986).

[9] A.E.P. Veldman and K. Rinzema, Playing with nonuniform grids, *J. Engng. Math* **26**, 119 (1991).

[10] F.H. Harlow and J.E. Welsh, Numerical calculation of time-dependent viscous incompressible flow of fluid with free surface, *Phys. Fluids* **8**, 2182 (1965).

[11] M. Antonopoulos-Domis, Large-eddy simulation of a passive scalar in isotropic turbulence, *J. Fluid Mech.* **104**, 55 (1981).

[12] J. Kim, P. Moin and R. Moser, Turbulence statistics in fully developed channel flow at low Reynolds number, *J. Fluid Mech.* **177**, 133 (1987).

[13] N. Gilbert and L. Kleiser, Turbulence model testing with the aid of direct numerical simulation results, in *Proc. Turb. Shear Flows 8*, Paper 26-1, Munich (1991).

[14] A. Kuroda, N. Kasagi and M. Hirata, Direct numerical simulation of turbulent plane Couette-Poisseuille flows: effect of mean shear rate on the near-wall turbulence structures, in: *Proc Turb. Shear Flows 9*, F. Durst *et al.* (eds.), Berlin: Springer-Verlag 241-257 (1995).

[15] H.P. Kreplin and H. Eckelmann, Behavior of the three fluctuating velocity components in the wall region of a turbulent channel flow, *Phys Fluids* **22**, 1233 (1979).

[16] R.W.C.P. Verstappen and A.E.P. Veldman, Direct numerical simulation of turbulence at lower costs, *J. Engng. Math.* **32**, 143 (1997).

[17] R.B. Dean, Reynolds number dependence of skin friction and other bulk flow variables in two-dimensional rectangular duct flow, *J. Fluids. Engng.* **100**, 215 (1978).

[18] T.J. Hanratty, L.G. Chorn and D.T. Hatziavramidis, Turbulent fluctuations in the viscous wall region for Newtonian and drag reducing fluids, *Phys Fluids* **20**, S112 (1977). *Int. J. Heat Mass Transfer* **24**, 1541 (1981).

Chapter 4

ANALYSIS AND CONTROL OF ERRORS IN THE NUMERICAL SIMULATION OF TURBULENCE

Sandip Ghosal

Department of Mechanical Engineering, Northwestern University, 2145 Sheridan Road, Evanston, IL 60208, USA

s-ghosal@northwestern.edu

Abstract Turbulent flows are characterised by a continuum of length and time scales, a feature that introduces some unique problems in relation to the analysis and control of errors in numerical simulations of turbulent flows. In direct numerical simulation (DNS) one attempts to fully resolve the flow field. The primary sources of error in DNS are the aliasing error, which arises due to the evaluation of the nonlinear term on a discrete grid in physical space, and, the truncation error, due to the discretization of the derivative operator. In addition there are the time-stepping errors on account of the temporal discretization. In Large Eddy Simulation (LES) only the large scale flow is computed whereas the collective effect of the small scales are modeled. In that case, in addition to the above three types of errors, there are commutation errors arising out of the averaging process used to derive the LES equations and sub-grid modeling errors that arise because in the absence of a systematic method, the subgrid stress is evaluated using an ad hoc closure model. Methods for analyzing and quantifying these errors in turbulence simulations are discussed. In some instances the analysis points to suitable methods of error control.

Keywords: Turbulence, LES, DNS, aliasing error, truncation error, sub-grid modeling

1. Introduction

Estimation and control of numerical errors is the central issue in all areas of science and engineering where we seek to use numerical simulation to make useful predictions. Not surprisingly, the subject has received a lot of attention from physical scientists and applied mathematicians and is a well established branch of Applied Mathematics. The numerical simulation of turbulence, how-

D. Drikakis and B.J. Geurts (eds.), Turbulent Flow Computation, 101–140.

ever, presents certain issues that are peculiar to that subject, and it is to this special area that this chapter is devoted.

In many areas of scientific computation, one has a set of equations for which the existence and uniqueness of solutions under given initial and boundary conditions are known, though it may not be possible to obtain the solution in analytical form. In such situations one can solve the problem numerically, and the accuracy of the solution is judged by its deviation from the exact solution, which, may be known perhaps from experiments. In turbulence, first, the existence of solutions to the equations themselves have not been established. Secondly, the computed solution certainly cannot be expected to be "close" to the real solution in the usual sense. The equations are nonlinear and the system is chaotic, so that even infinitesimal errors are amplified exponentially. To the "practitioner" numerical solutions are considered "successful" if statistical measures of interest (such as energy spectra) are correctly predicted or the simulation correctly predicts the time evolution of large scale structures (the path of a large hurricane for example) over a certain time period of interest. To a mathematician, perhaps, the numerical solution of such a chaotic system is "accurate" if it has an attractor that is "close" to the attractor of the original system with respect to some appropriate measure. However, issues of what the dimension of the attractor is for the Navier-Stokes system (if one does exist) remains unknown. Thus, even a precise formulation of the sense in which turbulence computations are "accurate" is frought with enormous difficulties. No attempt will be made here to answer these deep mathematical questions in computational turbulence. Instead, we adopt here a much more pedantic approach. We will attempt to quantify numerical errors by studying the power spectra of the "residuals" that appear on the right hand side of the Navier Stokes equations when it is approximated by a discrete system. The effect of this additional forcing on the dynamics of the system is much more difficult to quantify, and, will not be addressed here.

2. Sources of Errors & Their Nature

We begin first with a brief overview of the various sources of errors in turbulence computations and a qualitative discussion as to their nature.

2.1 Discrete Representation

Unlike real space and time, which are continuous, the computer, which has a finite memory, must operate on a discrete representation of it. Thus, the velocity components $u_i(x, y, z, t)$ $(i = 1, 2, 3)$ must be approximated on a finite basis:

$$u_i(x, y, z, t) = \sum_{n=1}^{\infty} c_n w_n \approx \sum_{n=1}^{N} c_n w_n \tag{4.1}$$

where the replacement $\infty \to N$ constitutes the process of discretization. The larger N is, the more accurate is the representation, and of course, the higher is the computational cost. In a spectral method, the "w_n" are certain simple analytical functions (such as sines, cosines Legendre polynomials etc.). In a finite element method, the "w_n" are an appropriate finite element basis, usually simple polynomials with compact support. The difference

$$e = \left| \sum_{n=1}^{\infty} c_n w_n - \sum_{n=1}^{N} c_n w_n \right| \tag{4.2}$$

is the error due to the discretization of the system. Naturally, one would prefer to choose the basis w_n in such a way that the error is a minimum for any given value of N. In this regards, spectral methods are the best, as they have the property of "exponential convergence". That is, for infinitely smooth functions, e decreases faster than any negative power of N as $N \to \infty$. Spectral methods however can be applied only for simple boundaries where boundary conditions can be satisfied in a straight forward way (such as, by requiring all basis elements to satisfy the boundary conditions individually). When this is not the case, the advantage of exponential convergence must be sacrificed in favor of the simplicity of implementation of finite difference and finite element methods. Functions which are discontinous (such as the density near a shock wave) produce undesirable oscillations (Gibbs oscillations) and may not be easily handled with spectral methods either.

2.2 Differentiation

In addition to the error inherent in the representation of an infinite system by a finite one, errors are introduced whenever exact mathematical operations are replaced by approximate numerical equivalents. The most common example is $\frac{d}{dx} \to \frac{\delta}{\delta x}$ the replacement of the differentiation operator d/dx by its numerical equivalent $\delta/\delta x$. The effect of such an approximation is easily quantified. For example, for the central difference scheme,

$$\frac{\delta}{\delta x} u = \frac{u(x+\Delta) - u(x-\Delta)}{2\Delta} \tag{4.3}$$

we have

$$\frac{\delta}{\delta x} u = \frac{u(x+\Delta) - u(x-\Delta)}{2\Delta} = u'(x) - \frac{\Delta^2}{6} u''' + \cdots \tag{4.4}$$

so that the error at leading order scales as the square of the grid spacing, Δ. That is, we have here a second order method.

The order of the method is not always the best indicator of its desirability. The "spectral resolution" could be more important. The concept is illustrated

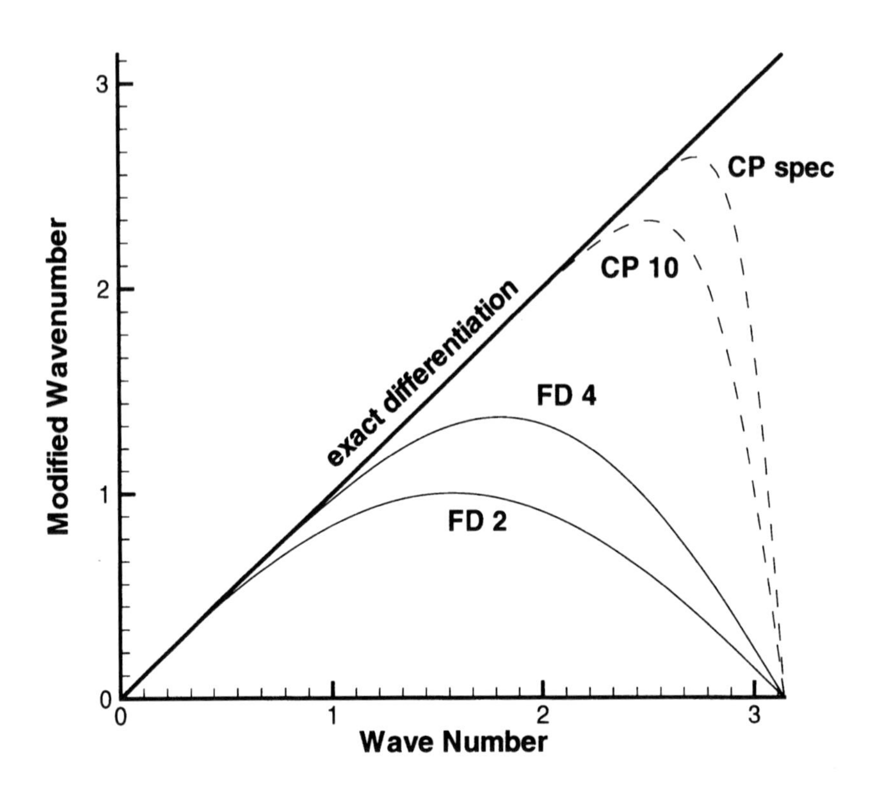

Figure 4.1. Modified wave numbers for central difference schemes (solid lines) of 2nd (FD 2) & 4th (FD 4) order, Lele's pentadiagonal compact schemes (dashed lines) of highest (10th order) formal accuracy (CP 10) & of 4th order formal accuracy with "spectral-like" resolution (CP spec).

most clearly by applying the operators d/dx and its approximation $\delta/\delta x$ to the function $\exp(ikx)$:

$$\frac{d}{dx}\exp(ikx) = ik\exp(ikx) \tag{4.5}$$

$$\begin{aligned}\frac{\delta}{\delta x}\exp(ikx) &= \frac{1}{2\Delta}\exp[ik(x+\Delta)] - \exp[ik(x-\Delta)] \\ &= i\frac{\sin(k\Delta)}{\Delta}\exp(ikx) \equiv ik'\exp(ikx)\end{aligned} \tag{4.6}$$

where

$$k' = \frac{\sin(k\Delta)}{\Delta} \tag{4.7}$$

is called the "modified wavenumber". The "modified wavenumber" is simply the numerical differentiation operator expressed in Fourier-space [1]. Clearly, as $\Delta \to 0, k' \to k$. An idea of how "close" k' is to k provides a sense of the accuracy of the approximate differentiation scheme. In the Fourier-spectral method, differentiation is actually performed by multiplying the Fourier-Transform of the dependent variable by the wavenumber k. Thus, the Fourier-spectral method has "infinite resolution". Figure 4.1 shows k' as a function of k for the second order central difference scheme, the fourth order central difference scheme, and, the Fourier-spectral scheme (corresponding to exact differentiation). Also shown are the modified wave numbers for two "compact schemes" designed by Lele [2]. Compact schemes combine the flexibility of finite difference methods while being "spectral like" in their accuracy. This combination makes them extremely suitable for turbulence simulations and a brief account of them is provided here (a complete discussion may be found in [2]).

2.2.1 Compact Schemes. The general pentadiagonal compact scheme for evaluating the first derivative f_i' of a function with values f_i at discrete grid points indexed by i, is defined through the relation:

$$\beta f_{i-2}' + \alpha f_{i-1}' + f_i' + \alpha f_{i+1}' + \beta f_{i+2}' = c\frac{f_{i+3} - f_{i-3}}{6\Delta} + b\frac{f_{i+2} - f_{i-2}}{4\Delta} + a\frac{f_{i+1} - f_{i-1}}{2\Delta} \tag{4.8}$$

Determination of f_i' in general involves solving a set of equations generated by letting i vary from $i = 1$ to $i = N$, or equivalently, inverting the matrix form of (4.8):

$$\mathbf{A}\mathbf{f}' = \mathbf{B} \tag{4.9}$$

which involves inverting the pentadiagonal matrix $\mathbf{A}$. The coefficients α, β, a, b and c are to be suitably chosen to ensure that f_i' so obtained is a "good" approximation to the true derivative of f at grid point i where the sense in which an approximation is "good" remains to be defined. An obvious way to define "good" is to require that the coefficients be chosen, such that, the Taylor expansions of the left and right hand sides about the grid point i match up to terms of order Δ^n. In this way, it is easy to show, that the choice

$$a = \frac{1}{3}(4 + 2\alpha - 16\beta + 5c) \tag{4.10}$$

$$b = \frac{1}{3}(-1 + 4\alpha + 22\beta - 8c) \tag{4.11}$$

results in an error term

$$\frac{4}{5}(-1 + 3\alpha - 12\beta + 10c)\Delta^4 f_i^{(5)}, \tag{4.12}$$

so that, we have a family of fourth order schemes parametrized by the three independent coefficients α, β and c. Here $f_i^{(n)}$ denotes the n th derivative of the function f at grid location 'i'. At this stage it may appear that we have achieved nothing that would compensate for the added cost (compared to a finite difference scheme) involved in inverting the pentadiagonal matrix $\mathbf{A}$. However, we now have the freedom of choosing the coefficients α, β and c in an optimum way. An obvious choice would be to make the remainder term (4.12), and, as many subsequent terms as possible identically zero. This uses up all of the remaining degrees of freedom:

$$\beta = \frac{43}{1798} \quad \alpha = \frac{334}{899} \quad a = \frac{1065}{1798} \quad b = \frac{1038}{899} \quad c = \frac{79}{1798} \tag{4.13}$$

and leads to a scheme that has the highest possible formal accuracy, namely, a tenth order scheme. The interesting insight provided by Lele is that achieving the highest order in terms of formal accuracy is not necessary the best way to "spend" the degrees of freedom afforded by the three floating parameters. Instead, one could require, that, the modified wave number k' for the scheme be as close to the ideal $k' = k$ provided by truly spectral schemes. There are many ways to define "close". One possibility is to require that $k' = k$ at three values (since there are three degrees of freedom) $k = k_1$, $k = k_2$ and $k = k_3$ where k_1, k_2 and k_3 are chosen so as to "straighten out" the modified wavenumber curve as much as possible, especially near the higher wavenumbers in the interval $k_{max} \geq k \geq 0$. The modified wavenumber for the general pentadiagonal compact scheme is easily found on substitution of $f = \exp(ikx)$ in (4.8) which gives

$$k' = \frac{a \sin k + (b/2)\sin(2k) + (c/3)\sin(3k)}{1 + 2\alpha\cos(k) + 2\beta\cos(2k)}. \tag{4.14}$$

The fourth-order "spectral-like" pentadiagonal compact scheme of Lele uses $k_1 = 2.2$, $k_2 = 2.3$, $k_3 = 2.4$ ($k_{max} = \pi$) to determine $\alpha = 0.5771439$, $\beta = 0.0896406$, $a = 1.3025166$, $b = 0.9935500$ and $c = 0.03750245$. Figure 4.1 shows the modified wavenumber for this scheme together with the modified wavenumber for the tenth order scheme discussed above. It is clear that the "spectral-like" scheme has a higher "resolving power' even though it has only fourth order formal accuracy. This distinction between "formal accuracy" and resolution is an extremely important one for turbulence simulations, and it is a matter to which we shall return later.

2.3 Aliasing

The aliasing error comes about due to the method of computation of the nonlinear term [3] in the Navier-Stokes (or LES) equations.

For illustration, let us suppose, that we are seeking a numerical solution to the one dimensional Burger's equation (a simple analog of the Navier Stokes equations in one dimension):

$$\frac{\partial u}{\partial t} + u\frac{\partial u}{\partial x} = \nu\frac{\partial^2 u}{\partial x^2}. \tag{4.15}$$

If we are seeking periodic solutions in x, and the domain is normalized so that $2\pi \geq x \geq 0$, we may seek solutions in the form

$$u = \sum_{k=-\infty}^{+\infty} \hat{u}_k \exp(ikx) \tag{4.16}$$

where k is an integer, and, the $\hat{u}_k$ are the Fourier coefficients of u. On substitution of (4.16) in (4.15) we find, that, the Fourier coefficients $\hat{u}_k$ must satisfy the ordinary differential equations:

$$\frac{d\hat{u}_k}{dt} = -ik \sum_{\substack{m,n \\ m+n=k}} \hat{u}_m\hat{u}_n - \nu k^2 \hat{u}_k \tag{4.17}$$

where m and n can take on any integer values subject to the constraint $m+n = k$. In a numerical simulation, one has information on only a finite set of modes $|k| \leq \frac{N}{2}$, where N is an even integer that defines the size of the computation. The resolved modes $\hat{u}_k$ ($|k| \leq N/2$) then evolve in time according to the equation

$$\frac{d\hat{u}_k}{dt} = -ik \sum_{\substack{|m|\&|n|\leq N/2 \\ m+n=k}} \hat{u}_m\hat{u}_n - ik \sum_{\substack{|m|or|n|>N/2 \\ m+n=k}} \hat{u}_m\hat{u}_n - \nu k^2 \hat{u}_k \tag{4.18}$$

in which, all of the terms are known exactly except for the second term on the right hand side. In DNS, it is simply set to zero. In LES, this term is approximated by a "subgrid model," and, the resulting error (the "subgrid modeling error") is the only source of inaccuracy, except for the time-stepping error introduced by the discretization of the d/dt operator. If equation (4.18) is integrated directly as a family of ordinary differential equations to determine the Fourier coefficients $\hat{u}_k$ ($|k| \leq N/2$), we are using a "Fourier-spectral method" which has no numerical errors associated with it except for (a) the subgrid modeling error associated with the replacement of the second term on the right hand side by a suitable approximation (b) the temporal discretization error.

The Fourier-spectral method however, is expensive, as, it involves N multiplications for each of the N modes, so that the operation count in evaluating the nonlinear term $\sim O(N^2)$. A cheaper method is the "Fourier pseudo-spectral" method, where the Fourier inverse is first computed to find the value of u at the $N+1$ grid points $x_j = (2\pi/N)j$, $j = 0, 1, 2, \ldots N$, where $u_j \equiv u(x_j)$, and

$u_0 = u_N$. The $\hat{u}_k$ and u_j may be readily interconverted through the "Discrete Fourier Transform" (DFT) and its inverse defined through the exact relations

$$\hat{u}_k = \frac{1}{N}\sum_{j=0}^{N-1} u_j \exp(-ikx_j) \tag{4.19}$$

$$u_j = \sum_{k=-N/2}^{N/2-1} \hat{u}_k \exp(ikx_j). \tag{4.20}$$

The DFT can be implemented in $O(N \log N)$ operations through an algorithm due to Cooley & Tukey [4]. Since the nonlinear operation $u\partial_x u = \partial_x(u^2/2)$ is compact in physical space, evaluation of u^2 in the whole domain requires only $O(N)$ operations. Therefore, the following method of computation of the k-th Fourier mode,

$$\frac{d\hat{u}_k}{dt} = -ik\widehat{(u_j^2)}_k - ik(SGS) - \nu k^2 \hat{u}_k, \tag{4.21}$$

where (SGS) is to be replaced by the relevant subgrid model (LES) or zero (DNS), involves only $O(N \log N)$ operations, a substantial cost saving if N is large (for example $N > 10^5$). Here u_j and $\widehat{(u_j^2)}_k$ are to be evaluated using the inverse and direct DFT respectively.

The saving in computational cost, however comes at a price. To see why this is so, we need to express the term ${u_j}^2$ in (4.21) in terms of the $\hat{u}_k$ using (4.20). Then we get

$$\frac{d\hat{u}_k}{dt} + \nu k^2 \hat{u}_k = -ik \sum_{\substack{|m|\&|n|\leq N/2 \\ m+n=k}} \hat{u}_m \hat{u}_n - ik \sum_{\substack{|m|\&|n|\leq N/2 \\ m+n=k\pm N}} \hat{u}_m \hat{u}_n - ik(SGS) \tag{4.22}$$

The terms

$$-ik \sum_{\substack{|m|\&|n|\leq N/2 \\ m+n=k\pm N}} \hat{u}_m \hat{u}_n$$

are not present in (4.18), they appear as a result of computing the nonlinear product on a discrete grid in physical space. This collection of spurious terms is called the "aliasing error". It appears not only in the Fourier pseudo-spectral method, but in any numerical method where the nonlinear term is computed by multiplying an approximate representation of the variable on a discrete grid in physical space. The name "aliasing error" hints at its origin. Nonlinear products create modes of higher frequency, and, when these can no longer be resolved on the grid, it becomes indistinguishable from, or, in other words, "aliases to" a lower frequency mode on the grid.

It is clear from the condition $m+n = k \pm N$ that the aliasing error is important only when the modes near the highest resolvable wavenumbers carry significant energy. Since this is certainly the case in LES and true to some extent in DNS as well, aliasing errors are extremely important in turbulence computations. The negative effect of aliasing errors was recognized early, when it was found that these spurious terms can lead to numerical instabilities [5].

Several methods for the control of aliasing errors have been devised. For example, the "3/2 rule" involves expanding the number of Fourier modes $N \to (3/2)N$ and "padding" the newly created modes with zeroes. The inverse of the DFT is then used to obtain the u_j in physical space and the nonlinear term computed by multiplication at each grid point. The result is then transformed back to Fourier space using the DFT and the number of modes is contracted back, $(3/2)N \to N$. This "dealiasing rule" completely removes the error associated with the quadratic nonlinearity in the Fourier pseudo-spectral method. The expansion to a $(3/2)N$ grid however is expensive. The method of "phase randomization" [2] greatly reduces the aliasing error without removing it completely. It however has the advantage that expansion to a larger set of modes and the associated rise in computational cost is avoided. Surprisingly, the truncation error introduced by the finite differencing operator itself plays a role in reducing aliasing errors! This is because, the $-ik$ (in the case of the Burger's equation, a more complicated expression involving wavevectors in the case of the Navier Stokes equations) premultiplying the aliasing error term gets replaced by $-ik'$, where k' is the modified wavenumber. From Figure 4.1 $k' \to 0$ as $k \to k_{max}$, so that the aliasing error term gets supressed. Therefore, control of aliasing error becomes more crucial with increasing accuracy of the method. The nonlinear term in the Burgers equation (4.15) can be written in an alternate form (the conservative form):

$$u\frac{\partial u}{\partial x} = \frac{\partial}{\partial x}\left(\frac{u^2}{2}\right)$$

Though these forms are completely equivalent, their discretization lead to completely different numerical schemes with distinct properties, including different aliasing errors. The effect on aliasing errors of various alternate forms for the nonlinear term in the Navier-Stokes equations was studied by Blaisdel *et al.* [7]. The skew symmetric form was found to have the least aliasing error.

2.4 Subgrid modeling

In LES, the computation is restricted to the large scales while the collective effect of all the small scales is modeled through a "subgrid" term. The separation of large and small scales is achieved through the formal procedure discussed in the next section of "filtering" the Navier Stokes equations. For a uniform filter size, this results in a set of equations for the filtered variables $(\bar{u}_i, \bar{p})$ that satisfy the Navier Stokes equations augmented with an additional "subgrid stress" term accounting for the collective effect of the small scales on the LES equations. For constant density flows, these equations are (we will use index notation with the Einstein summation convention throughout this chapter)

$$\frac{\partial \bar{u}_i}{\partial t} + \bar{u}_k \frac{\partial \bar{u}_k}{\partial x_i} = -\frac{\partial \tau_{ij}}{\partial x_j} + \nu \nabla^2 \bar{u}_i \tag{4.23}$$

$$\frac{\partial \bar{u}_i}{\partial x_i} = 0, \tag{4.24}$$

where $\tau_{ij} \equiv \overline{u_i u_j} - \bar{u}_i \bar{u}_j$ is the subgrid stress tensor. Since, in general, $\overline{u_i u_j}$ cannot be expressed in terms of the resolved fields $\bar{u}_i$, the equations are not closed, and therefore, cannot be solved without additional assumptions. In the absense of a complete theory of turbulence, one is forced to replace τ_{ij} with some model, a common example being the eddy-viscosity model

$$\tau_{ij} = \frac{1}{3}\delta_{ij}\tau_{kk} - \nu_t \left(\frac{\partial \bar{u}_i}{\partial x_j} + \frac{\partial \bar{u}_j}{\partial x_i} \right), \tag{4.25}$$

where ν_t is a constant (or variable) "eddy viscosity". Note that here the modeling is done only on the "traceless" part of the τ_{ij} tensor, since $(1/3)\tau_{kk}\delta_{ij}$ can be lumped with the pressure term as $p + (1/3)\tau_{kk}$, and the combination is determined by using the incompressibility condition. "Subgrid modeling" is a very large enterprise, and a wide variety of models $\tau_{ij} \to \tau_{ij}^{(M)}$ of various degrees of sophistications have been proposed in the literature. Without going into the details of various models, it suffices to say here that the modeling introduces an unknown error $\partial_j(\tau_{ij}^{(M)} - \tau_{ij})$ to the right hand side of the LES momentum equations.

The subgrid modeling error is very difficult to quantify or estimate. One approach, sometimes referred to as "apriori testing," consists of performing a high resolution DNS, filtering the data at some "cut-off" length 'Δ' and actually computing the τ_{ij} term[8, 9]. The computed result can then be compared with various models $\tau_{ij}^{(M)}$. An experimental version of the procedure also exists, where the τ_{ij} is obtained from high resolution experimental measurements rather than DNS [10, 11]. These and other tests generally show that large discrepencies exist between τ_{ij} and $\tau_{ij}^{(M)}$ when instantaneous pointwise comparisons are made between these quantities, at least for most commonly used subgrid models. Correlations between τ_{ij} and $\tau_{ij}^{(M)}$ exist only in some average sense, such as, when mean dissipation rates are compared:

$$\langle \tau_{ij} \bar{S}_{ij} \rangle \approx \langle \tau_{ij}^{(M)} \bar{S}_{ij} \rangle. \tag{4.26}$$

Here $\bar{S}_{ij} = (\partial_j \bar{u}_i + \partial_i \bar{u}_j)/2$ and $\langle \quad \rangle$ is some appropriate average.

2.5 Non-Commutation

The filtering operation $f \to \bar{f}$ may be defined formally by

$$\bar{f}(\mathbf{x}) = \int G\left(\frac{\mathbf{y}-\mathbf{x}}{\Delta}\right) f(\mathbf{y})\, d\mathbf{y} \tag{4.27}$$

where G is a smoothing operator or "filter kernel" such as (the Gaussian filter)

$$G(\mathbf{r}) = \frac{1}{\sqrt{2\alpha\pi}} \exp\left(-\frac{r^2}{2\alpha^2}\right) \tag{4.28}$$

where α is any positive number of order unity. A wide variety of filters have been discussed in the LES context [12], they all have the effect of "smoothing", that is removing or attenuating fluctuations on a length scale smaller than $\sim \Delta$. A filter kernel could be any unimodal function with a maximum at $x = 0$ and the properties

$$G(\pm\infty) = 0 \quad \int_{-\infty}^{+\infty} G(x)\, dx = 1 \quad \int_{-\infty}^{+\infty} x^2 G(x)\, dx \sim 1 \tag{4.29}$$

Equations (4.23)-(4.24) follow from the incompressible Navier-Stokes equations provided the following is true for any function $f(\mathbf{x})$:

$$\overline{\frac{\partial f}{\partial x_i}} = \frac{\partial \bar{f}}{\partial x_i}. \tag{4.30}$$

It is easily verified that (4.30) is true if the the filter width Δ in (4.27) is a constant but not otherwise.

It would appear, therefore, that it is to our advantage to treat 'Δ' as fixed. However, this introduces a problem (we work in one dimension for illustration) as may be seen in the following sequence of steps

$$\begin{aligned}
\overline{\frac{\partial f}{\partial x}} &= f(y)G\left(\frac{y-x}{\Delta}\right)\Big|_{y=a}^{y=b} - \int_a^b f(y)\frac{\partial}{\partial y}G\left(\frac{y-x}{\Delta}\right) dy \\
&= f(y)G\left(\frac{y-x}{\Delta}\right)\Big|_{y=a}^{y=b} + \int_a^b f(y)\frac{\partial}{\partial x}G\left(\frac{y-x}{\Delta}\right) dy \\
&= f(y)G\left(\frac{y-x}{\Delta}\right)\Big|_{y=a}^{y=b} + \frac{\partial}{\partial x}\int_a^b f(y)G\left(\frac{y-x}{\Delta}\right) dy. \\
&= f(y)G\left(\frac{y-x}{\Delta}\right)\Big|_{y=a}^{y=b} + \frac{\partial \bar{f}}{\partial x}.
\end{aligned} \tag{4.31}$$

The difficulty is with the boundary term on the right hand side, which involves the (unknown) value of f at the domain boundaries $x = a$ and $x = b$. For no slip boundary conditions and rigid boundaries, these terms would of course vanish for the velocity components, but not for the pressure or viscous terms.

There is another, and perhaps more important reason for considering a variable filter width. In LES, there is an implicit assumption that the filter width "Δ" divides the turbulent eddies into the large scales which are problem dependent and the small scales which may be represented by a *universal* subgrid model because they represent the small scale Kolmogorov cascade that is independent of the details of the large scale forcing field. Such a division presents no problem far from the boundary, but sufficiently close to boundaries such a separation is impossible. Detailed numerical solutions near walls [13] have shown that turbulence is dominated by organized coherent structures whose effect cannot be described by something as simple as an "eddy viscosity". Clearly, the scale 'Δ' below which turbulence is universal cannot be less than the normal distance to the wall.

A simple way to introduce a variable filter width that avoids the boundary terms in (4.31) is to define the filtering operation [14] as follows:

$$\bar{f}(x) = \frac{1}{\Delta}\int_a^b G\left(\frac{F(y) - F(x)}{\Delta}\right) f(y)F'(y)\, dy. \tag{4.32}$$

where F is any monotonically increasing function defined on $[a, b]$ such that $F(a) = -\infty$ and $F(b) = +\infty$. On going through the steps shown in (4.31) it is easily seen, that, the boundary terms now involve $F(a)$ or $F(b)$ in the argument of G so that they become $f(a \quad \text{or} \quad b)G(\pm\infty)$. But $G(\pm\infty) = 0$ by definition of the filter kernel G. If we use the approximation $F(y) - F(x) \approx (y - x)F'(x)$ and $F'(y) \approx F'(x)$ for $y \approx x$ in (4.32), then, it becomes

$$\bar{f}(x) = \frac{1}{\delta(x)}\int_a^b G\left(\frac{y - x}{\delta(x)}\right) f(y)\, dy. \tag{4.33}$$

where $\delta(x) \equiv \Delta/F'(x)$. Thus, definition (4.32) is essentially the same as what one would get by simply making Δ variable in (4.27), except it ensures that the boundary terms one gets on performing the steps (4.31) are exactly zero.

Introduction of the variable filter width in this way makes the unpleasant boundary terms vanish and is also consistent with the physical fact, that, close to the boundary there are no "universal scales" so that all scales must in principle be resolved. This follows, since $F(\pm\infty) = +\infty$ so that $\delta(x) \to 0$ as $x \to a$ or b. Thus, the kernel in (4.32) becomes a Dirac-delta function, so that the LES field $\bar{f}$ smoothly approaches the DNS field f at the boundary. However, by allowing a variable filter width we have now violated the condition (4.30), so that, (4.23) and (4.24) are no longer true.

The "commutation error," that is, the difference between the left and right hand sides of (4.30) may however be calculated, in terms of the derivatives of

$\bar{u}$ [14]

$$\overline{\frac{du}{dx}} = \frac{d\bar{u}}{dx} - \alpha\delta^2 \left(\frac{\delta'}{\delta}\right) \frac{d^2\bar{u}}{dx^2} + O(\delta^4) \tag{4.34}$$

where $\delta(x) = \Delta/F'(x)$ is the local filter-width introduced earlier and $\alpha = \int \zeta^2 G(\zeta)\, d\zeta$ is a numerical constant of order unity. The physical meaning of the additional terms may be understood by considering the example of a Gaussian wave packet $u(x)$ traveling from left to right according to the evolution equation

$$\frac{\partial u}{\partial t} + \frac{\partial u}{\partial x} = 0. \tag{4.35}$$

Suppose that $\delta(x)$ increases monotonically from left to right. Then, as the wave travels it would encounter ever increasing filter widths so that $\bar{u}$ would have a decreasing peak and broadening width as it propagates to the right, even though $u(x)$ travels unchanged in form. On filtering (4.35) and using (4.34) we get an equation for $\bar{u}$,

$$\frac{\partial \bar{u}}{\partial t} + \frac{\partial \bar{u}}{\partial x} = \nu \frac{\partial^2 \bar{u}}{\partial x^2} \tag{4.36}$$

where $\nu = \alpha\delta^2(\delta'/\delta)$ is positive. The term on the right hand side of (4.36) is a diffusion term which causes the necessary spreading of the wave packet $\bar{u}$.

These corrections accounting for the "commutation error" can be systematically calculated and incorporated in the basic LES equations for the filtered field. However, no consensus seems to have emerged in the community as to its significance. The inclusion of the corrections due to the commutation errors are seen to change the basic structure of the Navier Stokes equations, in particular, raising the order of the highest derivatives in the equation. Methods for dealing with this added complexity have not been developed. Van der Ven have proposed [15] a class of special filters G that would eliminate the "corrections" to any desired order in δ. These filters however do not satisfy the requirements of positivity, so that, the turbulent kinetic energy $k = \overline{u_i u_i} - \bar{u}_i \bar{u}_i$ is not guaranteed to be positive. Methods of numerical implementation of such filters in the context of dynamic models and aliasing error control have also been worked out [16, 17]. Perhaps LES should be formulated in a completely different way and we should not be attempting to derive the LES equations by filtering the Navier Stokes equations. Such alternate formulations have been proposed by various authors but none is completely satisfactory and there is no single approach that has found wide acceptance in the community.

2.6 Time Advancement

In turbulence computations, stability, and, the related issue of the size of time steps are usually of greater concern than the errors introduced by temporal discretization. Analysis of stability and accuracy of various temporal discretization schemes can be found in standard treatises [18, 19] and are not discussed here in detail.

3. Spectral Analysis of Errors

In section 2.2 the distinction between "formal accuracy" and "resolution" of a numerical differentiation method was pointed out. In problems involving a continuum of scales, neither of these are completely satisfactory in giving a sense of the magnitude of the errors. To see this, let us consider the truncation error of a second order central difference scheme applied to a function $u = u(x)$

$$\frac{\delta u}{\delta x} = \frac{du}{dx} + \frac{d^3u}{dx^3}\frac{\Delta^2}{6} + \cdots \tag{4.37}$$

In problems that do not involve a wide range of scales, one can reasonably define unique characteristic length and time scales. When the variables are normalized with these characteristic scales, u''' is of order unity. Thus, one may reasonably conclude that for the central difference scheme, the truncation error $\sim \Delta^2$. This is no longer true in turbulence or in other nonlinear problems characterized by a broad spectrum of scales. In LES or DNS one could write the above error estimate in Fourier-space as $-ik^3\Delta^2\hat{\mathbf{u}}(\mathbf{k})/6$, however, the magnitude of this error depends on which Fourier-modes one looks at, and the magnitude of $\hat{\mathbf{u}}(\mathbf{k})$ which depends on the turbulence spectrum. Therefore, it is difficult to judge, without a more precise analysis, what the true magnitude of the errors are. We now present a systematic method of obtaining such "error spectra" that give a better sense of the magnitude of numerical errors in turbulence simulations. The spectrum of the truncation and aliasing errors will be seen to depend not only on the numerical method, but on the turbulence spectrum as well. In order to be specific, the method will be illustrated with reference to an arbitrary finite difference scheme applied to isotropic turbulence in a cubical box.

3.1 Formulation

We consider isotropic turbulence in a cubical box, Ω, of side 'L' with periodic boundary conditions on the boundaries of Ω. The volume of the box will be denoted by $V = L^3$. We consider a finite difference method of solving the Navier-Stokes equations (4.23)-(4.24). In this method, the cubical domain is embedded with a rectangular grid of uniform spacing 'Δ' and the velocity is defined only at the N^3 nodes of the grid that we will denote by Ω_0. Here N is the number of nodes in each direction which is assumed to be an even integer. Clearly, $\Delta = L/(N-1)$. Equations (4.23)-(4.24) are now replaced by their finite difference approximations

$$\frac{\partial \bar{u}_i}{\partial t} = -\frac{\delta}{\delta x_j}(\bar{u}_i\bar{u}_j) - \frac{\delta \bar{P}}{\delta x_i} - \frac{\delta T^M_{ij}}{\delta x_j} + \nu\frac{\delta}{\delta x_k}\frac{\delta}{\delta x_k}\bar{u}_i, \tag{4.38}$$

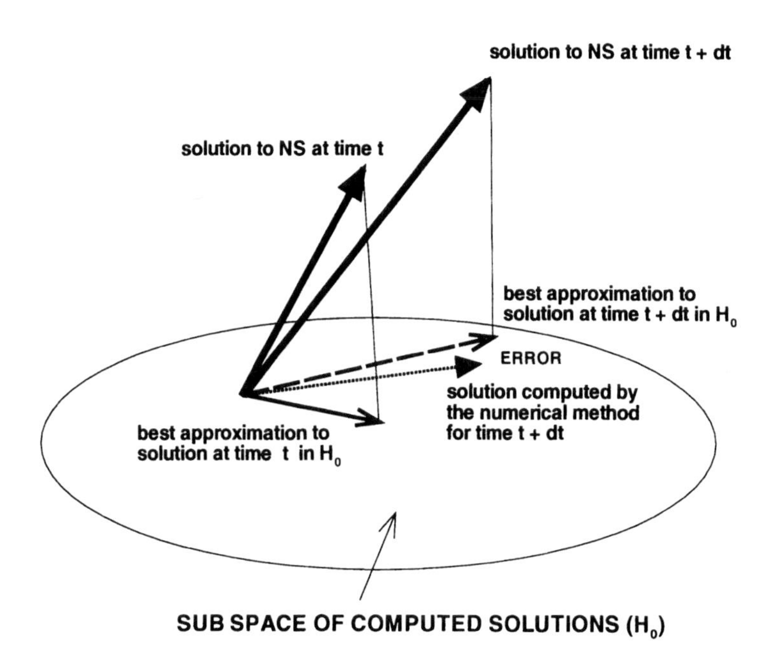

Figure 4.2. Geometric illustration of operator formulation of spectral error analysis

$$\frac{\delta \bar{u}_k}{\delta x_k} = 0 \tag{4.39}$$

where $\delta/\delta x_k$ represents a numerical differentiation scheme that approximates the continuous operator $\partial/\partial x_k$. In (4.38), T_{ij}^M is the discretized version of a "subgrid model" τ_{ij}^M that is specified in LES as a function (or functional) of the '$\bar{u}_i$' and its derivatives. In DNS, $\tau_{ij}^M = 0$. The first term on the right of (4.38) can be written in several alternative forms. These alternate expressions for the advective term, though equivalent in the continuum limit, result in distinct behavior of the finite discretized system. Equation (4.38) is known as the "divergence form" (or "conservative form"). The other most commonly used (see e.g. [7] and references therein) discretizations are the "skew-symmetric form", "advective form" (or "non-conservative form") and "vorticity form". For simplicity, we consider only the divergence form of the discretization (4.38).

We will also assume that the time integration of (4.38) is exact and suppress the time dependence of the dependent variable in our notation wherever convenient. This is done to keep the problem tractable and might be justified if time-stepping errors are negligible in comparison to errors due to finite-differencing.

Let $\mathcal{H}$ be the function space of vectors $\boldsymbol{\Psi} = (u_1, u_2, u_3)$ whose components u_1, u_2 and u_3 are square integrable in Ω. The norm is defined as

$$||\boldsymbol{\Psi}|| = \frac{1}{3}\int_\Omega d\mathbf{x}\left[(u_1)^2 + (u_2)^2 + (u_3)^2\right]. \tag{4.40}$$

The solutions of the Navier-Stokes equations (4.23)-(4.24) can be regarded as elements of $\mathcal{H}$ and the equations can be formally written as

$$\partial_t \boldsymbol{\Psi} = \mathcal{N}\boldsymbol{\Psi} \tag{4.41}$$

where $\mathcal{N}$ is a (nonlinear) operator defined in $\mathcal{H}$. If we define the set of wavevectors $\mathbf{k} = (2\pi n_1/L, 2\pi n_2/L, 2\pi n_3/L)$ where n_1, n_2, n_3 are integers (positive, negative or zero), then each component of $\boldsymbol{\Psi}$ can be expanded in terms of the orthogonal basis functions $\exp(i\mathbf{k}\cdot\mathbf{x})$. The vectors

$$(\ \exp(i\mathbf{k}_1\cdot\mathbf{x}),\ \exp(i\mathbf{k}_2\cdot\mathbf{x}),\ \exp(i\mathbf{k}_3\cdot\mathbf{x})\ )$$

where $\mathbf{k}_1$, $\mathbf{k}_2$ and $\mathbf{k}_3$ span the set of possible wavevectors therefore form a basis $\mathcal{B}$ of $\mathcal{H}$. Any element $(u_1, u_2, u_3) \in \mathcal{H}$ can be expanded in terms of this basis in the following way:

$$u_i(\mathbf{x}) = \frac{8\pi^3}{V}\sum_{\mathbf{k}} \hat{u}_i(\mathbf{k})\exp(i\mathbf{k}\cdot\mathbf{x}) \tag{4.42}$$

where

$$\hat{u}_i(\mathbf{k}) = \frac{1}{8\pi^3}\int u_i(\mathbf{x})\exp(-i\mathbf{k}\cdot\mathbf{x})\,d\mathbf{x}. \tag{4.43}$$

Throughout the rest of this chapter, if the range of summation is not explicitly indicated, any summation over the variable $\mathbf{k}$ is assumed to run over all allowed values of $\mathbf{k}$. Similarly, integration with respect to $\mathbf{k}$ would imply that the domain is $\mathbf{R}^3$ unless otherwise indicated. In physical space, all integrals with respect to the variable $\mathbf{x}$ are over Ω and all sums are over Ω_0 unless there is a subscript specifying an alternate domain. We would like to define the difference between the solution of the approximate equations (4.38)-(4.39) and the solution of the exact equations (4.23)-(4.24), as the "error". However, such a comparison is not possible, because, the exact solution is defined in a continuous region Ω whereas the approximate system has values only on the discrete set Ω_0. In order to compare the two, we need to either extend the solution defined on Ω_0 to Ω by means of an interpolation procedure, or, consider the values of the exact solution on the finite set Ω_0. We choose the former approach, as it is analytically more convenient to work in the continuum. Interpolation from Ω_0 to Ω may be done in an infinite number of ways. For simplicity, a trigonometric interpolation will be adopted. Thus, for each velocity component u_i we construct a function

$$\sum_{\mathbf{k}\in\omega_0} c_i(\mathbf{k}) \exp(i\mathbf{k}\cdot\mathbf{x}) \tag{4.44}$$

with the $c_i(\mathbf{k})$ chosen in such a way that (4.44) coincides with the computed u_i at each of the N^3 node points, where ω_0 denotes the set of wavevectors whose components are in the interval $(-\pi/\Delta, +\pi/\Delta)$. This is precisely, the definition of the discrete Fourier Transform [3], and it is known that the coefficients can be chosen uniquely and are given by

$$c_i(\mathbf{k}) = \left(\frac{\Delta}{L}\right)^3 {\sum_{\mathbf{x}\in\Omega_0}}' u_i(\mathbf{x}) \exp(-i\mathbf{k}\cdot\mathbf{x}) \tag{4.45}$$

where the prime over the summation sign indicates that grid points on the three planes $x = L, y = L$ and $z = L$ are to be excluded from the sum. If $\mathcal{H}_0$ denotes the subspace of $\mathcal{H}$ spanned by the basis functions with components $\exp(i\mathbf{k}\cdot\mathbf{x})$ where $\mathbf{k} \in \omega_0$, then it is clear from (4.45), that any vector $(u_1, u_2, u_3) \in \mathcal{H}_0$ has the following expansion in terms of the basis $\mathcal{B}$

$$u_i(\mathbf{x}) = \frac{8\pi^3}{V} \sum_{\mathbf{k}} \tilde{u}_i(\mathbf{k}) \exp(i\mathbf{k}\cdot\mathbf{x}) \tag{4.46}$$

(the factors of $8\pi^3/V$ is for later convenience) where

$$\tilde{u}_i(\mathbf{k}) = \begin{cases} \frac{\Delta^3}{8\pi^3} {\sum_{\mathbf{x}}}' u_i(\mathbf{x}) \exp(-i\mathbf{k}\cdot\mathbf{x}) & \text{if } \mathbf{k} \in \omega_0 \\ 0 & \text{otherwise.} \end{cases} \tag{4.47}$$

Thus, the elements of $\mathcal{H}_0$ can be expanded in terms of the basis B, just like the elements of $\mathcal{H}$, except, that, the expansion coefficients are the discrete Fourier Transforms $\tilde{u}_i$, instead of the Fourier Transforms $\hat{u}_i$ and they vanish if $\mathbf{k} \notin \omega_0$.

The solution $\mathbf{\Psi}_0$ of (4.38) can be considered to "live" in $\mathcal{H}_0$, and may be described formally by

$$\partial_t \mathbf{\Psi}_0 = \mathcal{N}_0 \mathbf{\Psi}_0 \tag{4.48}$$

where $\mathcal{N}_0$ is the "discretized Navier-Stokes operator" that maps elements of $\mathcal{H}_0$ to itself. Obviously, the discrete operator $\delta/\delta x_k$ in (4.38) must be extended in the obvious way to act on functions defined in Ω rather than Ω_0. For example, for a second order central difference scheme,

$$\frac{\delta}{\delta x_1} f(\mathbf{x}) = \frac{1}{2\Delta}[f(x_1 + \Delta, x_2, x_3) - f(x_1 - \Delta, x_2, x_3)] \tag{4.49}$$

where $\mathbf{x}$ can be any point in Ω not necessarily restricted to the nodes of the lattice, and, the finite-differences near boundary points are evaluated using the usual artifice of 'periodic extension' of the domain. Such an extension can be accomplished in an unambiguous way, since, as may be easily shown, the effect of applying the continuous version of the difference operator to the trigonometric extension of a function defined on a grid gives the same result as applying the difference operator first on the discrete lattice and then constructing its trigonometric extension to the continuous domain.

The problem of designing a good numerical method can be formulated as the problem of choosing the operator $\mathcal{N}_0$ in some optimal way. A good numerical method seeks to make the solution of the finite problem, $\mathbf{\Psi}_0$, as "close" as possible to the true solution $\mathbf{\Psi}$. Now, an element of $\mathcal{H}_0$ cannot approximate an element $\mathbf{\Psi} \in \mathcal{H}$ arbitrarily closely, but, there exists a unique element $\mathbf{\Psi}_0 \in \mathcal{H}_0$ that is closest to $\mathbf{\Psi}$ in the sense of $||\mathbf{\Psi}_0 - \mathbf{\Psi}||$ having the minimum possible value. This element is the projection of $\mathbf{\Psi}$ onto $\mathcal{H}_0$ and the operator $\mathcal{P}$ that maps $\mathbf{\Psi}$ to its projection is called the projection operator[20] corresponding to the subspace $\mathcal{H}_0$. In terms of the basis $\mathcal{B}$,

$$\begin{aligned}\mathcal{P}(u_1, u_2, u_3) &= \frac{8\pi^3}{V}\left(\sum_{\mathbf{k}\in\omega_0} \hat{u}_1(\mathbf{k}) \exp(i\mathbf{k}\cdot\mathbf{x})\,,\right. \\ &\left.\sum_{\mathbf{k}\in\omega_0} \hat{u}_2(\mathbf{k}) \exp(i\mathbf{k}\cdot\mathbf{x}),\ \sum_{\mathbf{k}\in\omega_0} \hat{u}_3(\mathbf{k}) \exp(i\mathbf{k}\cdot\mathbf{x})\right)\end{aligned} \tag{4.50}$$

where $\hat{u}_i(\mathbf{k})$ are given by (4.43). This is true, since the best approximation (with respect to the L_2 norm) of a function by a finite series of trigonometric polynomials is achieved when the expansion coefficients are the Fourier coefficients [21].

Clearly, the best one can hope for from the numerical method (4.48) is $\mathbf{\Psi}_0 = \mathcal{P}\mathbf{\Psi}$. From (4.41), the equation satisfied by $\mathcal{P}\mathbf{\Psi}$ is

$$\partial_t \mathcal{P}\mathbf{\Psi} = \mathcal{P}\mathcal{N}\mathbf{\Psi} \tag{4.51}$$

On subtracting (4.48) from (4.51), and, on adding and subtracting the term $\mathcal{N}_0\mathcal{P}\mathbf{\Psi}$ from the right hand side, one obtains[1]

$$\partial_t \mathbf{e} = \mathcal{N}_0\mathcal{P}\mathbf{\Psi} - \mathcal{N}_0\mathbf{\Psi}_0 + (\mathcal{P}\mathcal{N} - \mathcal{N}_0\mathcal{P})\mathbf{\Psi} \tag{4.52}$$

where $\mathbf{e} \equiv \mathcal{P}\mathbf{\Psi} - \mathbf{\Psi}_0$. As the true solution $\mathbf{\Psi}$ moves around in the space "$\mathcal{H}$", its "shadow," $\mathcal{P}\mathbf{\Psi}$ moves around in the subspace $\mathcal{H}_0$ (Figure 4.2). An "ideal" numerical method, "$\mathcal{N}_0$" is such, that, if initially $\mathbf{\Psi}_0 = \mathcal{P}\mathbf{\Psi}$, then $\mathbf{\Psi}_0$ should remain "locked" to $\mathcal{P}\mathbf{\Psi}$ for all subsequent times. The condition for this to happen is $\partial_t \mathbf{e} = 0$ if $\mathbf{e} = 0$. Equation (4.52) shows, that, when $\mathbf{e} = \mathcal{P}\mathbf{\Psi} - \mathbf{\Psi}_0 = 0$,

$$\partial_t \mathbf{e} = (\mathcal{P}\mathcal{N} - \mathcal{N}_0\mathcal{P})\mathbf{\Psi} \tag{4.53}$$

Therefore, the numerical method $\mathcal{N}_0$ is "ideal" if and only if

$$\mathcal{N}_0\mathcal{P} = \mathcal{P}\mathcal{N}. \tag{4.54}$$

For any $\mathbf{\Psi} \in \mathcal{H}$, we define the difference

$$\mathbf{E} \equiv (\mathcal{P}\mathcal{N} - \mathcal{N}_0\mathcal{P})\mathbf{\Psi} \tag{4.55}$$

as the "error", that is, the extent of the departure of the solution from the best possible one on the given grid. It will be seen that this general, though somewhat abstract definition, contains the usual "truncation error", "aliasing error", and "subgrid modeling error" discussed earlier.

3.2 Explicit Evaluation & Classification

In this analysis, we seek to characterize the magnitude of the error $\mathbf{E}$, not by a single number, but rather, by statistical properties such as its power spectral density. Such statistical measures can be precisely defined only in the limit where the wavevector can assume a continuum rather than a discrete set of values. In physical space this implies that we are considering the grid size Δ and some characteristic scale of turbulence λ fixed and taking the limit as the size of the box $L \to \infty$. In actual simulations, the box size, L, is of course finite. However, L is taken much larger than Δ or λ so that smooth power spectra can be defined and computed statistical quantities are not changed when the box size is further increased. This ensures that the computed quantities are indistinguishable from the ideal limit, $L \to \infty$. For the purpose of theoretical analysis it is advantageous to take the limit $L \to \infty$ first rather than at the end of the computation, thus, from now on we will assume that our lattice is infinite. Thus, in the Fourier-basis, the exact solution will be characterized by the continuous family of wave vectors $\mathbf{k} \in \mathbf{R}^3$ and the numerical solution will be characterized by the subset $\mathbf{k} \in \omega_0$ where $\omega_0 \equiv [-\pi/\Delta, \pi/\Delta] \times [-\pi/\Delta, \pi/\Delta] \times [-\pi/\Delta, \pi/\Delta]$. In the limit of infinite box size, equations (4.42) and (4.43) take the form

$$u_i(\mathbf{x}) = \int \hat{u}_i(\mathbf{k}) \exp(i\mathbf{k} \cdot \mathbf{x})\, d\mathbf{k} \tag{4.56}$$

and

$$\hat{u}_i(\mathbf{k}) = \frac{1}{8\pi^3} \int u_i(\mathbf{x}) \exp(-i\mathbf{k} \cdot \mathbf{x})\, d\mathbf{x}. \tag{4.57}$$

For elements of $\mathcal{H}_0$, the expansions (4.46) and (4.47) take the form

$$u_i(\mathbf{x}) = \int \tilde{u}_i(\mathbf{k}) \exp(i\mathbf{k} \cdot \mathbf{x}) d\mathbf{k} \tag{4.58}$$

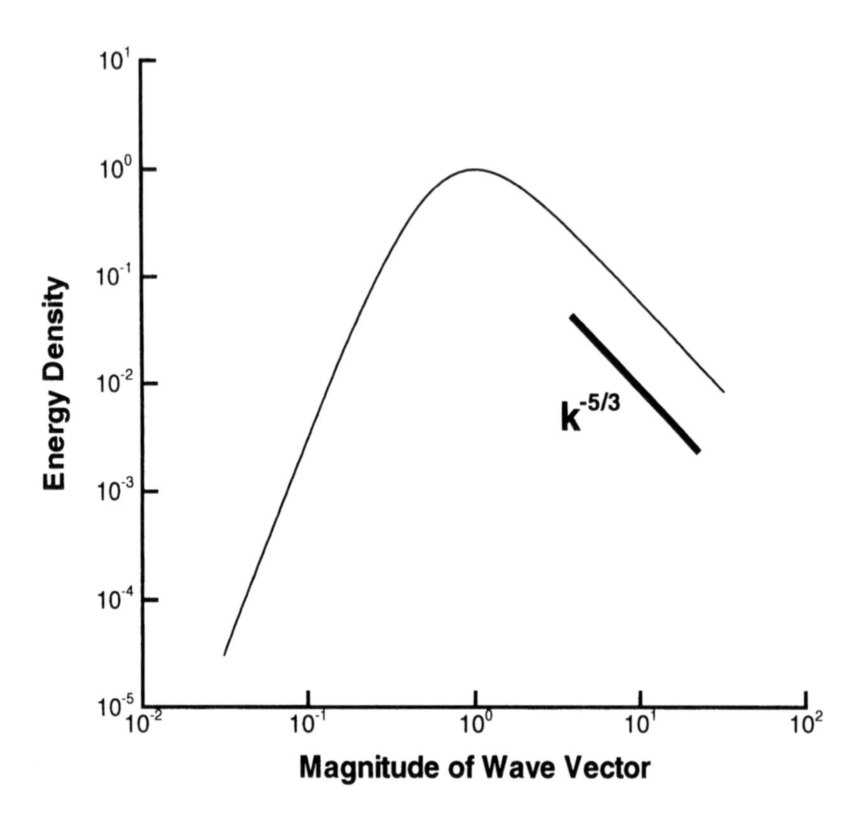

Figure 4.3. The Von Karman energy spectrum for the inertial range of fully developed turbulence.

and

$$\tilde{u}_i(\mathbf{k}) = \begin{cases} \frac{\Delta^3}{8\pi^3}\sum_{\mathbf{x}} u_i(\mathbf{x})\exp(-i\mathbf{k}\cdot\mathbf{x}) & \text{if } \mathbf{k}\in\omega_0 \\ 0 & \text{otherwise.} \end{cases} \tag{4.59}$$

The following useful identity is also readily derived by first proving it for a finite box, and then, taking the limit of infinite box size;

$$\frac{\Delta^3}{8\pi^3}\sum_{\mathbf{x}}\exp(i\mathbf{K}\cdot\mathbf{x}) = \sum_{\mathbf{a}\in\Lambda}\delta(\mathbf{K}-\mathbf{a}) \tag{4.60}$$

where 'δ' is the Dirac delta function, Λ is the set of wavevectors of the form $(2pk_m, 2qk_m, 2rk_m)$ where p, q and r are integers (positive, negative or zero), $k_m = \pi/\Delta$, and $\mathbf{K}$ is any vector (not necessarily restricted to ω_0)[2]. Equation

(4.60) is the discrete lattice equivalent of

$$\frac{1}{8\pi^3}\int d\mathbf{x}\exp(i\mathbf{K}\cdot\mathbf{x}) = \delta(\mathbf{K}) \tag{4.61}$$

and indeed reduces to (4.61) in the limit $\Delta \to 0$.

Let us denote the expansion coefficients of $\mathbf{E}$ in the Fourier-basis $\mathcal{B}$ by $(E_1(\mathbf{k}), E_2(\mathbf{k}), E_3(\mathbf{k}))$. To compute $E_i(\mathbf{k})$, we first evaluate $\mathcal{PN}$ in the basis $\mathcal{B}$. Clearly, the projection operator $\mathcal{P}$ has the effect of making all Fourier coefficients corresponding to wavevectors $\mathbf{k}$ outside the region ω_0 vanish while the remaining wavevectors are unaffected. Therefore,

$$\begin{aligned}(\mathcal{PN})_i(\mathbf{k}) &= H(\mathbf{k})\\ &\times\left[-iP_{imn}(\mathbf{k})\int\int d\mathbf{k}'d\mathbf{k}''\delta(\mathbf{k}'+\mathbf{k}''-\mathbf{k})\hat{u}_m(\mathbf{k}')\hat{u}_n(\mathbf{k}'') - \nu k^2\hat{u}_i(\mathbf{k})\right]\end{aligned} \tag{4.62}$$

where $H(\mathbf{k})$ is the unit step function defined by

$$H(\mathbf{k}) = \begin{cases} 1 & \text{if } \mathbf{k}\in\omega_0 \\ 0 & \text{otherwise.}\end{cases} \tag{4.63}$$

The right hand side of (4.62) for $\mathbf{k}\in\omega_0$ is simply the well known (see e.g. [3]) right hand side of the Navier-Stokes equation in Fourier-space. The tensor P_{imn} is defined by

$$P_{imn}(\mathbf{k}) = \begin{cases} (k_n P_{im} + k_m P_{in})/2 & \text{if } \mathbf{k}\neq 0, \\ 0 & \text{otherwise}\end{cases} \tag{4.64}$$

with $P_{ij} = \delta_{ij} - k_i k_j/k^2$. It is customary to write (4.62) in the form

$$\begin{aligned}(\mathcal{PN})_i(\mathbf{k}) &= -\nu k^2 H(\mathbf{k})\hat{u}_i(\mathbf{k}) - iP_{imn}(\mathbf{k})H(\mathbf{k})\\ &\times\left[\int_{\omega_0}\int_{\omega_0} d\mathbf{k}'d\mathbf{k}''\delta(\mathbf{k}'+\mathbf{k}''-\mathbf{k})\hat{u}_m(\mathbf{k}')\hat{u}_n(\mathbf{k}'') + \hat{\tau}_{mn}(\mathbf{k})\right]\end{aligned} \tag{4.65}$$

where

$$\hat{\tau}_{mn}(\mathbf{k}) = H(\mathbf{k})\left[\int\int - \int_{\omega_0}\int_{\omega_0}\right] d\mathbf{k}'d\mathbf{k}''\hat{u}_m(\mathbf{k}')\hat{u}_n(\mathbf{k}'')\delta(\mathbf{k}'+\mathbf{k}''-\mathbf{k}) \tag{4.66}$$

is called the subgrid stress.

Next we consider the term $\mathcal{N}_0\mathcal{P}$. Let U_i denote the result of using the projection operator $\mathcal{P}$ on the velocity field with components u_i. Then,

$$U_i(\mathbf{x}) = \int_{\omega_0}\hat{u}_i(\mathbf{k})\exp(i\mathbf{k}\cdot\mathbf{x})\,d\mathbf{k}. \tag{4.67}$$

In finite difference, pseudo-spectral and collocation methods the nonlinear term is constructed by multiplying together velocity components at each grid point. From equation (4.58) and (4.59), the extension into $\mathcal{H}_0$ of such a product is $f_{mn}(\mathbf{x})$ whose Fourier-coefficients are

$$\hat{f}_{mn}(\mathbf{k}) = \frac{\Delta^3}{8\pi^3} H(\mathbf{k}) \sum_{\mathbf{x}} U_m(\mathbf{x}) U_n(\mathbf{x}) \exp(-i\mathbf{k}\cdot\mathbf{x}). \tag{4.68}$$

When the expressions (4.67) for $U_i(\mathbf{x})$ are substituted into (4.68) we have

$$\hat{f}_{mn}(\mathbf{k}) = \frac{\Delta^3}{8\pi^3} H(\mathbf{k}) \sum_{\mathbf{x}} \int_{\omega_0}\int_{\omega_0} d\mathbf{k}' d\mathbf{k}'' \hat{u}_m(\mathbf{k}') \hat{u}_n(\mathbf{k}'') \exp[i(\mathbf{k}'+\mathbf{k}''-\mathbf{k})\cdot\mathbf{x}]. \tag{4.69}$$

The summation over the lattice points can be taken inside the integral signs and executed in accordance with (4.60):

$$\hat{f}_{mn}(\mathbf{k}) = H(\mathbf{k}) \sum_{\mathbf{a}\in\Lambda} \int_{\omega_0}\int_{\omega_0} d\mathbf{k}' d\mathbf{k}'' \hat{u}_m(\mathbf{k}') \hat{u}_n(\mathbf{k}'') \delta(\mathbf{k}'+\mathbf{k}''-\mathbf{k}-\mathbf{a}). \tag{4.70}$$

Thus, $\mathcal{N}_0\mathcal{P}$ differs from $\mathcal{P}\mathcal{N}$ in that the sum of $\hat{f}_{mn}(\mathbf{k})$ and the Fourier Transform of the subgrid force replaces the nonlinear term

$$H(\mathbf{k}) \int\int d\mathbf{k}' d\mathbf{k}'' \delta(\mathbf{k}' + \mathbf{k}'' - \mathbf{k}) \hat{u}_m(\mathbf{k}') \hat{u}_n(\mathbf{k}'')$$

in equation (4.62). Further, all expressions involving the wavevectors $\mathbf{k}$ in (4.62) should be replaced by the corresponding modified wavevectors $\tilde{\mathbf{k}}$, since, if the exact derivative operator $\partial/\partial x_k$ is replaced by the numerical differentiation $\delta/\delta x_k$, multiplication by wavevectors $\mathbf{k}$ in Fourier-space are replaced by multiplication by the corresponding modified wavevectors $\tilde{\mathbf{k}}$.

In order to complete the evaluation of $\mathcal{N}_0\mathcal{P}$ we need an explicit specification of the subgrid model τ_{ij}^M. In the absence of a theory of turbulence, it is impossible to quantify the error introduced by the subgrid modeling except through direct measurement or fully resolved numerical simulation. Therefore, in this analysis we will neglect the error due to subgrid modeling and work with "the ideal subgrid model"

$$\tau_{ij}^M = \tau_{ij}(\mathbf{x}, t) \tag{4.71}$$

where $\tau_{ij}(\mathbf{x}, t)$ is the exact subgrid stress that we assume is given as a function of position and time. Now we can write down the expression for $\mathcal{N}_0\mathcal{P}$:

$$(\mathcal{N}_0\mathcal{P})_i(\mathbf{k}) = -\nu\tilde{k}^2 H(\mathbf{k}) \hat{u}_i(\mathbf{k}) - iP_{imn}(\tilde{\mathbf{k}}) H(\mathbf{k}) \times \left[\sum_{\mathbf{a}\in\Lambda} \int_{\omega_0}\int_{\omega_0} d\mathbf{k}' d\mathbf{k}'' \delta(\mathbf{k}' + \mathbf{k}'' - \mathbf{k} - \mathbf{a}) \hat{u}_m(\mathbf{k}') \hat{u}_n(\mathbf{k}'') + \hat{\tau}_{mn}(\mathbf{k})\right] \tag{4.72}$$

On taking the difference of (4.62) and (4.71) we observe that the error may be written as

$$E_i(\mathbf{k}) = E_i^{(\mathrm{FD})}(\mathbf{k}) + E_i^{(\mathrm{alias})}(\mathbf{k}) \tag{4.73}$$

where

$$\begin{aligned} E_i^{(\mathrm{FD})}(\mathbf{k}) &= \nu(\tilde{k}^2 - k^2)H(\mathbf{k})\hat{u}_i(\mathbf{k}) + iH(\mathbf{k})\left[P_{imn}(\tilde{\mathbf{k}}) - P_{imn}(\mathbf{k})\right] \times \\ &\left[\int_{\omega_0}\int_{\omega_0} d\mathbf{k}' d\mathbf{k}'' \delta(\mathbf{k}' + \mathbf{k}'' - \mathbf{k})\hat{u}_m(\mathbf{k}')\hat{u}_n(\mathbf{k}'') + \hat{\tau}_{mn}(\mathbf{k})\right] \\ &= \nu(\tilde{k}^2 - k^2)H(\mathbf{k})\hat{u}_i(\mathbf{k}) + iH(\mathbf{k})\left[P_{imn}(\tilde{\mathbf{k}}) - P_{imn}(\mathbf{k})\right] \times \\ &\left[\int\int d\mathbf{k}' d\mathbf{k}'' \delta(\mathbf{k}' + \mathbf{k}'' - \mathbf{k})\hat{u}_m(\mathbf{k}')\hat{u}_n(\mathbf{k}'')\right] \end{aligned} \tag{4.74}$$

and

$$\begin{aligned} E_i^{(\mathrm{alias})}(\mathbf{k}) &= iP_{imn}(\tilde{\mathbf{k}})H(\mathbf{k}) \times \\ &\sum_{\mathbf{a}\in\Lambda_0}\int_{\omega_0}\int_{\omega_0} d\mathbf{k}' d\mathbf{k}'' \delta(\mathbf{k}' + \mathbf{k}'' - \mathbf{k} - \mathbf{a})\hat{u}_m(\mathbf{k}')\hat{u}_n(\mathbf{k}''). \end{aligned} \tag{4.75}$$

In (4.74), the "reciprocal lattice" Λ was replaced by the smaller set Λ_0 consisting of the vectors $(2pk_m, 2qk_m, 2rk_m)$ where p, q and r can independently take on the values 0 or ± 1 but excluding the case $p = q = r = 0$. The reason integer values of p, q and r with modulus greater than 1 are not included in Λ_0 is that the relation $\mathbf{a} = \mathbf{k}' + \mathbf{k}'' - \mathbf{k}$ cannot be satisfied for such values if $\mathbf{k}, \mathbf{k}', \mathbf{k}'' \in \omega_0$ and hence the delta function ensures that they do not contribute to the sum. The first term in (4.73) arises because of the inability of the finite-differencing operator, $\delta/\delta x_k$, to accurately compute the gradient of short-wavelength waves. We call this the "truncation error." It vanishes for a spectral method which can differentiate waves of all wavelengths exactly. The second term arises due to the method of computation of the nonlinear term by taking products in physical space on a discrete lattice. This, we identify with the "aliasing error."

3.2.1 Truncation Errors. We will now calculate the power spectrum of the truncation error. This may be defined as

$$\frac{\mathcal{E}^{(\mathrm{FD})}(k)}{4\pi k^2} = \lim_{V\to\infty} \frac{8\pi^3}{V} \left\{\langle E_i^{(\mathrm{FD})}(\mathbf{k}) E_i^{(\mathrm{FD})}(\mathbf{k})^*\rangle\right\}_{\Omega}, \tag{4.76}$$

$\{\ \}_{\Omega}$ denotes angular average in wave-number space over the surface of the sphere $|\mathbf{k}| = k$ and V is the volume of the physical box containing the fluid. The prefactor $8\pi^3/V$ in (4.76) is necessary because we would like the integral of the power spectra over the entire region ω_0 to give us the mean-square error rather than the square of the error summed over the infinite lattice (which of course would be infinite!). From (4.73), we have, for $\mathbf{k} \in \omega_0$,

$$\begin{aligned}
&\langle E_i^{\rm FD}(\mathbf{k}) E_i^{\rm FD}(\mathbf{k})^* \rangle = \\
&\Delta_{imn}(\mathbf{k}, \tilde{\mathbf{k}}) \Delta^*_{ipq}(\mathbf{k}, \tilde{\mathbf{k}}) \iint d\mathbf{k}' d\mathbf{k}'' \langle \hat{u}_m(\mathbf{k}') \hat{u}_n(\mathbf{k} - \mathbf{k}') \hat{u}^*_p(\mathbf{k}'') \hat{u}^*_q(\mathbf{k} - \mathbf{k}'') \rangle \\
&+2\nu \Im \left[i \Delta^*_{imn}(\mathbf{k}, \tilde{\mathbf{k}}) (\tilde{k}^2 - k^2) \int d\mathbf{k}' \langle \hat{u}^*_m(\mathbf{k}') \hat{u}^*_n(\mathbf{k} - \mathbf{k}') \hat{u}_i(\mathbf{k}) \rangle \right] \\
&+\nu^2 |\tilde{k}^2 - k^2|^2 \langle \hat{u}_i(\mathbf{k}) \hat{u}^*_i(\mathbf{k}) \rangle
\end{aligned} \tag{4.77}$$

where $\langle\ \ \rangle$ denotes ensemble average, $*$ denotes complex conjugate, $\Im$ denotes the imaginary part, and $\Delta_{imn}(\mathbf{k}, \tilde{\mathbf{k}}) \equiv P_{imn}(\tilde{\mathbf{k}}) - P_{imn}(\mathbf{k})$. The following two properties of the Δ_{imn} tensors follow immediately from the corresponding properties of P_{imn}; $\Delta_{imm} = 0$, $\Delta_{imn} = \Delta_{inm}$.

In order to make further analytical work possible with (4.77) we now introduce the "Millionshchikov hypothesis" [25] that in fully developed turbulence, the joint probability density function of any set of velocity components at arbitrary space-time points can be assumed to be joint-normal. The joint-normal hypothesis was originally evoked in turbulence in an attempt to close the hierarchy of equations for moments [3]. Though this did not succeed, the joint-normal hypothesis has been successfully used in other contexts. For example, Batchelor [26] used it with success to predict the pressure spectrum of isotropic turbulence. The joint-normal hypothesis implies in particular

$$\begin{aligned}
\langle u_i(\mathbf{x}_1) u_j(\mathbf{x}_2) u_k(\mathbf{x}_3) u_l(\mathbf{x}_4) \rangle &= \langle u_i(\mathbf{x}_1) u_j(\mathbf{x}_2) \rangle \langle u_k(\mathbf{x}_3) u_l(\mathbf{x}_4) \rangle \\
+\langle u_i(\mathbf{x}_1) u_k(\mathbf{x}_3) \rangle \langle u_j(\mathbf{x}_2) u_l(\mathbf{x}_4) \rangle &+ \langle u_i(\mathbf{x}_1) u_l(\mathbf{x}_4) \rangle \langle u_j(\mathbf{x}_2) u_k(\mathbf{x}_3) \rangle
\end{aligned} \tag{4.78}$$

and that all third order moments are zero. Here $\mathbf{u}(\mathbf{x}, t)$ is the true velocity field defined at all space time points. On taking the Fourier-transform (in infinite continuous space) of (4.78) and assuming the turbulence to be homogeneous, we have,

$$\begin{aligned}
\langle \hat{u}_i(\mathbf{k}_1) \hat{u}_j(\mathbf{k}_2) \hat{u}_k(\mathbf{k}_3) \hat{u}_l(\mathbf{k}_4) \rangle = \; & \delta(\mathbf{k}_1 + \mathbf{k}_2) \delta(\mathbf{k}_3 + \mathbf{k}_4) \Phi_{ij}(\mathbf{k}_2) \Phi_{kl}(\mathbf{k}_4) \\
& +\delta(\mathbf{k}_1 + \mathbf{k}_3) \delta(\mathbf{k}_2 + \mathbf{k}_4) \Phi_{ik}(\mathbf{k}_3) \Phi_{jl}(\mathbf{k}_4) \\
& +\delta(\mathbf{k}_1 + \mathbf{k}_4) \delta(\mathbf{k}_2 + \mathbf{k}_3) \Phi_{il}(\mathbf{k}_4) \Phi_{jk}(\mathbf{k}_3),
\end{aligned} \tag{4.79}$$

where Φ_{ij} is the Fourier-transform of the correlation tensor $R_{ij}(\mathbf{x}_2 - \mathbf{x}_1) \equiv \langle u_i(\mathbf{x}_1) u_j(\mathbf{x}_2) \rangle$. On substituting (4.78) into the first term in (4.77) we get a sum of three terms. It is readily seen that the first of these three terms is proportional to $\delta(\mathbf{k})$ which when combined with $8\pi^3/V$ from (4.76) gets replaced by the

Kronecker Delta operator, $\delta_{\mathbf{k}}$, which is unity if $\mathbf{k} = 0$, and zero otherwise [in accordance with the familiar rule $(V/8\pi^3)\delta_{\mathbf{k}} \to \delta(\mathbf{k})$ for passing to the continuum limit]. Since $\Delta_{ipq}(\mathbf{k}, \tilde{\mathbf{k}})$ vanishes if $\mathbf{k} = 0$, there is no contribution from this term. Further, since $\Delta_{ipq}(\mathbf{k}, \tilde{\mathbf{k}})$ is invariant with respect to an interchange of the last two indices, the second and third terms are equal. Thus, the total contribution is

$$2\Delta_{imn}(\mathbf{k}, \tilde{\mathbf{k}})\Delta^*_{ipq}(\mathbf{k}, \tilde{\mathbf{k}}) \int d\mathbf{k}' \Phi^*_{mp}(\mathbf{k}')\Phi^*_{nq}(\mathbf{k} - \mathbf{k}') \tag{4.80}$$

Here the factor $8\pi^3\delta(0)/V$, which in general is undefined, has been replaced with $\delta_{00} = 1$. This is in accordance with the prescribed procedure for taking the limit of an infinitely large box, viz. the step $L \to \infty$ should be performed last in the evaluation. The second term of (4.77) vanishes under the joint-normal hypothesis and the last term is easily shown to be

$$\frac{\int d\mathbf{x}}{8\pi^3}\,\nu^2|\tilde{k}^2 - k^2|^2\Phi_{ii}. \tag{4.81}$$

The first factor is eliminated by the $8\pi^3/V$ in (4.76) and we obtain finally

$$\begin{aligned}\mathcal{E}^{(\mathrm{FD})}(k) = \Big\{&8\pi k^2\Delta_{imn}(\mathbf{k}, \tilde{\mathbf{k}})\Delta_{ipq}{}^*(\mathbf{k}, \tilde{\mathbf{k}}) \int \Phi^*_{mp}(\mathbf{k}')\Phi^*_{nq}(\mathbf{k} - \mathbf{k}')\, d\mathbf{k}' \\ &+4\pi k^2\nu^2|\tilde{k}^2 - k^2|^2\Phi_{ii}(\mathbf{k})\Big\}_{\Omega}\end{aligned} \tag{4.82}$$

Equation (4.82) is the general result for homogeneous turbulence. If in addition, the turbulence is isotropic, Φ_{ij} simplifies [27] to

$$\Phi_{ij}(\mathbf{k}) = \frac{E(k)}{4\pi k^4}(k^2\delta_{ij} - k_ik_j) \tag{4.83}$$

where $E(k)$ is the three dimensional energy spectrum and δ_{ij} is the Kronecker-delta symbol. The integral in the first term of (4.82) may be written after substitution of (4.83) as

$$\begin{aligned}J_{mpnq}(\mathbf{k}) &\equiv 8\pi k^2 \int \Phi^*_{mp}(\mathbf{k}')\Phi^*_{nq}(\mathbf{k} - \mathbf{k}')\, d\mathbf{k}' \\ &= \frac{k^2}{2\pi}\int\!\!\int \frac{E(P)E(Q)}{P^4Q^4}\left[P^2Q^2\delta_{mp}\delta_{nq} - P_mP_pQ^2\delta_{nq} - Q_nQ_qP^2\delta_{mp}\right. \\ &\qquad \left.+P_mP_pQ_nQ_q\right]\quad \delta(\mathbf{P} + \mathbf{Q} - \mathbf{k})\, d\mathbf{P}d\mathbf{Q}.\end{aligned} \tag{4.84}$$

With some algebra, the integral may be expressed [22] as

$$\begin{aligned}J_{mpnq}(\mathbf{k}) &= F_1(k)\delta_{mp}\delta_{nq} + F_2(k)(\delta_{mn}\delta_{pq} + \delta_{pn}\delta_{mq}) \\ &+F_3(k)\left[\frac{k_mk_p}{k^2}\delta_{nq} + \frac{k_nk_q}{k^2}\delta_{mp}\right] + F_4(k)\frac{k_mk_pk_nk_q}{k^4}\end{aligned} \tag{4.85}$$

where

$$F_1(k) = \frac{1}{16}[7I_4 + 6I_3 - 2I_2 + 5I_1] \tag{4.86}$$

$$F_2(k) = \frac{1}{16}[-3I_4 + 2I_3 + 2I_2 - I_1] \tag{4.87}$$

$$F_3(k) = \frac{1}{16}[-15I_4 - 6I_3 + 2I_2 + 3I_1] \tag{4.88}$$

$$F_4(k) = \frac{1}{16}[45I_4 - 30I_3 - 6I_2 + 7I_1]. \tag{4.89}$$

The terms I_m are defined as

$$I_m = k\int_0^\infty d\xi \int_{|\xi-1|}^{\xi+1} d\eta\, E(k\xi)E(k\eta)W_m(\xi,\eta) \tag{4.90}$$

where the weights W_m are defined as follows:

$$W_1(\xi,\eta) = \frac{1}{\xi\eta} \tag{4.91}$$

$$W_2(\xi,\eta) = \frac{(1-\xi^2-\eta^2)^2}{4\xi^3\eta^3} \tag{4.92}$$

$$W_3(\xi,\eta) = \frac{(1+\xi^2-\eta^2)^2}{4\xi^3\eta} \tag{4.93}$$

$$W_4(\xi,\eta) = \frac{[1-(\xi^2-\eta^2)^2]^2}{16\xi^3\eta^3}. \tag{4.94}$$

Therefore, after substituting (4.84) in (4.82) and using the properties $\Delta_{imm} = 0$ and $\Delta_{imn} = \Delta_{inm}$, the following expression is obtained for the power spectrum of the truncation error:

$$\begin{aligned}
\mathcal{E}^{(\mathrm{FD})}(k) &= [F_1(k) + F_2(k)]\left\{\sum_{i,m,n} |\Delta_{imn}(\mathbf{k},\tilde{\mathbf{k}})|^2\right\}_\Omega \\
&+2F_3(k)\left\{\sum_{i,m,n,p} \frac{k_m k_p}{k^2}\Delta_{imn}(\mathbf{k},\tilde{\mathbf{k}})\Delta^*_{ipn}(\mathbf{k},\tilde{\mathbf{k}})\right\}_\Omega \\
&+F_4(k)\left\{\sum_{i,m,n,p,q} \frac{k_m k_p k_n k_q}{k^4}\Delta_{imn}(\mathbf{k},\tilde{\mathbf{k}})\Delta^*_{ipq}(\mathbf{k},\tilde{\mathbf{k}})\right\}_\Omega \\
&+2\nu^2 E(k)\left\{|\tilde{k}^2 - k^2|^2\right\}_\Omega
\end{aligned} \tag{4.95}$$

In equation (4.95), the functions $F_1(k)$, $F_2(k)$, $F_3(k)$ and $F_4(k)$ are known once the energy spectrum is specified. They are not affected by the choice of numerical schemes. On the other hand, the coefficients of these functions in (4.95) depend *only* on the numerical method (through the dependence of $\tilde{\mathbf{k}}$ on $\mathbf{k}$) and is independent of the energy spectrum. Thus, given a specific numerical scheme and energy spectrum, equation (4.95) can be used to compute the power spectrum of the truncation error. This is done in the § 3.3

3.2.2 Aliasing errors. The power spectrum of the aliasing-error,

$$\frac{\mathcal{E}^{(\mathrm{alias})}(k)}{4\pi k^2} = \lim_{V\to\infty} \frac{8\pi^3}{V} \left\{ \langle E_i^{(\mathrm{alias})}(\mathbf{k}) E_i^{(\mathrm{alias})}(\mathbf{k})^* \rangle \right\}_\Omega, \tag{4.96}$$

may be calculated similarly. From (4.74) one obtains, for $\mathbf{k} \in \omega_0$,

$$\begin{aligned} \langle E_i^{\mathrm{alias}}(\mathbf{k}) E_i^{\mathrm{alias}}(\mathbf{k})^* \rangle &= P_{imn}(\tilde{\mathbf{k}}) P^*_{ipq}(\tilde{\mathbf{k}}) \\ \times \sum_{\mathbf{a},\mathbf{a}'\in\Lambda_0} &\int_{\omega_0}\int_{\omega_0}\int_{\omega_0}\int_{\omega_0} d\mathbf{k}_1 d\mathbf{k}_2 d\mathbf{k}_3 d\mathbf{k}_4 \langle \hat{u}_m(\mathbf{k}_1)\hat{u}_n(\mathbf{k}_2)\hat{u}^*_p(\mathbf{k}_3)\hat{u}^*_q(\mathbf{k}_4)\rangle \\ \times\, &\delta(\mathbf{k}+\mathbf{a}-\mathbf{k}_1-\mathbf{k}_2)\,\delta(\mathbf{k}+\mathbf{a}'-\mathbf{k}_3-\mathbf{k}_4). \end{aligned} \tag{4.97}$$

On applying the joint-normal hypothesis, in analogy to the derivation of (4.95), we get

$$\begin{aligned} \lim_{V\to\infty} \frac{8\pi^3}{V} \langle E_i^{\mathrm{alias}}(\mathbf{k}) E_i^{\mathrm{alias}}(\mathbf{k})^* \rangle &= 2 \sum_{\mathbf{a}\in\Lambda_0} P_{imn}(\tilde{\mathbf{k}}) P^*_{ipq}(\tilde{\mathbf{k}}) \\ \int_{\omega_0}\int_{\omega_0} d\mathbf{k}' d\mathbf{k}'' \delta(\mathbf{k}'+\mathbf{k}''-\mathbf{k}-\mathbf{a}) &\Phi^*_{mp}(\mathbf{k}')\Phi^*_{nq}(\mathbf{k}''). \end{aligned} \tag{4.98}$$

Thus,

$$\begin{aligned} \mathcal{E}^{(\mathrm{alias})}(k) = 8\pi k^2 \sum_{\mathbf{a}\in\Lambda_0} &\left\{ P_{imn}(\tilde{\mathbf{k}}) P^*_{ipq}(\tilde{\mathbf{k}}) \times \right. \\ \int_{\omega_0}\int_{\omega_0} d\mathbf{k}' d\mathbf{k}'' \Phi^*_{mp}(\mathbf{k}') &\left. \Phi^*_{nq}(\mathbf{k}'')\delta(\mathbf{k}+\mathbf{a}-\mathbf{k}'-\mathbf{k}'') \right\}_\Omega. \end{aligned} \tag{4.99}$$

The integral in (4.98) is difficult to handle analytically because integration over the cubical region ω_0 destroys the spherical symmetry of the problem that was exploited in the computation of J_{mpnq} in the last section. In order to make analytical progress, the following approximation is introduced. The region ω_0, which is a cube in $\mathbf{k}$-space, is replaced by the largest sphere contained in it. Clearly, this procedure can be implemented simply by removing the suffix 'ω_0' from the integral signs in (4.98) and replacing the energy spectrum $E(k)$ by

$$E^{\mathrm{min}}(k) = \begin{cases} E(k) & \text{if } k < k_m \\ 0 & \text{otherwise.} \end{cases} \tag{4.100}$$

The superscript 'min' indicates that this procedure underestimates the true aliasing error by failing to take account of the contribution of modes close to the eight corners of the cube. An alternative method that overestimates the error can be provided by replacing the cube by the smallest sphere that contains it. To obtain this estimate one needs to use in place of $E^{\min}$ the following spectrum;

$$E^{\max}(k) = \begin{cases} E(k) & \text{if } k < \sqrt{3}k_m \\ 0 & \text{otherwise.} \end{cases} \tag{4.101}$$

The true aliasing error is then expected to lie between these two bounds. With the approximation so described, the integral in (4.98) may be extended to the entire wave space. Thus, one obtains

$$\mathcal{E}^{(\text{alias})}(k) = \sum_{\mathbf{a}\in\Lambda_0} \left\{ P_{imn}(\tilde{\mathbf{k}}) P^*_{ipq}(\tilde{\mathbf{k}}) J_{mpnq}(\mathbf{k}+\mathbf{a}) \right\}_\Omega . \tag{4.102}$$

Substitution of the expression for J_{mpnq} used earlier, gives

$$\mathcal{E}^{(\text{alias})}(k) = \sum_{\mathbf{a}\in\Lambda_0} \left\{ [F_1(K) + F_2(K)] \sum_{i,m,n} |P_{imn}(\tilde{\mathbf{k}})|^2 \right.$$
$$\left. +2F_3(K) \sum_{i,m,n,p} \frac{K_m K_p}{K^2} P_{imn}(\tilde{\mathbf{k}}) P^*_{ipn}(\tilde{\mathbf{k}}) \right\}$$
$$\left. + \quad F_4(K) \sum_{i,m,n,p,q} \frac{K_m K_p K_n K_q}{K^4} P_{imn}(\tilde{\mathbf{k}}) P^*_{ipq}(\tilde{\mathbf{k}}) \right\}_\Omega \tag{4.103}$$

where $\mathbf{K} = \mathbf{k}+\mathbf{a}$. Note that in this case the $F_i(K)$ does depend on the direction of $\mathbf{k}$ so that the $F_i(K)$ cannot be extracted from the $\{\quad\}_\Omega$ operation. Though the summation over the set Λ_0 consists of $3^3 - 1 = 26$ terms, for a cubical box one only needs to evaluate 3 terms due to symmetry. Indeed, the full set of "aliasing modes", $\mathbf{a} \in \Lambda_0$, fall into three classes [2]:

$$3D \left\{ (\pm 2k_m, \pm 2k_m, \pm 2k_m) \right. \quad 2D \begin{cases} (\pm 2k_m, \pm 2k_m, 0) \\ (\pm 2k_m, 0, \pm 2k_m) \\ (0, \pm 2k_m, \pm 2k_m) \end{cases} \quad 1D \begin{cases} (\pm 2k_m, 0, 0) \\ (0, \pm 2k_m, 0) \\ (0, 0, \pm 2k_m). \end{cases} \tag{4.104}$$

By symmetry, all the contributions within each class are equal. Therefore,

$$\mathcal{E}^{(\text{alias})}(k) = 6\mathcal{E}^{(\text{alias})}_{1D}(k) + 12\mathcal{E}^{(\text{alias})}_{2D}(k) + 8\mathcal{E}^{(\text{alias})}_{3D}(k) \tag{4.105}$$

where $\mathcal{E}_{1D}^{(\text{alias})}(k)$ is the contribution from any one of the 1D modes, $\mathcal{E}_{2D}^{(\text{alias})}(k)$ is the contribution from any one of the 2D modes, and $\mathcal{E}_{3D}^{(\text{alias})}(k)$ is the contribution from any one of the 3D modes respectively.

If the modified wave-number $\tilde{\mathbf{k}}$ of a numerical method and the energy spectrum of the turbulence, $E(k)$ are known, (4.102) may be evaluated numerically using either $E^{\min}(k)$ or $E^{\max}(k)$ (defined in (4.100) and (4.101)) to get the lower or upper bound for the aliasing error respectively.

3.2.3 Subgrid and total contributions. For the purpose of comparison with the numerical errors, we will calculate the power spectrum of the exact subgrid force and the total nonlinear term. The total nonlinear term $\mathbf{N}$ and the (exact) subgrid force $\mathbf{S}$ can be readily written down in terms of the Fourier-basis:

$$N_i(\mathbf{k}) = -iP_{imn}(\mathbf{k}) \int\int d\mathbf{k}'\, d\mathbf{k}''\delta(\mathbf{k}' + \mathbf{k}'' - \mathbf{k})\hat{u}_m(\mathbf{k}')\hat{u}_n(\mathbf{k}''), \quad (4.106)$$

and

$$S_i(\mathbf{k}) = -iP_{imn}(\mathbf{k})H(\mathbf{k}) \times$$
$$\left(\int\int - \int_{\omega_0}\int_{\omega_0}\right) d\mathbf{k}' d\mathbf{k}''\delta(\mathbf{k}' + \mathbf{k}'' - \mathbf{k})\hat{u}_m(\mathbf{k}')\hat{u}_n(\mathbf{k}''). \quad (4.107)$$

The power-spectra are defined as

$$\frac{\mathcal{S}(k)}{4\pi k^2} = \lim_{V\to\infty} \frac{8\pi^3}{V} \{\langle S_i(\mathbf{k})S_i(\mathbf{k})^*\rangle\}_\Omega \quad (4.108)$$
$$\frac{\mathcal{N}(k)}{4\pi k^2} = \lim_{V\to\infty} \frac{8\pi^3}{V} \{\langle N_i(\mathbf{k})N_i(\mathbf{k})^*\rangle\}_\Omega \quad (4.109)$$

where $\{\ \}_\Omega$ as usual denotes angular average over the sphere $|\mathbf{k}| = k$.

The evaluation of (4.109) is similar to the calculation of $\mathcal{E}^{\text{FD}}(k)$ in § 3.2.1. One only needs to replace 'Δ_{imn}' in (4.95) by '$-P_{imn}$' and drop the last term involving the viscosity. The resulting expressions can be further simplified using the following properties of the P_{imn} tensor:

$$P_{imn}P_{imn} = k^2, \quad (4.110)$$
$$k_m k_p P_{imn}P_{ipn} = \frac{1}{4}k^4 P_{in}P_{in} = \frac{k^4}{2}, \quad (4.111)$$
$$k_m k_n P_{imn} = 0. \quad (4.112)$$

Thus,

$$\mathcal{N}(k) = k^2[F_1(k) + F_2(k) + F_3(k)] \quad (4.113)$$

where $F_1(k)$, $F_2(k)$ and $F_3(k)$ are as defined in § 3.2.1.

The computation of $\mathcal{S}(k)$ once again requires us to restrict the $\mathbf{k}$ space integration to a cubical domain which makes it difficult to handle the integrals analytically. This difficulty is dealt with in precisely the same manner as was done in the computation of the aliasing error. The cubical domain in $\mathbf{k}$ space is replaced by a spherical region of appropriate size. This is completely equivalent to replacing the energy spectrum $E(k)$ by a pseudo-spectrum as in § 3.2.2. With this modification, the calculation is exactly identical to that just presented for the nonlinear term. Thus, one obtains

$$\mathcal{S}(k) = k^2[F_1(k) + F_2(k) + F_3(k)]. \tag{4.114}$$

where in the evaluation of the functions F_i, the "pseudo-spectrum"

$$E^{\min}(k) = \begin{cases} 0 & \text{if } k < \sqrt{3}k_m \\ E(k) & \text{otherwise,} \end{cases} \tag{4.115}$$

or

$$E^{\max}(k) = \begin{cases} 0 & \text{if } k < k_m \\ E(k) & \text{otherwise,} \end{cases} \tag{4.116}$$

should be used in place of $E(k)$ to obtain the lower and upper bound respectively.

3.3 Conclusion and Recommendations

The results established in the previous sections will now be applied to derive quantitative measures of errors in LES. We focus on LES, since, due to the significant amplitude of the highest resolved modes, the effect of numerical errors are much more critical in LES than in DNS. However, the results presented above are general and may be used to obtain error spectra for DNS as well as LES. The difference lies only in the choice of the energy spectrum $E(k)$ (in DNS the dissipation range must be included in $k_{max} \geq k \geq 0$) and in whether the subgrid force or the molecular viscosity term is neglected.

In large-eddy simulation the grid spacing Δ is typically much larger than the Kolmogorov length so that molecular viscosity plays a negligible role. Therefore 'ν' is set to zero throughout this section. For the energy spectrum we assume the "Von-Karman form"

$$E(k) = \frac{ak^4}{(b + k^2)^{17/6}} \tag{4.117}$$

where the constants $a = 2.682$ and $b = 0.417$ are chosen so that the maximum of $E(k)$ occurs at $k = 1$ and the maximum value $E(1) = 1$. This can always be ensured by a proper choice of length and time-scales. The Von-Karman spectrum has the property $E(k) \sim k^4$ as $k \to 0$ and $E(k) \sim k^{-5/3}$ as $k \to \infty$

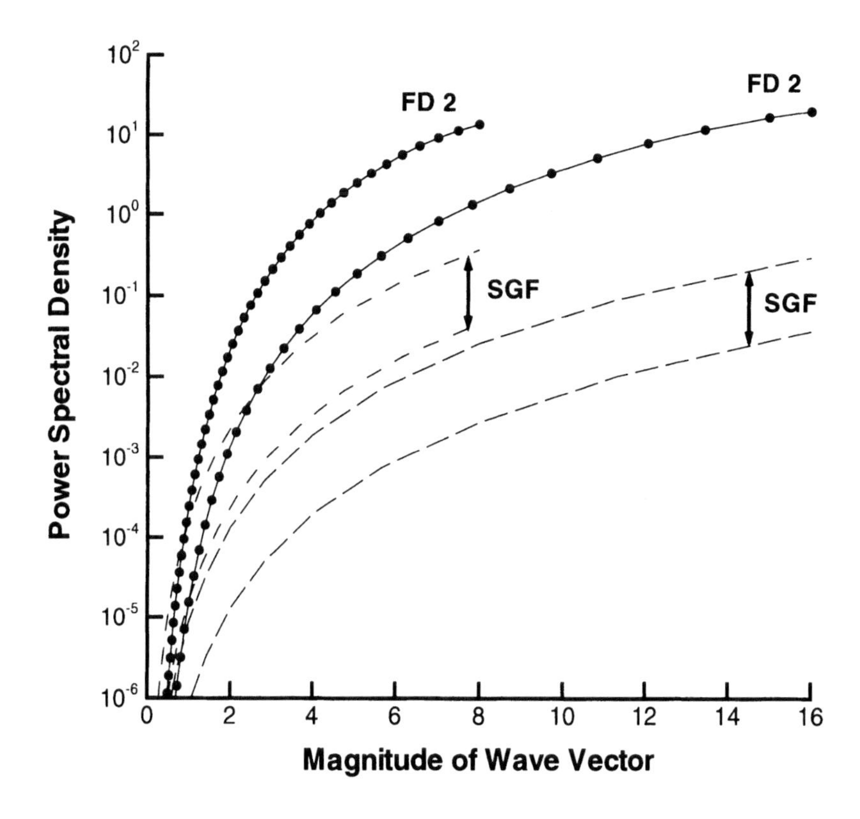

Figure 4.4. Spectral distribution of truncation error $\mathcal{E}^{(FD)}(k)$ for the 2nd order central difference scheme (FD 2), at two resolutions $k_{max} = 8$ and $k_{max} = 16$. Dashed lines indicate upper and lower bounds for the subgrid force (SGF).

and is a fair representation of inertial range turbulence. A plot of this spectrum is shown in Figure 4.3.

A natural way of defining an "acceptable" level of numerical error is to require that the error in the numerical method be much smaller than the subgrid force throughout all, or at least most, of the resolved wavenumber range. Our analytical expression for the power spectra of errors allows us to evaluate various numerical methods against this criterion. Let us first consider a low order method. As a representative example, we choose the second order central-difference scheme implemented with the nonlinear term in the divergence form. Such a scheme is characterized by the modified wavenumber $\tilde{k}_i = \sin(k_i\Delta)/\Delta$ ($i = 1$, 2 or 3). Equation (4.95) is used to compute the power spectrum of the truncation error $\mathcal{E}^{(\mathrm{FD})}(k)$. These results are compared to the power spectra of the upper and lower bounds of the subgrid force computed using (4.114).

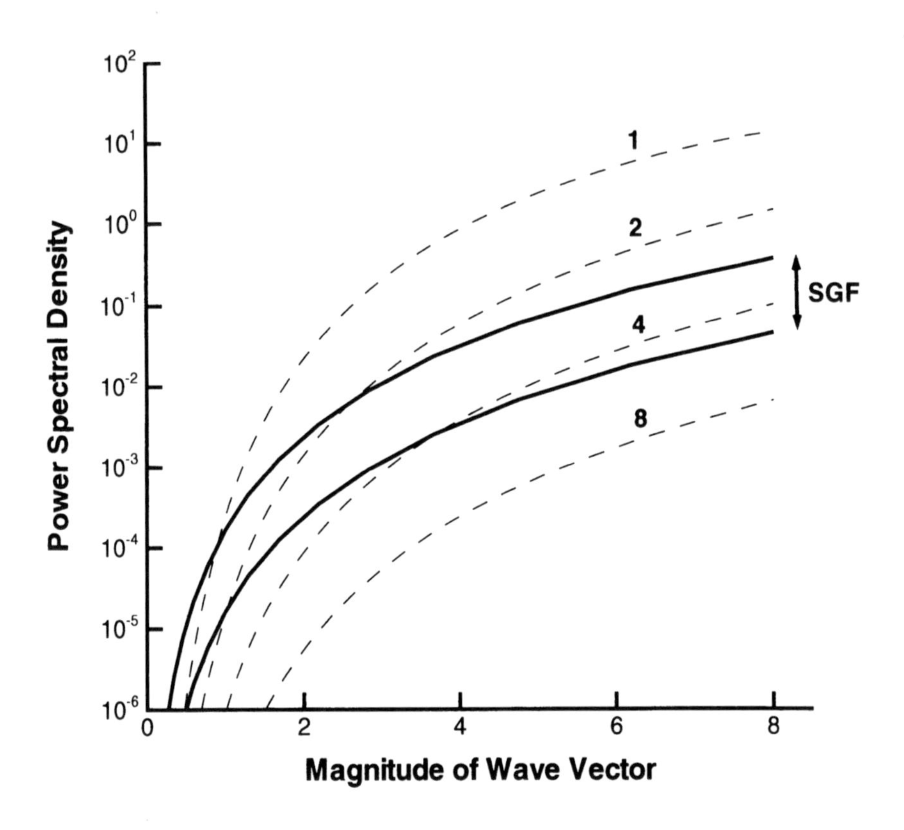

Figure 4.5. Spectral distribution of truncation error $\mathcal{E}^{(FD)}(k)$ for the 2nd order central difference schemes, for several values of filter to grid ratio indicated near respective graph (dashed line). Upper and lower bounds for the subgrid force (solid line) are shown for comparison (SGF).

The results are plotted in Figure 4.4 for $k_m = 8$ and 16. They have the same qualitative appearance for other values of k_m. The truncation error as well as the subgrid force increase monotonically with k. Note that the ordinate is on a logarithmic scale, on a linear scale the curve has the classical "cusp like" appearance. The significant feature is that, the truncation error is substantially larger than the subgrid force over the entire wavenumber range!

Figure 4.4 also illustrates the point that the error in a low order scheme cannot be reduced to an acceptable level by sufficiently refining the grid. This is because, in an LES, the "filter width" is usually identified with the "grid size", as a result, "grid refinement" does not quite mean the same thing in LES as it does in DNS. If the grid (and filter) is refined, that is, k_m is increased, in LES more Fourier modes enter the computation. Thus, we are not refining the grid for a problem of fixed size, but rather *enlarging* the problem with each

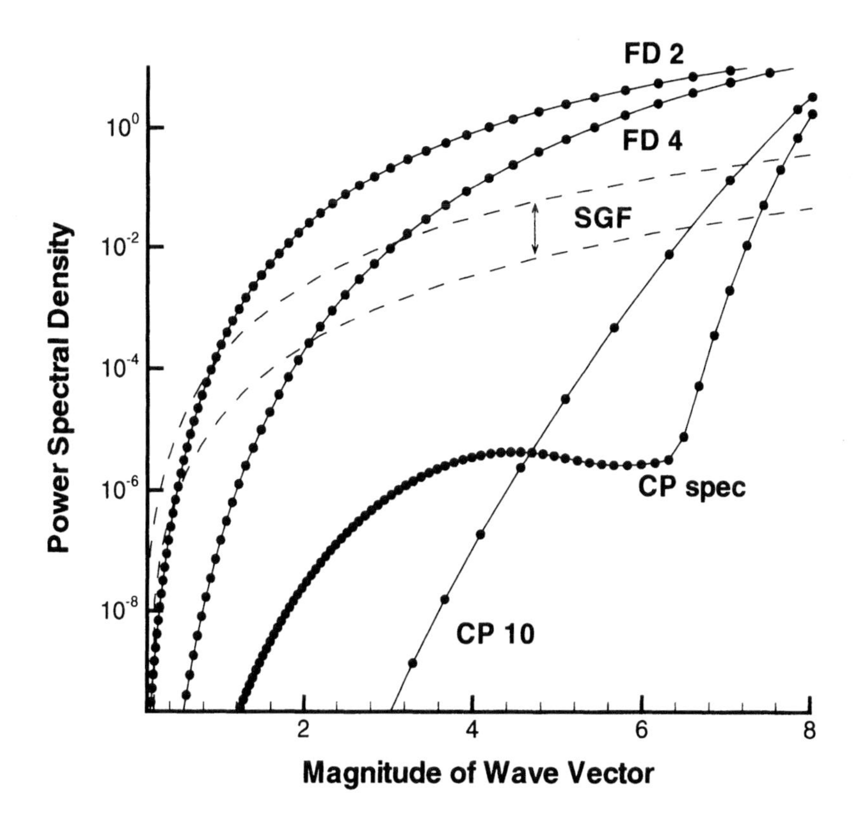

Figure 4.6. Spectral distribution of truncation error $\mathcal{E}^{(FD)}(k)$ for the 2nd (FD 2) & 4th (FD 4) order central difference schemes, Lele's pentadiagonal 10th order compact scheme (CP 10) & "spectral-like" compact scheme (CP spec) with formal 4th order accuracy. Dashed lines indicate upper and lower bounds for the subgrid force (SGF).

grid refinement! It is therefore not surprising that such "grid refinement" does not lead to any improvement in the accuracy of the computation. Figure 4.4 shows that though both the subgrid force and the truncation error decreases on increasing k_m, the truncation error continues to dominate the subgrid contribution irrespective of the resolution. This state of affairs continue if we keep refining the grid, until, ultimately, the resolution is fine enough, that we reach the dissipation range of the spectrum, at which stage, the truncation error as well as the subgrid force would rapidly drop to negligible levels. Such high resolutions are of course irrelevant to this discussion, as they would correspond not to an LES, but to a highly resolved DNS.

The lack of improvement in numerical accuracy on refinement of the grid, is clearly a result of a misapplication of the concept of "grid refinement" in LES.

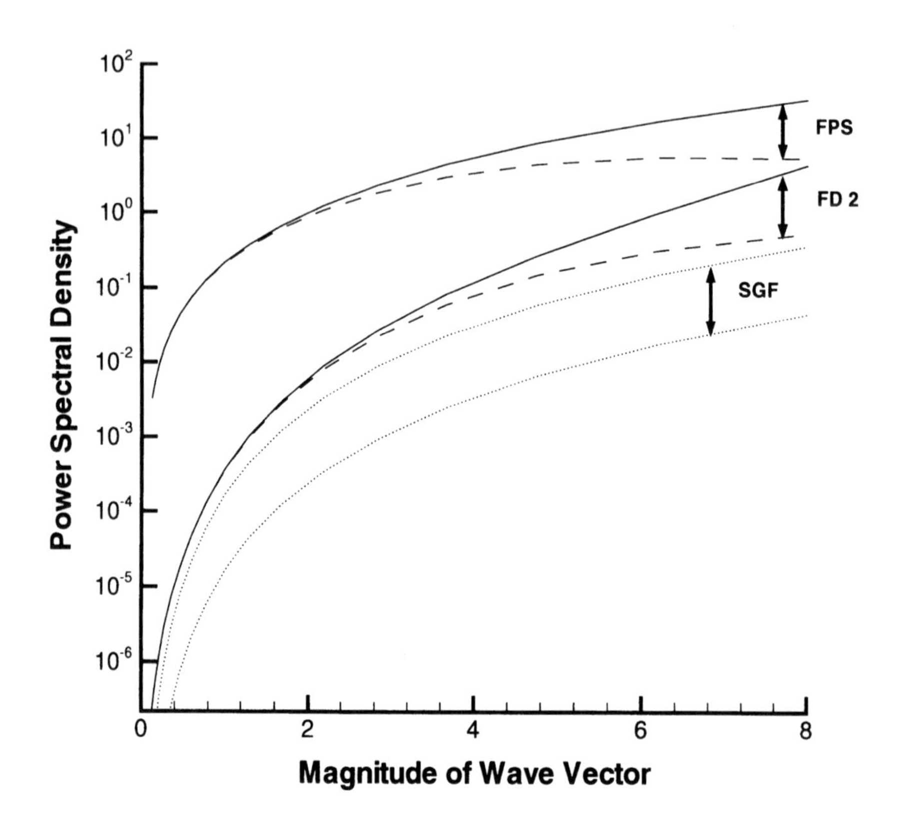

Figure 4.7. Upper (solid) and lower (dashed) bounds of the power spectrum of the aliasing error $\mathcal{E}^{(alias)}(k)$ with the 2nd order central difference scheme (FD2) and the Fourier Pseudo-Spectral method (FPS). The upper and lower bounds for the subgrid force (SGF) are also shown (dotted) for comparison

The "filter size", Δ in an LES represents the level of smoothing that *we* wish to apply to the turbulent field. The resulting field $(\bar{u}_i, \bar{p})$ is expected to have spatial variations on a scale Δ, and they evolve according to the LES equations (4.23) and (4.24). The grid spacing, h, on the other hand is a property of the numerical method that we adopt in order to solve (4.23) and (4.24). In practical terms, 'Δ' is the "thing" that goes into the Smagorinsky model of eddy viscosity

$$\tau_{ij} - (1/3)\delta_{ij}\tau_{kk} = -2C\Delta^2|\bar{S}|\bar{S}_{ij}$$

and in the "test filter" $\hat{\Delta} = 2\Delta$ of the "dynamic model" [28, 29, 30], while h defines the computational nodes (e.g. $x_i = h(i-1)$, $i = 1, \cdots N$). "Grid refinement" should ideally mean reducing h while keeping Δ the *same*. For a fixed numerical scheme (second order central difference in this example) how much does one need to refine the grid in order to achieve acceptable numerical accuracy in LES? The answer is provided by Figure 4.5 where the truncation error is evaluated using (4.95) for $h = \Delta$, $h = \Delta/2$, $h = \Delta/4$ and $h = \Delta/8$ keeping $\Delta = \pi/k_m$ fixed. Since Δ is fixed, the upper and lower bounds of the subgrid force evaluated using (4.114) is the same in all four cases. It is seen that in order to keep the truncation error much lower than the subgrid force throughout the wavenumber range of interest, we must have $\Delta/h \geq 8$. This corresponds to an $8^3 = 64$ fold increase in the number of nodes in a 3D computation in comparison to the (grossly inaccurate) computation with $h = \Delta$. Numerical simulation of turbulence is an enterprise where one is constantly pushing the limits of available computer speed and memory, and such massive increase in computational size is an unaffordable luxury! Therefore, in order to achieve LES with reliable numerical accuracy, we must look elsewhere, namely towards numerical methods of high resolving power considered next.

Figure 4.6 shows the truncation error evaluated using (4.95) for a second and fourth order central difference scheme as well as the tenth order and "spectral like" pentadiagonal compact schemes of Lele discussed earlier. The resolution is kept fixed at $h = \Delta = \pi/k_m$ ($k_m = 8$) which corresponds to a barely resolved computation. Increasing the accuracy of the numerical scheme is seen to lead to a dramatic improvement in the accuracy of the computation. The curves corresponding to the tenth order compact scheme and the spectral like compact scheme (both involve the inversion of a pentadiagonal matrix) are of particular interest. The "spectral like" scheme is seen to have lower levels of error near the limits of resolution $k \approx k_m$ compared to the tenth order scheme. The tenth order scheme has smaller errors in the low wavenumber range, however, this is largely irrelevant as the error levels with either scheme are very small at low k, and it is the accuracy in the neighborhood of k_m that is the limiting factor. A method of order n in formal accuracy has an error $\sim (k\Delta)^n = (k/k_m)^n$. A large n therefore leads to a faster decrease in the error as $k \to 0$. However, we are more interested in the behavior of the error as $(k/k_m) \to 1$ and the order of the method does not tell us anything about this all important regime. The truncation error in the limit $(k/k_m) \to 1$ is determined by the "resolving power" that is, the closeness of the modified wavenumber k' to k even as $k \to k_m$. The "spectral like" pentadiagonal compact scheme performs better than the corresponding tenth order scheme because it has higher resolving power, even though it is only fourth order in formal accuracy. In summary, methods of high resolving power are essential for efficient and accurate numerical simulation of the LES equations.

Figure 4.7 shows the corresponding comparison for the aliasing error. The upper and lower bounds for the aliasing error are computed using (4.105), for the pseudo-spectral and second order central difference schemes. The results are compared to the subgrid force for $h = \Delta = \pi/k_m$ (with $k_m = 8$). The aliasing error is seen to be substantially larger than the subgrid force . In general, increasing the order of a scheme has a relatively weak effect on the aliasing error and the effect is primarily in the high wavenumber region. This effect is in fact in the "reverse" direction compared to truncation errors, that is, the lowest pair of curves which correspond to a second-order scheme have the smallest aliasing error and the highest pair corresponding to an undealiased pseudo-spectral method have the largest. The aliasing errors for schemes that are intermediate between these in terms of resolution fall somewhere in the middle, and have been omitted for clarity. The effect is, of course, quite easy to understand. In the one dimensional problem, the aliasing part of the nonlinear term is multiplied by the modified wavenumber which approaches zero at the cut-off so that the aliasing error is also reduced to zero at k_m. In the three dimensional problem a similar situation applies except that the power spectrum does not actually go to zero on account of the averaging over wavenumber shells. However, the aliasing error is reduced at high wavenumbers for low order methods such as the second order central-difference schemes.

This suggests that together with a numerical method of high resolving power, some method of aliasing error control should be adopted in LES. The simplest such measure would be to adopt the "symmetric form" for the nonlinear term as suggested by Blaisdell [7]. Full dealiasing of quadratic nonlinearities can be achieved by using the familiar "$3/2$ rule". These can be adapted for physical space implementation using "compact filters" designed to mimic very closely a "Fourier cut-off" filter in wave space [31].

The early success of LES computations can be attributed to advances in the area of subgrid modeling coupled with the development and use of highly accurate spectral methods [32]. This was possible because "bench mark flows" of simple geometry were primarily the focus of these studies. It is only relatively recently that LES methods have been applied to domains of complex shapes where finite difference and finite element methods have distinct advantages[33]. Such efforts have led to renewed interest in the issue of numerical errors in LES and their control. The need for stringent error control methods in LES that is apparent from this analysis have been underlined in several recent studies.

The earliest "alarm call" of the possibility of large truncation errors in LES appears to be due to Vreman *et al.* [34]. This contribution appears not to have been noticed until recently, though the authors conclusions were stated in no uncertain terms[3]! In this early study, the authors 'filtered' a DNS database for a 2D compressible mixing layer and explicitly computed the (exact) subgrid force and the truncation errors that would result on discretization of the derivative operators. They presented their results in terms of the L_2 norm of the error rather than the power spectra that we have used here. With $\Delta = h$, and a second order method, the truncation error was about four times larger than the subgrid force! Subsequent studies by the authors with 3D computations confirmed these initial findings [35, 36]. Independently of the numerical simulation results of Vreman *et al.* theoretical analysis demonstrating the dominance of truncation and aliasing errors over the subgrid force was presented[4] by this author[22]. Subsequently, a comprehensive numerical study of the problem was undertaken by Kravchenko and Moin[37] using an ingenious artifice to make a spectral code for channel flow "mimic" a finite difference method of any order. This was achieved by using the modified wavenumber of the scheme in the step of multiplication by wavenumbers to compute derivatives in the streamwise and spanwise directions. A highly accurate B-spline method was used in the wall normal direction, so that truncation errors in the wall normal direction could be considered negligible. The second order central difference scheme gave poor agreement with the "law of the wall," moreover, the presence or absence of the subgrid model did not have any significant effect on the results. On the other hand, both the fully spectral method and the sixth order compact schemes gave excellent agreement with the law of the wall. The agreement became much worse when the subgrid model was turned off, suggesting that the subgrid force was having an effect and was no longer dominated by the truncation error. These conclusions are in full accord with the analysis presented in the last section. An identical conclusion is supported by a recent study due to Kosović *et al.* [38] of isotropic turbulence in a compressible medium using LES with the "stretched vortex model". Results of energy decay and spectra were in good agreement with high resolution DNS with the tenth order compact scheme but poor agreement was obtained on switching to a second order scheme. Very recently, the simulations were repeated using the pentadiagonal "spectral like" compact scheme of Lele which was found to give better agreement with DNS data for the temporal decay of energy than either the pentadiagonal scheme with highest (10th order) formal accuracy or the fourth order tridiagonal compact scheme, even though the spectral like scheme has only fourth order formal accuracy[39]. This observation is in agreement with the analysis presented in section (3).

In summary, it may be concluded, that, as the domain of application of LES moves from simple "building block" flows to more practical engineering problems, great care must be taken in order to assure the accuracy and reliability of the computed results. It is essential that the filter size be kept at least somewhat larger than the grid size and efforts at controlling aliasing errors be adopted. The use of the symmetric form of the nonlinear term and compact approximations to Fourier cut-off filters could potentially provide good control over aliasing errors. High resolution schemes (such as the compact schemes) are essential in order to keep truncation errors to acceptable levels. The role of "commutation errors" and possible remedies however remain unclear at this stage. Alternate formulations of LES that avoid these problems are a promising direction for future research[40, 41, 42, 43].

Notes

1. The first two terms on the right hand side were incorrectly combined as $\mathcal{N}_0 \mathbf{e}$ in [22]. Though the conclusions are not changed as a result, this step is not permissible since $\mathcal{N}_0$ is not a linear operator. The author is grateful to Professor T.J.R. Hughes for pointing out this error in the earlier formulation.

2. This relation is familiar in solid state physics [23] where the set Λ goes by the name "Reciprocal Lattice"

3. "In this paper we use DNS results in order to calculate not only the magnitude of the subgrid terms but also the magnitude of the discretization errors. It will be shown that for both second- and fourth-order-accurate spatial discretizations, the discretization errors dominate the subgrid terms, in case $\Delta = h$."

4. the author is grateful to Professor Vreman for bringing [34, 35, 36] to his attention

References

[1] R. Vichnevetsky and J. Bowles, *Fourier analysis of numerical approximations of hyperbolic equations*, SIAM, Philadelphia, (1982).

[2] S. Lele, "Compact Finite Difference Schemes with Spectral Like Resolution," *J. Comp. Phys.* **103**, 16 (1992).

[3] C. Canuto, M.Y. Hussaini, A. Quarteroni, T.A. Zang "Spectral Methods in Fluid Dynamics," Berlin: Springer (1988).

[4] J.W. Cooley and J.W. Tukey, "An algorithm for the machine calculation of complex Fourier Series," *Math. Comput.* **19**, 297 (1965).

[5] N.A. Phillips "An example of nonlinear computational instability," In *The Atmosphere and Sea in Motion* ed. B. Bolin, Rockefeller Inst. Press, New York (1959).

[6] R. Rogallo, NASA Tech. Memo. TM81315 (1981).

[7] G.A. Blaisdell, E.T. Spyropoulos and J.H. Qin "The effect of the formulation of nonlinear terms on aliasing errors in spectral methods" *Appl. Numer. Math.* **20**, 1 (1996).

[8] J.A. Domaradzki, R.W. Metcalfe, R.S. Rogallo and J.J. Riley "Analysis of subgrid-scale eddy viscosity with use of results from direct numerical simulations," *Phy. Rev. Lett.* **58**(6), 547 (1987).

[9] R.M. Kerr, J.A. Domaradzki, and G. Barbier "Small-scale properties of nonlinear interactions and subgrid-scale energy transfer in isotropic turbulence," *Phys. Fluids* **8**, 197 (1996).

[10] S. Cerutti and C. Meneveau "Statistics of filtered velocity in grid and wake turbulence," *Phys. Fluids* **12**(5), 1143 (2000).

[11] C. Meneveau and J. Katz "Conditional subgrid force and dissipation in locally isotropic and rapidly strained turbulence," *Phys. Fluids* **11**(8), 2317 (1999).

[12] A.A. Aldama "Filtering Techniques for Turbulent Flow Simulation" In *Lecture Notes in Engineering*, **56**, Springer-Verlag, New York/Berlin (1990).

[13] J. Kim, P. Moin and R. Moser "Turbulence statistics in fully-developed channel flow at low Reynolds number" *J. Fluid Mech.*, **177**, 133 (1987).

[14] S. Ghosal and P. Moin, "The Basic Equations for the Large Eddy Simulation of Turbulent Flows in Complex Geometry," *J. Comp. Phys.* **118**, 24 (1995).

[15] H. van der Ven, "A Family of Large Eddy Simulation (LES) filters with nonuniform filter widths," *Phys. Fluids* **7**(5), 1171 (1995).

[16] O.V. Vasilyev, T.S. Lund and P. Moin "A General Class of Commutative Filters for LES in Complex Geometries", *J. Comp. Phys.* **146**(1), 82 (1998).

[17] T.S. Lund "Discrete filters for LES" *CTR Annu. Res. Briefs* pg.83 (1997).

[18] H. Lomax, "Finite difference methods for fluid dynamics" Lecture notes (Stanford University).

[19] W. Gear, *Numerical initial value problems in ordinary differential equations*, Prentice-Hall, New Jersey, (1971).

[20] G. Helmberg *Introduction to Spectral Theory in Hilbert Spaces* North Holland, Amsterdam-London (1969).

[21] R.V. Churchill *Fourier Series and Boundary Value Problems* McGraw-Hill, New York (1963).

[22] S. Ghosal "An analysis of numerical errors in large eddy simulations of turbulence," *J. Comp. Phys.* **125**, 187 (1996).

[23] W. Jones and N. March, *Theoretical solid state physics, Vol.1: Perfect lattices in equilibrium*, Wiley-Interscience, London, (1973).

[24] M. Lesieur, *Turbulence in fluids*, Kluwer Academic Publishers, Dordrecht, The Netherlands, (1987).

[25] A. Monin and A. Yaglom, *Statistical Fluid Mechanics, Vol. 2* The MIT Press, Cambridge, (1979).

[26] G.K. Batchelor, "Pressure fluctuations in isotropic turbulence", *Proc. Camb. Phil. Soc.* **47**, 359 (1951).

[27] G. Batchelor, *The theory of homogeneous turbulence*, Cambridge Univ. Press, Cambridge, England, (1953).

[28] M. Germano, U. Piomelli, P. Moin and W.H. Cabot "A dynamic subgrid-scale eddy viscosity model," *Phys. Fluids A* **3**, 1760 (1991).

[29] S. Ghosal, T.S. Lund, P. Moin and K. Akselvoll "A dynamic localization model for the large eddy simulation of turbulent flow," *J. Fluid Mech.* **286**, 229 (1995).

[30] C. Meneveau, T.S. Lund and W.H. Cabot "A lagrangian dynamic subgrid-scale model of turbulence," *J. Fluid Mech.* **319**, 353 (1996).

[31] O.V. Vasilyev, T.S. Lund and P. Moin "A general class of commutative filters for LES in complex geometries," *J. Comp. Phys.* **146**(1), 82 (1998).

[32] M. Lesieur and O. Metais "New trends in large-eddy simulations of turbulence" *Annu. Rev. Fluid Mech.* **28**, 45 (1996).

[33] U. Piomelli "Large-eddy simulation: achievements and challenges" *Progress in Aerospace Sciences* **35**, 335 (1999).

[34] B. Vreman, B. Geurts and H. Kuerten "Discretization error dominance over subgrid terms in large eddy simulation of compressible shear layers in 2D" *Comm. in Num. methods Eng.* **10**, 785 (1994).

[35] B. Vreman, B. Geurts and H. Kuerten "A priori tests of large eddy simulation of the compressible plane mixing layer" *J. Eng. Math.* **29**, 299 (1995).

[36] B. Vreman, B. Geurts and H. Kuerten "Comparison of numerical schemes in large eddy simulation of the temporal mixing layer" *Int. J. Num. methods in Fluids* **29**, 299 (1996).

[37] A.G. Kravchenko and P. Moin "On the effect of numerical errors in large eddy simulation of turbulent flows" *J. Comp. Phys.* **131**, 310 (1997).

[38] B. Kosović, D.I. Pullin and R. Samtaney "Subgrid-scale modeling for Large-eddy simulations of compressible turbulence" *(preprint)* (2001).

[39] D.I. Pullin (private communication).

[40] T.J.R. Hughes, L. Mazzei and A.A. Oberai "The multiscale formulation of large eddy simulation: Decay of homogeneous isotropic turbulence" *Phys. Fluids* **13**(2), 505 (2001).

[41] T.J.R. Hughes, A.A. Oberai and L. Mazzei "Large eddy simulation of turbulent channel flows by the variational multiscale method" *Phys. Fluids* **13**(6), 1784 (2001).

[42] D. Drikakis, O.P. Iliev and D.P. Vassileva "A Nonlinear Multigrid Method for the Three-Dimensional Incompressible Navier-Stokes Equations" *J. Comp. Phys.* **146**, 301 (1998).

[43] J.A. Langford "Toward ideal large eddy simulation" *Thesis*, Univ. Illinois at Urbana Champaign (2000).

Chapter 5

DESIGNING ADAPTIVE LOW-DISSIPATIVE HIGH ORDER SCHEMES FOR LONG-TIME INTEGRATIONS

H. C. Yee
NASA Ames Research Center, Moffett Field, CA 94035, USA
yee@nas.nasa.gov

B. Sjögreen
Department of Numerical Analysis and Computer Sciences,
KTH, 100 44 Stockholm, Sweden
bjorns@nada.kth.se

Abstract A general framework for the design of adaptive low-dissipative high order schmes is presented. It encompasses a rather complete treatment of the numerical approach based on four integrated design criteria: (1) For stability considerations, condition the governing equations before the application of the appropriate numerical scheme whenever it is possible. (2) For consistency, compatible schemes that possess stability properties, including physical and numerical boundary condition treatments, similar to those of the discrete analogue of the continuum are preferred. (3) For the minimization of numerical dissipation contamination, efficient and adaptive numerical dissipation control to further improve nonlinear stability and accuracy should be used. (4) For practical considerations, the numerical approach should be efficient and applicable to general geometries, and an efficient and reliable dynamic grid adaptation should be used if necessary. These design criteria are, in general, very useful to a wide spectrum of flow simulations. However, the demand on the overall numerical approach for nonlinear stability and accuracy is much more stringent for long-time integration of complex multiscale viscous shock/shear/turbulence/acoustics interactions and numerical combustion. Robust classical numerical methods for less complex flow physics are not suitable or practical for such applications. The present approach is designed expressly to address such flow problems, especially unsteady flows. The minimization of employing very fine grids to overcome the production of spurious numerical solutions and/or instability due to under-resolved grids is also sought [79, 17]. The incremental studies to illustrate the performance of

D. Drikakis and B.J. Geurts (eds.), Turbulent Flow Computation, 141-198.

the approach are summarized. Extensive testing and full implementation of the approach is forthcoming. The results shown so far are very encouraging.

Keywords: Low-Dissipative Schemes, Adaptive Numerical Dissipation/Filer Controls, High Order Finite Difference Methods, Linear and Nonlinear Instabilities, Skew-Symmetric Form, Entropy Splitting, Summation-by-Parts, Integration-by-Parts, Wavelets, Multi-Resolution Wavelets, linear and nonlinear filters.

1. Introduction

Classical stability and convergence theory are based on linear and local linearized analysis as the time steps and grid spacings approach zero. This theory offers no guarantee for nonlinear stability and convergence to the correct solution of the nonlinear governing equations. The use of numerical dissipation has been the key mechanism in combating numerical instabilities. Recent nonlinear stability analysis based on energy norm estimate [24] offers stability improvement in combating nonlinear instability for long-time integrations and/or simple smooth flows. These recent developments offer two major sources of stabilizing mechanisms, namely, from the governing equation level and from the numerical scheme level. Employing a nonlinear stable form of the governing equation in conjunction with appropriate nonlinear schemes for initial-boundary-value problems (IBVPs) is one way of minimizing the use of numerical dissipations [56, 55]. On the other hand, even with the recent development, when employing finite time steps and finite grid spacings in the long-time integration of multiscale complex nonlinear fluid flows, nonlinear instability, although greatly reduced, still occurs and the use of a numerical-dissipation/filter is unavoidable.

Aside from acting as a post-processor step, most filters serve as some form of numerical dissipation. Without loss of generality, "numerical-dissipation/filter" is, hereafter, referred to as "numerical dissipation" unless otherwise stated. Proper control of the numerical dissipation to accurately resolve all relevant multiscales of complex flow problems while still maintaining nonlinear stability and efficiency for long-time numerical integrations poses a great challenge to the design of numerical methods. The required type and amount of numerical dissipation are not only physical problem dependent, but also vary from one flow region to another. This is particularly true for unsteady high-speed shock/shear/boundary-layer/turbulence/acoustics interactions and/or combustion problems, since the dynamics of the nonlinear effect of these flows are not well-understood [79], while long-time integrations of these flows have already stretched the limit of the currently available supercomputers and the existing numerical methods. It is of paramount importance to have proper control of the type and amount of numerical dissipation in regions where it is needed but nowhere else. Inappropriate type and/or amount can be detrimental to the integrity of the computed solution even with extensive grid refinement.

The present work is a sequel to [82, 83, 63, 65]. It is an expanded version of [84]. The objective here is to propose a rather complete treatment of the numerical approach based on the four integrated design criteria (1)-(4) stated in the abstract. The key emphasis here is to describe and illustrate with examples on an adaptive procedure employing appropriate sensors to switch on the desired numerical dissipation where needed, and leave the rest of the region free of numerical dissipation contamination, while at the same time improving nonlinear stability of the entire numerical process for long-time numerical integration of the complex multiscale problems in question. These sensors are capable of distinguishing shocks/shears from turbulent fluctuations and/or spurious high-frequency oscillations for a full spectrum of flow speeds and Reynolds numbers. The minimization of employing very fine grids to overcome the production of spurious numerical solutions and/or instability due to under-resolved grids is sought [17]. It was shown in [56, 20, 83, 63] that conditioning the governing equations via the so called entropy splitting of the inviscid flux derivatives [83] can improve the over all stability of the numerical approach for smooth flows. Therefore, the same shock/shear detector that is designed to switch on the shock/shear numerical dissipation can be used to switch off the entropy splitting form of the inviscid flux derivative in the vicinity the discontinuous regions to further improve nonlinear stability and minimize the use of numerical dissipation. The rest of the sensors, in conjunction with the local flow speed and Reynolds number, can also be used to adaptively determine the appropriate entropy splitting parameter for each flow type/region. These sensors are readily available as an improvement over existing grid adaptation indicators [20]. If applied correctly, the proposed adaptive numerical dissipation control is scheme independent, and can be a stand alone option for many of the favorite schemes used in the literature.

Outline: A brief summary of linear and nonlinear stability and the logistics of advocating design criteria (1)-(4) for a complete numerical approach are discussed in Sections 2 – 4. Adaptive numerical dissipation controls for high order schemes are discussed in Section 5. Some representative examples to illustrate the performance of the approach are given in Section 6.

2. Conditioning of the Governing Equations

Traditionally, conditioning the governing partial differential equations (PDEs) usually referred to rewriting the governing equations in an equivalent set of PDEs in order to prove the stability and/or well-posedness of the PDEs. When numerical methods are used to solve PDEs that are nonlinear, it is well-known that different equivalent forms of the governing equations might exhibit different numerical stability, accuracy and/or spurious computed solutions, even for problems containing no shock/shear discontinuities. There are many conditioned forms of the governing equations proposed in the literature. Different conditioned forms of the nonlinear convection fluxes and the viscous fluxes have been proposed for the incompressible and compressible Navier-Stokes equations. Here we concentrate on the convection terms of these equations and mention a few conditioning forms which are precursors of the so called entropy splitting of the compressible Euler equations [83]. If a method-of-lines approach is used to discretize these equations, the entropy splitting reduces to the splitting of the convection flux derivatives. For the viscous terms, we only adapt the method of preventing odd and even decoupling on all of our numerical experiments whenever it is applicable [61].

The splitting of the nonlinear convection terms (for both the compressible and incompressible Navier-Stokes equations) into a conservative part and a non-conservative part has been known for a long time. In the DNS, LES and atmospheric science simulation literature, it is referred to as the skew-symmetric form of the momentum equations [4, 30, 5, 86]. It consists of the mean average of the conservative and non-conservative (convective form [86]) part of the momentum equations. The spatial difference operator is then applied to the split form. From the numerical analysis standpoint, the Hirt and Zalesak's ZIP scheme [27, 85] is equivalent to applying central schemes to the non-conservative momentum equations (convective form of the momentum equations). MacCormack [39] proposed the use of the skew-symmetric form for problems other than DNS and LES. Harten [25] and Tadmor [73] discussed the symmetric form of the Euler equations and skew-adjoint form of hyperbolic conservation laws, respectively. Although the derivation in these works is different, the ultimate goal of using the split form is almost identical. This goal is to improve nonlinear stability, minimize spurious high-frequency oscillations, and enhance robustness of the numerical computations. The canonical splitting used by Olsson & Oliger [56] is a mathematical tool to prove the existence of a generalized energy estimate for a symmetrizable system of conservation laws. For the thermally perfect gas compressible Euler equations, the transformation consists of a convex entropy function that satisfies a mathematical entropy condition. The mathematical entropy function, in this case, can be a function of the physical entropy. Therefore, the resulting splitting was referred to as **entropy splitting** by Yee et al. [83]. The entropy splitting can be viewed as a more general form than its precursors which makes possible the L^2 stability proof of the nonlinear Euler equations with physical boundary conditions (BCs) included. The following subsections which were part of [65], provide more details.

2.1 Introduction to skew-symmetric splitting

Consider a variable coefficient linear hyperbolic system

$$U_t + A(x)U_x = 0 \quad a < x < b, \tag{5.1}$$

where U is a vector and the matrix $A(x)$ is symmetric. Define the scalar product and norm,

$$(U,V) = \int_a^b U(x)^T V(x)\,dx, \quad ||U||^2 = (U,U). \tag{5.2}$$

It is possible to obtain an energy estimate for the solution by integration-by-parts. To do this, write the system in skew-symmetric form,

$$U_t + \frac{1}{2}(A(x)U)_x + \frac{1}{2}A(x)U_x - \frac{1}{2}A_xU = 0. \tag{5.3}$$

Start from

$$\frac{1}{2}\frac{d}{dt}||U||^2 = (U,U_t) = -\frac{1}{2}[(U,(A(x)U)_x) + (U,A(x)U_x) + (U,A_xU)] \tag{5.4}$$

and perform the integration-by-parts

$$(U,(A(x)U)_x) = -(U_x, A(x)U) + [U^TAU]_a^b = -(U,AU_x) + [U^TAU]_a^b, \tag{5.5}$$

where the last equality follows from the symmetry of A. This gives the energy norm estimate

$$\frac{1}{2}\frac{d}{dt}||U||^2 = -\frac{1}{2}([U^TAU]_a^b - (A_xU,U)), \tag{5.6}$$

which is a standard result that has been known for a long time. It can be found in many textbooks on PDEs, e.g., [19].

For semi-discrete difference approximations, the same idea can be used. Introduce the grid points $x_j = a+(j-1)h, j = 1,2,\ldots,N$ on the interval $[a,b]$, with uniform grid spacing $h = (b-a)/(N-1)$. Apply a spatial discretization to the skew-symmetric form

$$\frac{dU_j(t)}{dt} = -\frac{1}{2}A(x_j)DU_j - \frac{1}{2}D(A(x_j)U_j) + \frac{1}{2}DA(x_j)U_j \tag{5.7}$$

where D is a finite difference operator, approximating the spatial derivative. We will obtain an estimate in a discrete scalar product,

$$(U,V)_h = h\sum_{i,j=1}^{N} \sigma_{i,j}U_iV_j, \tag{5.8}$$

in the same way as for the continuous case. Here $\sigma_{i,j}$ a positive definite matrix (identity matrix for the L^2 norm). The estimate becomes

$$\begin{aligned}\tfrac{1}{2}\tfrac{d}{dt}||U||_h^2 = & -\tfrac{1}{2}((U, ADU)_h + (U, D(AU))_h) + \tfrac{1}{2}(U, D(A)U)_h = \\ & -\tfrac{1}{2}(U_N^T A_N U_N - U_1^T A_1 U_1) + \tfrac{1}{2}(U, D(A)U)_h,\end{aligned} \tag{5.9}$$

where we now assume that the difference operator has the summation-by-parts (SBP) property,

$$(U, DV)_h = -(DU, V)_h + U_N V_N - U_1 V_1 \tag{5.10}$$

with respect to the discrete scalar product. The SBP property here is the discrete analogue of the integration-by-parts energy norm property. One simple SBP operator is given by

$$\begin{aligned} DU_1 &= D_+ U_1 \\ DU_j &= D_0 U_j, \quad j = 2, 3, \ldots, N-1 \\ DU_N &= D_- U_N \end{aligned} \tag{5.11}$$

for the scalar product

$$(U, V)_h = \frac{h}{2} U_1 V_1 + h \sum_{j=2}^{N-1} U_j V_j + \frac{h}{2} U_N V_N. \tag{5.12}$$

Here we define $D_0 U_j = (U_{j+1} - U_{j-1})/(2h)$, $D_+ = (U_{j+1} - U_j)/h$, $D_- U_j = (U_j - U_{j-1})/h$. Higher order accurate SBP operators can be found; see [70]. For periodic problems, the SBP property is usually easy to verify. In this case the boundary terms disappear.

The crucial point is the splitting of the convective term,

$$AU_x = \frac{1}{2} AU_x + \frac{1}{2}(AU)_x - \frac{1}{2} A_x U, \tag{5.13}$$

into one conservative and one non-conservative parts. The difference approximation is applied to the split form. The skew-symmetric splitting for difference approximations has also been known for a long time. It was used in [30], and [31] to prove estimates for the Fourier method. See also [47], where SBP is proved for the Fourier method and a fourth-order difference method, when the boundaries are periodic.

Although this L^2 estimate does not give uniform boundedness of the solution, it has turned out in practical computations that methods based on skew-symmetric splitting perform much better for long-time integrations than un-split methods.

Actually, for a symmetric hyperbolic nonlinear system with $A(x)$ replaced by $A(U(x))$, similar skew-symmetric splitting and energy norm can be obtained. One of the earliest works on skew-symmetric splitting is Arakawa [4], where a splitting was derived for the 2D Euler equations for incompressible fluid flow in vorticity stream function formulation,

$$\omega_t = \psi_y \omega_x - \psi_x \omega_y. \tag{5.14}$$

Here ω is the vorticity and ψ is the stream function, such that the velocity (u, v) is $(-\psi_y, \psi_x)$. In [4] it is shown that the approximation

$$\begin{aligned} \frac{d\omega_{i,j}(t)}{dt} = \; & \tfrac{1}{3}(D_y\psi_{i,j}D_x\omega_{i,j} - D_x\psi_{i,j}D_y\omega_{i,j}) + \\ & \tfrac{1}{3}(D_x(\omega_{i,j}D_y\psi_{i,j}) - D_y(\omega_{i,j}D_x\psi_{i,j})) + \\ & \tfrac{1}{3}(D_y(\psi_{i,j}D_x\omega_{i,j}) - D_x(\psi_{i,j}D_y\omega_{i,j})) \end{aligned} \tag{5.15}$$

leads to the estimates

$$\begin{aligned} & \tfrac{d}{dt}||\omega||^2 = (\omega, \omega_t)_h = 0 \\ & (\psi, \omega_t)_h = 0. \end{aligned} \tag{5.16}$$

Here it is assumed that boundary terms are equal to zero (homogeneous). D_x and D_y denote finite difference operators acting in the x- and y-direction respectively. The second estimate is related to the conservation of kinetic energy,

$$\frac{1}{2}\frac{d}{dt}(||\psi_y||^2 + ||\psi_x||^2) = -(\psi, \omega_t). \tag{5.17}$$

The proof of the estimates only involves pairwise cancellation of terms according to the rule,

$$(u, D(uv))_h + (u, vDu)_h = 0, \tag{5.18}$$

which holds for zero boundary data, if D satisfies (5.10). In [4], the operator (5.11) is used. Note that the use of ω and ψ in this section pertains to the vorticicity formulation symbols. In later sections, the same symbols will have different meanings.

In [30], the inviscid Burgers' equation,

$$u_t + (u^2/2)_x = 0 \tag{5.19}$$

with the quadratic flux derivative split into $\frac{1}{3}uu_x + \frac{2}{3}(u^2/2)x$, is approximated as

$$\frac{du_j}{dt} = -\frac{1}{3}(u_j Du_j + Du_j^2). \tag{5.20}$$

For this approximation, we obtain

$$\frac{1}{2}\frac{d}{dt}||u||_h^2 = -\frac{1}{3}((u, uDu)_h + (u, Du^2)_h) = 0 \tag{5.21}$$

by using SBP on the last term. Again, boundary data is assumed to be zero. The split form $(Du^2 + uDu)/3$ is also used in [58].

It is important to note that the split form is non-dissipative in the sense that the highest grid frequency, $u_j = (-1)^j$, has derivative zero, and thus cannot be seen or smoothed by the time integration. In addition, the difference operator applied to the split form should be done with D on the non-conservative term, and $-D^T$, the negative adjoint of D, on the conservative term. However, since even order centered difference operators are anti-symmetric, we write D in both places. For odd orders of accuracy, the approximation should be done analogously to the first order example,

$$\frac{du_j}{dt} = -\frac{1}{3}(u_j D_+ u_j + D_- u_j^2). \tag{5.22}$$

2.2 Skew-Symmetric Splitting for Incompressible Fluid Flow

For a 3-D incompressible Navier-Stokes equations of the form,

$$\begin{aligned} \mathbf{u}_t + (\mathbf{u}^T\nabla)\mathbf{u} &= -\nabla p + \nu\triangle\mathbf{u}, \\ div\, \mathbf{u} &= 0, \end{aligned} \tag{5.23}$$

skew-symmetric splitting can be applied on the convective terms to estimate the kinetic energy, $\mathbf{u}^T\mathbf{u}$. Here the velocity vector is $\mathbf{u} = (u_1, u_2, u_3)^T$, the pressure p and the viscosity coefficient is ν. In [86] and in [26] the three forms

$$\begin{aligned} &(\mathbf{u}^T\nabla)\mathbf{u} && \text{(convective)} \\ &\tfrac{1}{2}((\mathbf{u}^T\nabla)\mathbf{u} + div(\mathbf{u}\mathbf{u}^T)) && \text{(skew-symmetric)} \\ &\tfrac{1}{2}(\nabla(\mathbf{u}^T\mathbf{u}) + \mathbf{u}\times\nabla\times\mathbf{u}) && \text{(rotational)} \end{aligned} \tag{5.24}$$

for the nonlinear terms are studied. They are equivalent to each other before the application of the numerical methods. Although the skew-symmetric and rotational forms are not in conservative form, they lead to conservation of kinetic energy that is important for long-time integration. For the inviscid case $\nu = 0$, when using the skew-symmetric form, we can estimate the kinetic energy as,

$$\begin{aligned} \tfrac{1}{2}\tfrac{d}{dt}(||\sqrt{\mathbf{u}^T\mathbf{u}}||^2) &= (u_1, (u_1)_t) + (u_2, (u_2)_t) + (u_3, (u_3)_t) = \\ &-(u_i, u_j\partial_j u_i) - (u_i, \partial_j(u_i u_j)) - (u_i, \partial_i p) = \\ &-(u_i, u_j\partial_j u_i) + (\partial_j u_i, u_i u_j) + (\partial_i u_i, p) = 0, \end{aligned} \tag{5.25}$$

where the summation convention is used. We assume that boundary velocities are zero. Integration-by-parts used above shows that the three pressure components are equal to $(div(\mathbf{u}), p)$, which is zero since the divergence is zero.

In order to carry out the same estimate for a difference approximation, we should use the skew-symmetric form for the convective terms, and discretize by a difference operator having the SBP property. The convective terms then disappear from the estimate directly, without use of the divergence condition, just as for the PDE. To eliminate the pressure term, it is enough that the pressure derivatives and the divergence condition are discretized by the same difference operator (or by operators that are negative adjoints, in the case of odd order of accuracy). We end up with

$$\frac{1}{2}\frac{d}{dt}||\mathbf{u}_{i,j,k}^T\mathbf{u}_{i,j,k}||_h^2 \leq 0 \tag{5.26}$$

for the difference approximation in a discrete norm. Note that the inequality (2.26) should really be an equality in the setting it is proved. Perhaps it is possible to keep the viscous terms, and make it an inequality.

The discrete estimate can also be derived from the discretized rotational form (as in [47]), but then the discretized divergence condition must be used to eliminate certain convective terms. For this to be possible, it is necessary that the divergence condition is discretized by the same operator as used for the other convection terms. Results from using the skew-symmetric form are compared with results from the rotational form in a turbulence simulation in [86]. The skew-symmetric form is found to give more accurate results. It is recommended that the pressure equation and the divergence condition should be discretized by the same SBP satisfying operator (or SBP operator for ease of reference), so that we can eliminate the term $(u_1, p_x) + (u_2, p_y) + (u_3, p_z)$.

2.3 Skew-Symmetric Splitting for Compressible Fluid Flows

Consider the equations of inviscid compressible fluid flow in one space dimension

$$\begin{pmatrix} \rho \\ \rho u \\ e \end{pmatrix}_t + \begin{pmatrix} \rho u \\ \rho u^2 + p \\ u(e+p) \end{pmatrix}_x = \begin{pmatrix} 0 \\ 0 \\ 0 \end{pmatrix}, \tag{5.27}$$

with ρ, u, e and p, the density, velocity, total energy per unit volume and pressure, respectively. In [5], a skew-symmetric splitting of the convective terms in momentum equation is used. This splitting was originally presented in [16]. The discretization in [5] is made for a more general equation, but for the simple equation (5.27), it becomes

$$\frac{d}{dt}\begin{pmatrix}\rho_j(t)\\(\rho u)_j(t)\\e_j(t)\end{pmatrix}+\begin{pmatrix}D(\rho_j u_j)\\\frac{1}{2}D(\rho_j u_j^2)+\frac{1}{2}\rho_j u_j Du_j+\frac{1}{2}u_j D(\rho_j u_j)+Dp_j\\D(u_j(e_j+p_j))\end{pmatrix}=\begin{pmatrix}0\\0\\0\end{pmatrix}\tag{5.28}$$

where we now only consider semi-discrete approximations. The skew-symmetric splitting of the convective term in the momentum equation makes it possible to estimate

$$\frac{d}{dt}h\sum_{j=1}^{N}\rho_j u_j^2=-h\sum_{j=1}^{N}u_j Dp_j+\text{boundary terms},\tag{5.29}$$

which is the discrete analogue of the estimate,

$$\frac{d}{dt}||\rho u^2||^2=-(u,p_x)+\text{ boundary terms},\tag{5.30}$$

obtained from the PDE.

2.4 Canonical (Entropy) Splitting for Systems of Conservation Laws

The skew-symmetric splitting for the nonlinear incompressible and compressible Navier-Stokes equations discussed above only involve the nonlinear convective terms of the momentum equation, and not the entire inviscid flux derivatives of the PDEs. Actually, for a general nonlinear system of conservation laws,

$$U_t+F(U)_x=0,\quad a<x<b,\ 0<t,\tag{5.31}$$

we can perform a skew-symmetric splitting of the entire inviscid flux vector derivative $F(U)_x$, if

1 $A(U)=\partial F/\partial U$ is symmetric.

2 F is homogeneous, i.e., $F(\lambda U)=\lambda^\beta F(U)$, with $\beta\neq -1$.

It is possible to show that

$$A(U)U=\beta F(U),\tag{5.32}$$

by differentiating the homogeneity relation with respect to λ, and setting $\lambda=1$. Define the splitting as

$$U_t+\frac{1}{1+\beta}A(U)U_x+\frac{\beta}{1+\beta}F(U)_x=0.\tag{5.33}$$

We then can show that

$$\begin{aligned}\frac{1}{2}\frac{d}{dt}\ ||U||^2 = (U, U_t) = -\tfrac{1}{1+\beta}(U, A(U)U_x) - \tfrac{\beta}{1+\beta}(U, F(U)_x) = \\ -\tfrac{1}{1+\beta}(U, A(U)U_x) + \tfrac{\beta}{1+\beta}(U_x, F(U)) - \tfrac{\beta}{1+\beta}[U^T F(U)]_a^b = \\ -\tfrac{1}{1+\beta}[U^T AU]_a^b.\end{aligned} \tag{5.34}$$

The two inner products disappear due to (5.32), and we are left with an estimate of the norm of the solution in terms of the solution on the boundary, a so called generalized energy estimate [56]. Olsson & Oliger [56] used the term generalized energy estimate because the norm will, in general, depend on an entropy vector W, but not the gradient of U in symmetrizable conservation laws.

In [56] the splitting is first done without the homogeneity assumptions. In that case, the entropy flux function $F^E(U)$

$$F^E(U) = \int_0^1 F(\theta U)\, d\theta \tag{5.35}$$

is introduced. It follows that

$$F_U^E(U)U = -F^E + F, \tag{5.36}$$

so that

$$F_x = (F_U^E U)_x + F_U^E U_x. \tag{5.37}$$

If F_U is symmetric, then F_U^E is symmetric too, and we obtain,

$$\begin{aligned}(U, F_x) &= (U, (F_U^E U)_x) + (U, F_U^E U_x) = \\ &-(U_x, F_U^E U) + (U, F_U^E U_x) + [U F_U^E U]_a^b = [U^T F_U^E U]_a^b.\end{aligned} \tag{5.38}$$

If the flux function F is homogeneous, F^E is just a scalar times F, and we recover the splitting (5.33).

In practice, the Jacobian matrix $A(U)$ is not symmetric, especially for more than 1-D. However, in many cases a symmetrizing variable transformation is available. The estimate for the homogeneous case above and symmetrizing transformations were given in [73]. In [56] the analysis is extended to non-homogeneous problems, and BCs are discussed in greater detail using the so called canonical splitting of the symmetrizable conservation laws. Formulas for symmetrizing the nonlinear compressible Euler equations are given in [25], and the corresponding analytical form for the canonical splitting of the perfect gas compressible case is given in [20]. It was further extended to thermally perfect gases and to 3-D generalized coordinates that preserve freestream in [83, 77].

Let the symmetrizing vector W be related to U via a transformation,

$$U = U(W). \tag{5.39}$$

It can be proved that the existence of a symmetrizing transformation is equivalent to the existence of an entropy function, $E(U)$, with E_{UU} positive definite. The entropy function is a function such that

$$E_t + F_x^E = 0 \tag{5.40}$$

is an additional conservation law, obtained by multiplying the original conservation law by $E_U(U)$. F^E is the entropy flux, related to the entropy by

$$(F_U^E)^T = E_U^T F_U, \tag{5.41}$$

an equation which is overdetermined, and therefore does not have a solution for all systems of conservation laws. Written in terms of the entropy function, (the inverse of) the symmetrizing change of variables (5.39) is defined by $W = E_U(U)$. The change of variables $\partial U/\partial W$ is symmetric and positive definite, and the new Jacobian $\partial F/\partial W$ is symmetric. If furthermore U and F are homogeneous in W of degree β, which is the case for the thermally perfect gas Euler equations for any $\beta \neq -1$, the formulas become simple. In that case we insert the change of variables into the conservation law and obtain

$$U_W W_t + F_W W_x = 0, \tag{5.42}$$

and define the split form of the flux derivative [56]

$$U_t + \frac{\beta}{1+\beta} F_x + \frac{1}{1+\beta} F_W W_x = 0, \tag{5.43}$$

with β a splitting parameter ($\beta = \infty$ recovers the original conservative form). Here $\beta \neq -1$ and, for a perfect gas, $\beta > 0$ or $\beta < \frac{\gamma}{1-\gamma}$. The theory only gives the range of β and does not give any guidelines on how to choose β for the particular flow. The vectors F_W and W can be cast as functions of the primitive variables (ρ, u, p) and β. From the study of [83], β is highly problem dependent. Multiplying the above equation by W and integrating gives

$$-(1+\beta)(W, U_t) = \beta(W, F_x) + (W, F_W W_x) = \beta(W, F_x) + (F_W W, W_x). \tag{5.44}$$

Integration-by-parts in space gives

$$(1+\beta)(W, U_t) = -[W^T F_W W]_a^b. \tag{5.45}$$

We thus obtain the estimate

$$\begin{aligned} \tfrac{d}{dt}(W, U_W W) = \quad & (W_t, U_W W) + (W, (U_W W)_t) = (U_t, W) + \beta(W, U_t) = \\ & (1+\beta)(W, U_t) = -[W^T F_W W]_a^b. \end{aligned} \tag{5.46}$$

In order to have an energy estimate, the boundary term $[W^T F_W W]_a^b$ should be of the sign that makes the time derivative of the norm negative. For stability the entropy norm $(W, U_W W)$ should be bounded.

It is noted that the energy estimate can be shown to be identical with

$$\frac{d}{dt}\int E(U) = [F^E]_a^b \tag{5.47}$$

obtained by integrating the entropy equation (5.40) in space. It follows that, $WU_W W = (1+\beta)E(U)$ [73]. We can show that for the thermally perfect gas compressible Euler equations the mathematical entropy function can be a function of the physical entropy. Therefore, the resulting splitting was referred to as **entropy splitting** by Yee et al. [83].

3. Discrete Analogue of the Continuum

Standard stability guidelines for finite difference methods in solving nonlinear fluid flow equations are based on a linearized stability analysis. The linear stability criterion is applied to the frozen nonlinear problem at each time step and grid point. Most often the numerical BC (or boundary scheme), if needed, is not part of the stability analysis. The preceding section summarizes the historical perspective of entropy splitting of the fluid flow equation related to stable spatial finite discretization without paying attention to numerical BC. This section expands on stable finite difference methods that have a discrete analogue of the conditioned governing equations IBVPs. For ease of reference, "scheme" or more precisely "interior scheme" here generally refers to spatial difference schemes for the interior grid points of the computational domain, whereas "boundary scheme" is the numerical boundary difference operator for grid points near the boundaries. However, without loss of generality, we also adopt the conventional terminology of denoting "scheme" interchangeably as either the "combined interior and boundary scheme" or just the "interior scheme" within the context of the discussion.

The only tool needed to derive the norm estimates presented in the preceding section was integration-by-parts. One main point in this section is that the same norm estimates can be made for a semi-discrete difference approximations, if the differential operators are approximated by difference operators having the SBP property. Examples of this were shown in (5.16) and (5.21). Of course, the other estimates in the preceding section can be carried over to a semi-discrete approximation by use of SBP difference operators.

The discussion is divided into linearly stable and nonlinearly stable difference methods. It is important to point out that when solving the Navier-Stokes equations with complex viscous shock, shear-layer, and boundary layer and/or chemical reaction interactions, even after incorporating tools from recent developments, the finite difference methods considered, although more rigorous than standard algorithms, are only linearly stable in a strict sense.

3.1 Linearly Stable Difference Methods

The most basic linear stability criterion is to investigate the behavior of the difference method when applied to a problem of constant coefficients and periodic boundaries. The Fourier symbol of the operator should be bounded. For higher than first-order methods, a complication is introduced by the numerical boundary treatment. Norm estimates, or normal mode analysis are normally employed (See Gustafsson et al. [24]). With these methods it is possible to prove stability for linear IBVPs. Difference operators having the SBP property with numerical BCs included have recently received some attention. See Strand, Olsson, and Nordström & Carpenter [70, 53, 54, 55, 51]. As already discussed in Section 2, the idea with these operators is to have the property

$$(DU_j, V_j)_h = -(U_j, DV_j)_h + V_N U_N - V_1 U_1, \tag{5.1}$$

where D is a difference operator approximating d/dx, including the accompanied boundary scheme. Typically D is a standard centered operator in the interior of the computational domain, and has a special one-sided form near boundaries. The discrete scalar product is defined by (5.12), and is weighted by a positive definite matrix, σ. For the standard L^2-norm, σ is the identity matrix. In [53, 70], formulas for the norm and boundary modifications of D are given which ensure the SBP property for operators up to order of accuracy eight. SBP satisfying numerical BCs are very different from the traditional numerical boundary treatment. For example, for a sixth-order central interior scheme, the SBP satisfying boundary schemes involved the modification of the central scheme at least 6 points from the boundary. The coefficients of these SBP boundary schemes are rational and irrational fractions. The coefficients of the boundary scheme are determined together with the weights, σ, in the scalar product, so that for each operator there is a particular scalar product in which the SBP property holds. With the SBP property, norm estimates of the difference approximation can be accomplished as the discrete analogue of the continuous energy estimate of the corresponding IBVP of the PDE.

3.2 Nonlinearly Stable Difference Methods

When using a linearly stable method on a nonlinear problem, nonlinear instabilities can appear. Instabilities can appear already for a linear problem with variable coefficients. For variable coefficient problems, it can be proved that numerical dissipation of not too high order will make the method stable. From a theorem by Strang it follows that a finite difference approximation of a nonlinear problem is stable, if the variable coefficient linearized approximation is stable, and the solution and the difference scheme are smooth functions. This is one reason for using numerical dissipation in practical flow simulation [24].

Before 1994, rigorous stability estimates for accurate and appropriate boundary schemes associated with fourth-order or higher spatial interior schemes were the major stumbling block in the theoretical development of combined interior and boundary schemes for **nonlinear** systems of conservation laws. Olsson [55] proved that an energy estimate can be established for second-order central schemes. To obtain a rigorous energy estimate for high order central schemes, one must apply the scheme to the split form of the inviscid governing equation. A discrete analogue of the continuum using a semi discrete approach can be written as

$$\frac{dU_j(t)}{dt} = -\frac{\beta}{1+\beta} DF(U_j) - \frac{1}{1+\beta} F_W(U_j) DW_j. \tag{5.2}$$

Here, D is a difference operator, having the SBP property [55, 70]. The estimate

$$\frac{d}{dt}(W, U_W W)_h = -W_N^T F_W(W_N) W_N + W_1^T F_W(W_1) W_1 \tag{5.3}$$

in the discrete scalar product follows in the same way as for the PDE with indices 1 and N the end points of the computational domain, and h the grid spacing. Here the SBP satisfying difference operator, for example, consists of central difference interior operators of even order together with the corresponding numerical boundary operators that obey the discrete energy estimate. See Olsson and Strand for forms of the SBP boundary operators [55, 70]. As noted in Section 2, if odd order of the spatial discretizations are used, the difference operator D in (5.2) should be modified. In this case, D should be employed on the non-conservative term, and $-D^T$, the negative adjoint of D on the conservative term, i.e.,

$$\frac{dU_j(t)}{dt} = -\frac{\beta}{1+\beta}(-D^T) F(U_j) - \frac{1}{1+\beta} F_W(U_j) DW_j. \tag{5.4}$$

3.3 SBP Difference Operators and Full Discretization

There are two additional difficulties when applying the above semi-discrete SBP spatial discretization methods to realistic problems.

- How to impose given physical BCs without destroying the SBP property. For example, assume that we are given a boundary value $u(x_1) = g$ at the leftmost grid point of the domain at $j = 1$. Applying the SBP operator at $j = 2, 3, \ldots, N$, and imposing $u_1 = g$ would not lead to an estimate, since the one sided operator that should have been applied at

$j = 1$ is not applied. This in turn leads to additional non-zero terms in the scalar product (e.g., from the operator at $j = 2$) which should have been canceled by terms from the operator at $j = 1$.

- How to discretize in time. The semi-discrete estimates that show the norm decreases, will not necessarily lead to a decrease of the norm in a time discretized approximation. In many practical cases we do obtain a small increase in the norm for the time discrete problem. However, the stability is still greatly improved by use of entropy splitting and SBP operators, when compared with more standard schemes.

Several methods showing how to impose the physical BC have been proposed to overcome the first difficulty. Examples are the projection method [53] and the penalty method called "simultaneous approximation term" (SAT) [10, 51]. For comparison of these methods, see [69, 48, 29]. The methods given in [10, 51] and [53] are based on linear properties and cannot be trivially generalized to the nonlinear Navier-Stokes equations, except for certain special cases. One such special case where the nonlinear case is covered by the theory involves imposing velocity zero on solid walls, where the simple approach of setting the velocity to zero after each time step coincides with the projection method in [53].

In addition, when time-dependent physical boundaries are involved, an additional complication arises, especially for multi-stage Runge-Kutta methods. If the time-dependent physical BC is not imposed correctly, the overall order of accuracy of the scheme cannot be maintained. Some systematic remedies are proposed but are rather complicated for variable coefficients and even more complicated for nonlinear problems. See [10, 8, 29].

For the full discretization of the problem, we should discretize in time in such a way that the discrete energy estimate also holds. The obvious solution would be to discretize in time in a skew-symmetric way, in a manner similar to the spatial discretization, e.g.,

$$\begin{aligned} \tfrac{\beta}{1+\beta} D_t U_j^n &+ \tfrac{1}{1+\beta} U_W(W_j^n) D_t W_j^n = \\ &- \tfrac{\beta}{1+\beta} DF(U_j^n) - \tfrac{1}{1+\beta} F_W(W_j^n) D W_j^n, \end{aligned} \tag{5.5}$$

where D_t is a difference operator acting in the time direction. However, it turns out that the SBP property of the time difference quotient leads to a problem which is coupled implicitly in the time direction. To solve it we have to solve a nonlinear system of equations for all time levels in the same system, leading to an impractically large computational effort. Furthermore, numerical experiments shown in Sjögreen & Yee [65] indicated that a bounded L^2 entropy norm $(W, U_W W)_h$ does not necessarily guarantee a well behaved numerical solution for long-time integrations. In other words, L^2 stability does not necessarily guarantee an accurate solution. In practical computations, the classical Runge-Kutta time discretizations using the method-of-lines approach (which we used for our numerical experiments) works well, but we have not been able to prove a time discrete entropy estimate for this method. In addition, numerical experiments shown in [65] indicate that the time discrete problem does not have a decreasing entropy norm for all values of β. Numerical experiments in Yee et al. [83, 61] also indicate the wide variations of the β value for a full spectrum of flow problems. For example, if a constant β is used for problems containing shock waves, a very large value of β is needed. Otherwise divergent solutions or wrong shock location and/or shock strengths are obtained. In view of these findings, employing a constant β (within the allowable range of β) throughout the entire computational domain appears not to be the best approach unless the flow problem is a simple smooth flow. Studies in [83, 63] indicate that the split form of the inviscid flux derivatives does help in minimizing the use of numerical dissipation. What is needed is adaptive control of the β parameter from one flow region to another as well as from one physical problem to another. We caution that if the adaptation is not handled correctly, an abrupt switching of the β can introduce spurious jumps in the numerical solutions. See [65] for the discussion.

In our computer code for the numerical experiment, we have implemented the sixth-order SBP operators by the projection and SAT methods given in [10, 51, 53]. They both perform satisfactorily, and no big difference in performance has been observed between them. See Section 6 for a 3-D compressible channel flow computation. We note that the majority of the physical boundaries of our viscous models are not time dependent, and the loss of spatial accuracy due to the multistage Runge-Kutta method is not a major concern.

4. Adaptive Numerical Dissipation Control

This section discusses the need for adaptive numerical dissipation controls in addition to conditioning the governing equations. An advanced numerical dissipation model for multiscale complex viscous flows is described.

The linear and nonlinear numerical dissipation (not filter) presently available is either built into the numerical scheme or added to the existing scheme. The built-in numerical dissipation schemes are, e.g., upwind, flux corrected transport (FCT), total variation diminishing (TVD), essentially non-oscillatory (ENO), weighted ENO (WENO), and hybrid schemes (e.g., those that switch between spectral and high-order shock-capturing schemes). The built-in nonlinear numerical dissipation in TVD, ENO and WENO schemes was designed to capture accurately discontinuities and high gradient flows while hoping to maintain the order of accuracy of the scheme away from discontinuities. These schemes have been shown to work well in a variety of rapidly developing shock-shock interactions that do not involve multiscale physics or long-time wave propagations. For multiscale physics that require low-dispersive errors, the amount of numerical dissipation built into these schemes is not optimal. In addition, analogous SBP theory for these schemes is not available. Moreover, they are more computationally expensive than standard high order centered schemes, and have severe limitations on the order of accuracy in the vicinity of the discontinuities and steep gradient regions. The inaccuracy of the numerical solutions can contaminate the entire flow field downstream [14]. Although, the amount of numerical dissipation is less than linear numerical dissipations, when applied to convection portions of viscous flows, it conflicts with the physical viscosity and can wash out the true physical steep gradient and/or turbulent structures. Aside from this fact, viscous reacting flows are even more difficult to simulate than non-reacting viscous flows. In the presence of numerical dissipations, even what is believed to be the optimal amount for non-reacting flows might have detrimental effects, e.g., wrong speeds of propagation and/or spurious traveling waves [36, 33, 34].

There exist different specialized linear and nonlinear filters to post process the numerical solution after the completion of a full time step of the numerical integration. Since they are post processors, the physical viscosity, if it exists, is taken into consideration. The main design principle of linear filters is to improve nonlinear stability, to stabilize under-resolved grids [17] and to de-alias smooth flows, while the design principle of nonlinear filters is to improve nonlinear stability as well as accuracy near discontinuities. When discontinuities are present in the solution, linear filtering and/or entropy splitting might not be helpful or not applicable. The nonconservative terms of the entropy splitting might lead to inconsistent behavior at shocks/shears [83, 63, 65]. See, for example, [22, 17, 18, 76] for forms of linear filters, and see [82, 83, 63] for forms of nonlinear filters. The use of the linear filter concept for smooth and/or turbulent flows has been employed for over two decades [76, 2, 37, 18]. For direct numerical simulation (DNS) and large eddy simulation (LES), there are additional variants of the linear filter approach. It was shown in Fischer & Mullen [17] that adding an appropriate filter can stabilize the Galerkin-based spectral element method in convection-dominated problems. The Fischer & Mullen numerical example illustrates the added benefit of the high-order linear filter. See Section 5.5 for a discussion.

For the last decade, CPU intensive high order schemes with built-in nonlinear dissipation have been gaining in popularity in DNS and LES for long-time integration of shock-turbulence interactions. Aside from the aforementioned short-coming of these built-in nonlinear dissipation high order schemes, their flow sensing mechanism is not sophisticated enough to clearly distinguish shocks/shears from turbulent fluctuations and/or spurious high-frequency oscillations. In [82, 83, 63] it was shown that these built-in numerical dissipations are more dissipative and less accurate than the nonlinear filter approach of [82, 83, 63] with a similar order of accuracy. It was also shown that these nonlinear filters can also suppress spurious high-frequency oscillations. However, a subsequent study of Sjögreen & Yee [65] showed that the high order linear filter can sustain longer time integration more accurately than the nonlinear filter for low speed smooth flows. In other words, for the numerical examples that were examined in [65], the high order linear filter can remove spurious high-frequency oscillation producing nonlinear instability better than the second-order nonlinear filter. Higher than third-order nonlinear filters might be able to improve their performance or might outperform the high order linear filters in combating spurious high-frequency oscillations at the expense of more CPU time and added complexity near the computational boundaries. These findings prompted the design of switching on and off or blending of different filters to obtain the optimal accuracy of high order spatial difference operators as proposed in Yee et al. and Sjögreen & Yee [83, 63]. The missing link of what was proposed in [83, 63] is an efficient, automated and reliable set of appropriate sensors that are capable of distinguishing key features of the flow for a full spectrum of flow speeds and Reynolds numbers.

We propose to enhance the conditioning of the equations with an advanced numerical dissipation model, which includes nonlinear sensors to detect shocks, shears and other small scale features, and spurious oscillation instability due to under-resolved grids. Furthermore, we will use the detector to switch off the entropy splitting at shocks/shears and adjust the entropy splitting parameter with the aid of the local Mach number and Reynolds number in smooth regions as discussed earlier. The advanced numerical dissipation model can be used: (Option I) as part of the scheme, (Option II) as an adaptive filter control after the completion of a full time step of the numerical integration or (Option III) as a combination of Options I and II. For example, we can combine high order nonlinear dissipation (with sensor control) using Option I and nonlinear filter (with a different sensor control) using Option II.

The numerical experiments we have conducted so far concentrate on an adaptive procedure that can distinguish three major computed flow features to signal the correct type and amount of numerical dissipation needed in addition to controlling the entropy splitting parameter. The major flow features and numerical instability are (a) shocks/shears , (b) turbulent fluctuations, and (c) spurious high-frequency oscillations. It is important to not damp out the turbulent fluctuations. The procedure can be extended if additional refinement or classification of flow types and the required type of numerical dissipation is needed. There exist different detection mechanisms in the literature for the above three features. These detectors are not mutually exclusive and/or are too expensive for practical applications. We believe that the multiresolution wavelet approach proposed in Sjögreen & Yee [63] is capable of detecting all of these flow features, resulting in three distinct sensors. If chosen properly, one multiresolution wavelet basis function might be able to detect all three features. For an optimum choice, one might have to use more than one type of wavelet basis functions but at the expense of an increase in CPU requirements. Some incremental studies into the use of entropy splitting and the application of these sensors were illustrated in [83, 63, 64, 66, 65, 50]. The next section summarizes the development of adaptive filters for a special class of high order discretizations.

5. High Order Filter Finite Difference Methods

The adaptive numerical dissipation controls discussed in the preceding section are scheme independent. This section applies the adaptive ideas to high order central schemes. We first summarize the high order nonlinear filter schemes that were developed for shock/shear capturing. We then extend these filtering ideas to include more complex flow structures by blending more than one filter and sensing tool.

5.1 ACM and Wavelet Filter Schemes for Discontinuity Capturing

An alternative to linear filter and/or standard shock-capturing schemes for viscous multiscale and long-time wave propagation computations is the ACM (artificial compression method) and wavelet filter schemes described in [82, 63]. A high order centered base scheme together with the nonlinear dissipative portion of a shock-capturing scheme, activated by an ACM or wavelet sensor is used as the filter. Often an entropy split form of the inviscid flux derivatives is used. The idea of the ACM filter scheme is to have the spatially higher non-dissipative scheme activated at all times and to add the full strength, efficient and accurate numerical dissipation only at the shock layers and steep gradients. Thus, it is necessary to have good detectors which flag the layers, and not the oscillatory turbulent parts of the flow field. While minimizing the use of numerical dissipation away from discontinuities and steep gradients, the ACM filter scheme consists of tuning parameters that are physical problem dependent. To minimize the tuning of parameters, new sensors with improved detection properties were proposed in Sjögreen & Yee [63]. The new sensors are derived from utilizing appropriate non-orthogonal wavelet basis functions, and they can be used to completely switch off the extra numerical dissipation outside shock layers. The non-dissipative spatial base scheme of arbitrarily high order of accuracy can be maintained without compromising its stability at all parts of the domain where the solution is smooth. The corresponding scheme is referred to as the wavelet filter scheme. This nonlinear filter approach is particularly important for multiscale viscous flows. The procedure takes the physical viscosity and the reacting terms into consideration since only non-dissipative high order schemes are used as the base scheme. In other words, numerical dissipations based on the convection terms are used to filter the numerical solution at the completion of the full step of the time integration at regions where the physical viscosity is inadequate to stabilize the high frequency oscillations due to the non-dissipative nature of the base scheme.

The method applied to the 2-D conservation law where U is the conservative vector and F and G are the inviscid fluxes,

$$U_t + F(U)_x + G(U)_y = 0, \tag{5.1}$$

can be described as taking, e.g., one full time step by a Runge-Kutta method on the semi discrete system without or with entropy splitting, respectively, by

$$\frac{dU_{j,k}}{dt} = -D_J F(U_{j,k}) - D_K G(U_{j,k}), \tag{5.2}$$

$$\begin{array}{rcl} \frac{dU_{j,k}}{dt} & = & -\frac{\beta}{1+\beta}[D_J F(U_{j,k}) + D_K G(U_{j,k})] \\ - & & \frac{1}{1+\beta}[F_W(U_{j,k}) D_J W_{j,k} + G_W(U_{j,k}) D_K W_{j,k}], \end{array} \tag{5.3}$$

where D_J and D_K are high order finite difference operators, acting in the j- and k-direction respectively. They can be the SBP satisfying higher-order difference operators (e.g., sixth-order central scheme with SBP boundary schemes). We consider here a rectangular grid with grid spacing Δx and Δy and time step Δt. Denote a full Runge-Kutta step by

$$\widetilde{U}_{j,k}^{n+1} = RK(U_{j,k}^n). \tag{5.4}$$

After the completion of a full Runge-Kutta step, a filter (post processing) step is applied leading to

$$U_{j,k}^{n+1} = \widetilde{U}_{j,k}^{n+1} - \lambda_x(\widetilde{F}_{j+1/2,k} - \widetilde{F}_{j-1/2,k}) - \lambda_y(\widetilde{G}_{j,k+1/2} - \widetilde{G}_{j,k-1/2}) \tag{5.5}$$

with $\lambda_x = \Delta t/\Delta x$ and $\lambda_y = \Delta t/\Delta y$. The filter numerical fluxes $\widetilde{F}_{j+1/2,k}$ and $\widetilde{G}_{j,k+1/2}$ act in the j- and k- coordinate directions respectively, and are evaluated on the function $\widetilde{U}^{n+1}$. For example,

$$\widetilde{F}_{j+1/2,k} = \frac{1}{2} R_{j+1/2} \Phi_{j+1/2}^* \tag{5.6}$$

where $R_{j+1/2}$ is the right eigenvector matrix of the Jacobian of the inviscid flux F evaluated at Roe's average state with the k index suppressed. The lth element of the filter flux $\Phi_{j+1/2}^*$ in the x-direction $({\phi^l}_{j+1/2})^*$ is a product of a sensor

$\omega^l_{j+1/2}$ and a nonlinear dissipation $\phi^l_{j+1/2}$, $l = 1, 2, 3, 4$. With the omission of the k index, it is of the form

$$(\phi^l{}_{j+1/2})^* = \omega^l_{j+1/2}\phi^l_{j+1/2}. \tag{5.7}$$

For the ACM sensor, $\omega^l_{j+1/2}$ is a product of a physical dependent sensor coefficient and a gradient like detector. The nonlinear numerical dissipation $\phi^l_{j+1/2}$ can be obtained, from the dissipative portion of a TVD, ENO or WENO scheme. For example, the numerical flux $H_{j+1/2,k}$ of a second- or third-order TVD, ENO or WENO scheme can be written

$$H_{j+1/2,k} = \frac{1}{2}(F_{j,k} + F_{j+1,k}) + \frac{1}{2}R_{j+1/2}\Phi_{j+1/2}, \tag{5.8}$$

with the first two terms corresponding to the flux average of a centered difference and $\Phi_{j+1/2}$ with elements $\phi^l_{j+1/2}$ being the numerical dissipation portion of the scheme.

For all the numerical experiments, the numerical dissipation portion of the Harten-Yee scheme is used. It has the form for the j-direction

$$\phi^l_{j+1/2} = \frac{1}{2}Q(a^l_{j+1/2})(g^l_{j+1} + g^l_j) - Q(a^l_{j+1/2} + \gamma^l_{j+1/2})\tilde{\alpha}^l_{j+1/2} \tag{5.9}$$

with $Q(x) = \sqrt{x^2 + \epsilon^2}$, the entropy satisfying remedy for the scheme with entropy correction parameter ϵ (not to be confused with the entropy splitting parameter). $a^l_{j+1/2}$ is the lth characteristic speed evaluated at the Roe's average state in the j-direction. $\gamma^l_{j+1/2}$ is the modified characteristic speed and g^l_j is a slope limiter which is a function of $\tilde{\alpha}^l_{j\pm1/2}$, the jump in the characteristic variable in the x-direction.

A form of the ACM sensor $\omega^l_{j+1/2}$ proposed in [82] is

$$\omega^l_{j+1/2} = \kappa \max(\theta^l_j, \theta^l_{j+1}) \tag{5.10}$$

where

$$\theta_j = \left| \frac{|\tilde{\alpha}^l_{j+1/2}| - |\tilde{\alpha}^l_{j-1/2}|}{|\tilde{\alpha}^l_{j+1/2}| + |\tilde{\alpha}^l_{j-1/2}|} \right|. \tag{5.11}$$

See [82, 83] for details. It was shown in [63] that the method can be improved by letting the sensor $\omega^l_{j+1/2}$ be based instead on a regularity estimate obtained from the wavelet coefficients of the solution. The wavelet analysis gives an estimate of the so called local Lipschitz exponent α. The dissipation is switched on for low α values, and switched off when α becomes large [63]. The wavelet analysis is more general and can be used to detect other features besides shocks/shears. The following gives a more detailed explanation.

5.2 Wavelet Sensor for Multiscale Flow Physics

Wavelets were originally developed for feature extraction in image processing and for data compression. It is well known that the regularity of a function can be determined from its wavelet coefficients [13, 46, 41] far better than from its Fourier coefficients. By computing wavelet coefficients (with a suitable set of wavelet basis functions), we obtain very precise information about the regularity of the function in question. This information is obtained just by analyzing a given grid function. No information about the particular problem which is solved is used. Thus, wavelet detectors are general, problem independent, and rest on a solid mathematical foundation.

As of the 1990's, wavelets have been a new class of basis functions that are finding use in analyzing and interpreting turbulence data from experiments. They also are used for analyzing the structure of turbulence from numerical data obtained from DNS or LES. See Farge [15] and Perrier et al. [57]. There are several ways to introduce wavelets. One standard way is through the continuous wavelet transform and another is through multiresolution analysis, hereafter, referred to as wavelet based multiresolution analysis. Mallet and collaborators [41, 42, 43, 44, 45, 46] established important wavelet theory through multiresolution analysis. See references [72, 71] for an introduction to the concept of multiresolution analysis. Recently, wavelet based multiresolution analysis has been used for grid adaptation (Gerritsen & Olsson [20]), and to replace existing basis functions in constructing more accurate finite element methods. Here we utilize wavelet based multiresolution analysis to adaptively control the amount of numerical dissipation.

The wavelet sensor estimates the Lipschitz exponent of a grid function f_j (e.g., the density and pressure). The Lipschitz exponent at a point x is defined as the largest α satisfying

$$\sup_{h\neq 0} \frac{|f(x+h)-f(x)|}{h^{\alpha}} \leq C, \tag{5.12}$$

and this gives information about the regularity of the function f, where small α means poor regularity. For a C^1 wavelet function ψ with compact support, α can be estimated from the wavelet coefficients, defined as

$$w_{m,j} = < f, \psi_{m,j} > = \int f(x)\psi_{m,j}(x)dx, \tag{5.13}$$

where

$$\psi_{m,j} = 2^m \psi\Big(\frac{x-j}{2^m}\Big) \tag{5.14}$$

is the wavelet function $\psi_{m,j}$ on scale m located at the point j in space. This definition gives a so called redundant wavelet, which gives (under a few technical assumptions on ψ) a non-orthogonal basis for L^2. Theorem 9.2.2 in [13] states that if ψ is C^1 and has compact support, and if the wavelet coefficients $\max_j |w_{m,j}|$ in a neighborhood of j_0 decay as $2^{m\alpha}$ as the scale is refined, then the grid function f_j has Lipschitz exponent α at j_0. In practical computations, we have a smallest scale determined by the grid size. We evaluate $w_{m,j}$ on this scale, m_0, and a few coarser scales, m_0+1, m_0+2, and estimate the Lipschitz exponent at the point j_0 by a least square fit to the line [63]

$$\max_{j \text{ near } j_0} \log_2 |w_{m,j}| = m\alpha + c. \tag{5.15}$$

Proper selection of the wavelet ψ is very important for an accurate detection of the flow features. The result in [46, 45], which is used in [20], gives the condition that $\psi(x)$ should be the kth-derivative of a smooth function $\eta(x)$ with the property

$$\eta(x) > 0, \quad \int \eta(x)\,dx = 1, \quad \lim_{x\to\pm\infty} \eta^{(k)}(x) = 0. \tag{5.16}$$

Then the result is valid for $0 < \alpha < k$. A continuous function $f(x)$ has a Lipschitz exponent $\alpha > 0$. A bounded discontinuity (shock) has $\alpha = 0$, and a Dirac function (local oscillation) has $\alpha = -1$. Large values of k can be used in turbulent flow so that large vortices or vortex sheets can be detected. Although the theorem above does not hold for α negative, a useful upper bound on α can be obtained from the wavelet coefficient estimate.

For the numerical experiments, the wavelet coefficient $w_{m,j}$ is computed numerically by a recursive procedure, which is a second-order B-spline wavelet or a modification of Harten's multi-resolution scheme [63]. We can express the algorithm as follows. Introduce the grid operators

$$\begin{aligned} Af_j &= \textstyle\sum_{k=-p}^{q} d_k f_{j+k} \\ Df_j &= \textstyle\sum_{k=-p}^{q} c_k f_{j+k} \end{aligned} \tag{5.17}$$

and its mth level expanded versions

$$\begin{aligned} A_m f_j &= \textstyle\sum_{k=-p}^{q} d_k f_{j+2^m k} \\ D_m f_j &= \textstyle\sum_{k=-p}^{q} c_k f_{j+2^m k}, \end{aligned} \tag{5.18}$$

where the integers p and q and the coefficients c_k and d_k are related to the chosen $\psi(x)$ and $\phi(x)$, and can be determined from them. Here $\phi(x)$ is the so called scaling function of the multiresolution wavelets.

The mth level of wavelet coefficients can be written as

$$w_{m,j} = \langle f, \psi_{m,j} \rangle = D_{m-1}A_{m-2}A_{m-3}\ldots A_0 f_j, \quad m = 1, 2, \ldots. \tag{5.19}$$

Once the coefficients d_k and c_k are determined, the computation is a very standard application of grid operators. In practice, we only use $m_0 = 3$ to 5. To be able to compute up to the boundary, we use one sided versions of the given operators. This seems to work well in practice, although it is not covered by the wavelet framework described above.

5.2.1 Detectors from the B-Spline Wavelet Basis Function.

Developing the best suited wavelets that can characterize all of the flow features might involve the switching or blending of more than one mother wavelet $\psi(x)$ and scaling function $\phi(x)$, especially if one needs to distinguish turbulent fluctuations from shock/shear and/or spurious high frequency oscillations. The mother wavelet function used in [20] and described in detail in [46] meets some of our requirements. It is obtained from second order B-splines.

$$\psi(x) = \begin{cases} 0 & x > 1 \\ -2(x-1)^2 & 1/2 < x < 1 \\ -4x(1-x) + 2x^2 & 0 < x < 1/2 \\ -4x(1+x) - 2x^2 & -1/2 < x < 0 \\ 2(x+1)^2 & -1 < x < -1/2 \\ 0 & x < -1 \end{cases} . \tag{5.20}$$

For this wavelet (5.20), there exists a scaling function, given by

$$\phi(x) = \begin{cases} 0 & x > 2 \\ \frac{1}{2}(x-2)^2 & 1 < x < 2 \\ -(x-1/2)^2 + 3/4 & 0 < x < 1 \\ \frac{1}{2}(x+1)^2 & -1 < x < 0 \\ 0 & x < -1 \end{cases} . \tag{5.21}$$

The normalization is such that the integral of the scaling function above is equal to one. The functions above are standard, and can be found in [13]. The scaling function differs by a shift from the scaling function used in [20], but the important relations

$$\begin{aligned} \phi(x) &= \tfrac{1}{4}\phi(2x+1) + \tfrac{3}{4}\phi(2x) + \tfrac{3}{4}\phi(2x-1) + \tfrac{1}{4}\phi(2x-2) \\ \psi(x) &= \phi(2x+1) - \phi(2x) \end{aligned} \tag{5.22}$$

hold, and give the grid operators

$$\begin{aligned} Af_j &= (f_{j-1} + 3f_j + 3f_{j+1} + f_{j+2})/8, \quad j = 2, \ldots, N-2 \\ Df_j &= (f_{j-1} - f_j)/2 \quad j = 2, \ldots, N. \end{aligned} \tag{5.23}$$

Note that this wavelet stencil is not symmetric. In general, the wavelet coefficients involve points from $p2^{m_0-1}$ to $-q2^{m_0-1}$, giving a stencil of totally $(p+q)2^{m_0-1} + 1$ points.

5.2.2 Detectors from the Redundant Form of Harten Multiresolution Wavelet. For the redundant form of Harten multiresolution wavelet there is more than one choice for the interpolation function. See Sjögreen [62] for a discussion. The exact form of the method for the computations in this article is

$$\begin{aligned} Af_j &= (f_{j-1} + f_{j+1})/2 \quad j = 2, \ldots, N-1 \\ Df_j &= f_j - Af_j \quad j = 2, \ldots, N-1. \end{aligned} \tag{5.24}$$

The above choice was made in order to have a simple and efficient method. The stencil is narrower than for the B-spline formulas that were given previously. With the formula above we also get a symmetric stencil, which is more natural if the other parts of the computation, such as difference approximations of PDEs are done by symmetric formulas. Furthermore, symmetry makes periodic BCs somewhat easier to implement. Note that the absence of symmetry for either the scaling function or the wavelet can lead to phase distortion. This can be shown to be important in signal processing applications.

5.2.3 Multi-Dimensional Wavelets. The computation of multi-dimensional wavelets is quite expensive, especially in 3-D. A simple minded efficient way is to evaluate the wavelet coefficients dimension-by-dimension. This means that we get two set of wavelet coefficients $w^x_{m,j}(y)$ and $w^y_{m,k}(x)$, where now (j, k) is the position and m is the scale. The precise definition is

$$\begin{aligned} w^x_{m,j}(y) &= \int f(x,y)\psi_{m,j}(x)\,dx \\ w^y_{m,k}(x) &= \int f(x,y)\psi_{m,k}(y)\,dy. \end{aligned} \tag{5.25}$$

Thus, the dimension-by-dimension approach involved only terms evaluated as finite differences in the x-direction and terms which are evaluated in the y-direction. We then use the $w^x_{m,j}(y)$ coefficients for the x-direction computation, and the y-coefficients for the y-direction computation.

5.2.4 Shock/Shear Wavelet Sensor. For the numerical experiments presented in the next section the wavelet sensor is obtained by computing a vector of the approximated Lipschitz exponent of a **chosen vector function to be sensed** with a suitable multiresolution non-orthogonal wavelet basis function. Here, “vectors or variables to be sensed” means the represented vectors or variables that are suitable for the extraction of the desired flow physics. The variables to be sensed can be the density, the combination of density and pressure, the characteristic variables, the jumps in the characteristic variables $\tilde{\alpha}^l_{j+1/2}$, or the entropy variable vector W ([20, 83]).

For example, if the characteristic variables are the chosen vector to be sensed by the wavelet approach, the sensor $\mathcal{S}^l_{j+1/2}$ can be defined as

$$\mathcal{S}^l_{j+1/2} = \tau(\alpha^l_{j+1/2}), \tag{5.26}$$

where $\alpha^l_{j+1/2}$ is the estimated Lipschitz exponent of the lth characteristic component with $l = 1, 2, 3, 4$, the four characteristic waves. $\tau(\alpha)$ is a sensing function which decreases from $\tau(0) = 1$ to $\tau(1) = 0$ (for the aforementioned type of wavelets). Note that the lth component of the estimated Lipschitz exponent $\alpha^l_{j+1/2}$ is not to be confused with the jump in the lth characteristic variables $\widetilde{\alpha}^l_{j+1/2}$ in Section 5.1. We use $\mathcal{S}^l_{j+1/2}$ as the sensor to distinguish it from the ACM sensor $\omega^l_{j+1/2}$ in Section 5.1. Noted that the k index is omitted (for the 2-D case) for simplicity.

If we instead base the exponent estimate on point centered quantities, we will use the sensor function

$$\mathcal{S}^l_{j+1/2} = \max(\tau(\alpha^l_j), \tau(\alpha^l_{j+1})). \tag{5.27}$$

If the exponent estimate is based on other quantities than the characteristic variables, (e.g., density and pressure), we use the switch

$$\mathcal{S}_{j+1/2} = \max_l \mathcal{S}^l_{j+1/2}, \tag{5.28}$$

where the maximum is taken over all components of the waves used in the estimate. In this case, the switch is the same for all characteristic fields.

The function $\tau(\alpha)$ should be such that $\tau(0) = 1$, and $\tau(1) = 0$. Three options considered are

$$\begin{aligned} \tau(\alpha) &= \begin{cases} 1 & \alpha < \alpha_0 \\ 0 & \alpha \geq \alpha_0 \end{cases} \\ \tau(\alpha) &= \tfrac{1}{2} + \tfrac{1}{\pi} \arctan K(\alpha_0 - \alpha) \\ \tau(\alpha) &= \max\{0, \min[1, (\alpha - 1)/(\alpha_0 - 1)]\}. \end{aligned} \tag{5.29}$$

Here, α_0 is a cut off exponent to be chosen. For the arctan function the values 0 and 1 are not attained, but we take the constant K large enough so that the function is close to zero for $\alpha > 1$, and close to one for $\alpha < 0$. We have tried values for K in the interval $[200, 500]$. Alternatively, one can integrate the actual α value into the sensor function instead of using the same amount of numerical dissipation at the cut off exponent.

After some experimentation we have found that switching on the dissipation at the grid points where $\alpha < 0.5$ works well, i.e.,

$$\tau(\alpha) = \begin{cases} 1 & \alpha < 0.5 \\ 0 & \alpha \geq 0.5 \end{cases}. \tag{5.30}$$

In fact the method does not seem to be very sensitive to the exact value of cut off α_0, (for $0.4 \leq \alpha_0 \leq 0.6$) for all the test cases considered. Furthermore, the same cut off value, 0.5, works well for all problems we have tried in Section 6 (except for the vortex convection case, where $\alpha_0 = 0.0$ is used in conjunction with entropy splitting [83]). Experiments with smoothed step functions do not give very different results.

5.3 Test Example of the Shock/Shear Wavelet Detectors

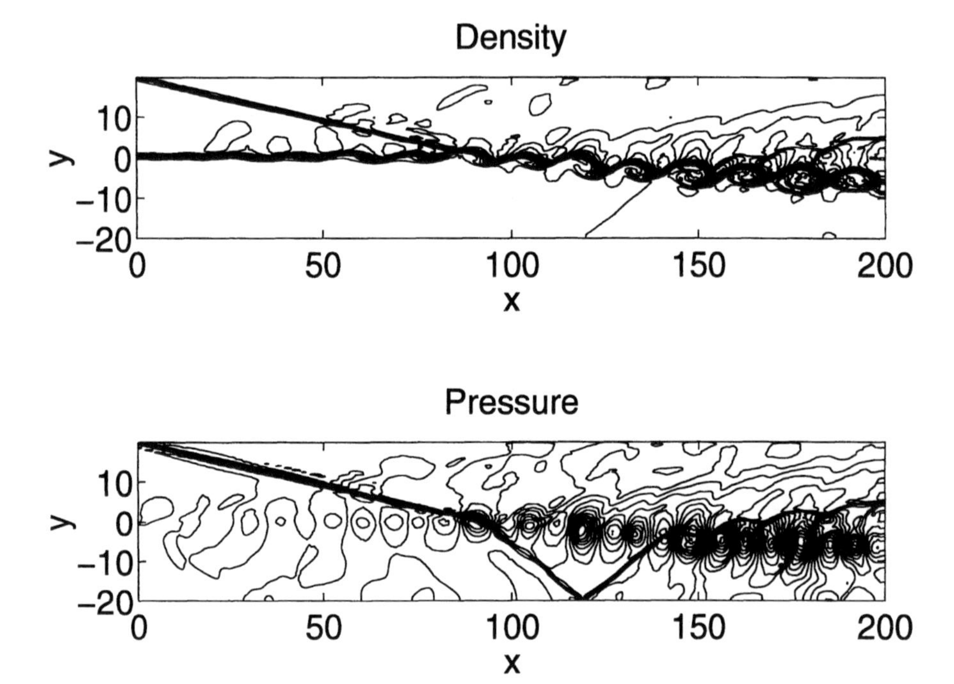

Figure 5.1. 2-D Testing discrete function, (density and pressure contours at $t = 120$).

This section shows the performance of the wavelet sensor using the dimension-by-dimension approach for a 2-D complex flow structure. It is important to note that the illustration involves only the feature extraction capability of the wavelet sensor on a given grid function. No dynamic behavior was involved (i.e., the numerical scheme is not part of the analysis). Figure 5.1 shows the computed density and pressure contours from a precomputed numerical simulation at $t = 120$ with $\Delta t = 0.12$ to be used as the two-dimensional discrete functions to be analyzed by the wavelet algorithm. The discrete functions represent a numerical data of a shock from the upper left corner, impinging on a horizontal shear layer in the middle of the domain. The shock is reflected from the lower wall boundary. For more details about the problem, see Yee et al. [82, 83].

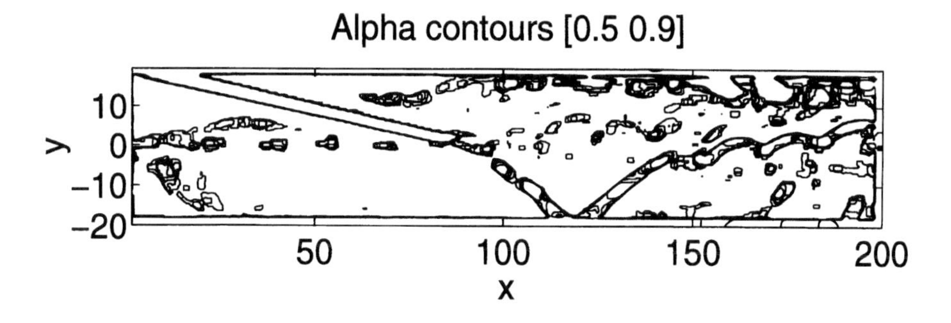

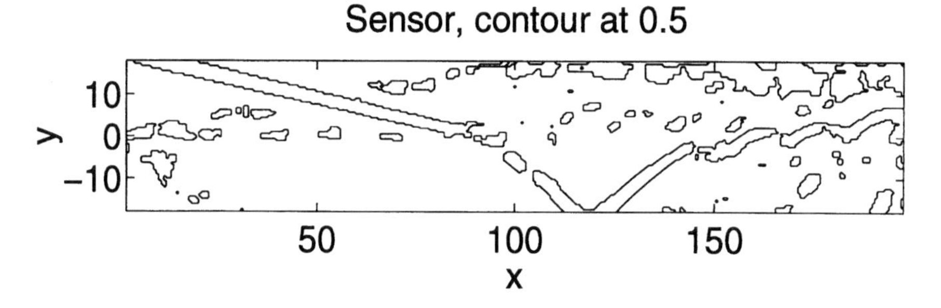

Figure 5.2. Top: α contours $0.5 \leq \alpha \leq 0.9$; Bottom: sensor contour at $\alpha = 0.5$. by the RH-wavelet.

Figure 5.2 shows contours of the estimated Lipschitz exponent α for the function in Fig. 5.1. The value α was computed here from three levels ($m_0 = 3$) of the wavelet algorithm, using the wavelet coefficient

$$w_{m,j,k} = \sqrt{(w^x_{m,j,k})^2 + (w^y_{m,j,k})^2}, \tag{5.31}$$

where the one dimensional coefficients were computed by the multiresolution operators (5.24) in each coordinate direction. The coefficients were computed for the pressure. The top figure in Fig. 5.2 shows α contours on levels from 0.5 to 0.9. The lower figure shows the corresponding sensor, a function which is one for $\alpha < 0.5$ and zero otherwise. The wavelet sensor clearly captures the shock and the shear layer. The low α at the upper boundary to the right is probably due to mildly unstable BCs at the upper boundary.

5.4 Blending of Different Filters

The nonlinear filters for the ACM or wavelet shock/shear capturing nonlinear filter might not be sufficient for (a) time-marching to steady state and (b) spurious high frequency oscillations due to insufficient grid resolution and nonlinear instability away from discontinuities, especially for turbulent and large-eddy simulations. This Section discusses the blending of other filters with these shock/shear filters.

In classical CFD codes, a second order accurate base method is used together with two constant strength linear numerical dissipation terms. One linear fourth-order dissipation is used everywhere except near shocks/shears/steep-gradients to remove nonlinear instabilities. It does not affect the second order accuracy of the base scheme. The second dissipation term is a second-order linear dissipation, which affects the order of accuracy, but is only switched on near discontinuities, and/or steep unresolved gradients using a gradient sensor. The sensor used cannot distinguish the different flow features distinctly and is not accurate enough for turbulent statistics and long-time acoustic computations, unless extreme grid refinement is employed.

In analogy with the aforementioned classical methods, a more advanced numerical dissipation model with improved flow feature extraction sensors for high order central schemes is proposed. Here, we consider a dissipation model with two parts. One part is a nonlinear filter ([82]) and the second part is a high order linear numerical dissipation term modified at boundaries to become a semi-bounded operator, see [67, 65]. The wavelet dissipation control sensor developed in [63] is used as the flow feature detector.

5.4.1 Time-Marching to Steady State. For time-marching to steady state one usually needs to add fourth-order linear dissipation to a second-order spatial differencing scheme (Beam and Warming (1976)). For the present schemes using characteristic filters, in addition to the nonlinear shock/shear filter, one might need to add a sixth-order linear dissipation to a fourth-order spatial base scheme and an eighth-order linear dissipation to a sixth-order spatial base scheme in regions away from shocks for stability and convergence. Let L_d be such an additional filter operator. The two ways of incorporating the L_d operator are options I and II discussed in Section 4.

To minimize the amount of dissipation due to L_d in the vicinity of shock waves, there should be a switching mechanism κ_d to turn off the L_d operator in the vicinity of shock waves. The L_d operator can be applied to the conservative, primitive or characteristic variables. The simplest form is to apply L_d to the conservative variables. Alternatively, since all of the work in computing the average states and the characteristic variables is done for the shock-capturing filter operator, one can apply the L_d operator to the characteristic variables. In this case, the switching mechanism k_d can be a vector so that it is more in tune with the nonlinear shock detector using the approximate Riemann solver. For example, one can set $\kappa = 0$ for the linearly degenerate fields and blend a small amount of κ_d to remove spurious noise generated by the lack of nonlinear filters. This blending of the nonlinear shock/shear filter with the L_d operator can be applied to time-accurate computations as well.

5.4.2 Suppression of Spurious High Frequency Oscillations. The nonlinear shock/shear filters might not be able to remove spurious high frequency oscillations effectively unless sufficient fine grid points are used. For the suppression of unphysical high frequency oscillations due to insufficient grid resolution and nonlinear instability away from discontinuities, higher-order spectral-like filters (Vichnevetsky (1974), Lele (1994), Alpert (1981), Visbal and Gaitonde (1998), Gaitonde and Visbal (1999)) might be needed at the locations where the value of the shock/shear sensor is very small or zero. If spectral-like filters are needed, a proper blending of nonlinear shock/shear filters with spectral-like filters should be applied. In this case, we can use the same procedures as the time-marching to the steady state except the L_d operator should be replaced with the spectral-like filters (for compact central schemes).

5.4.3 An Adaptive Numerical Dissipation Model for Shock-Turbulence Interactions. Assume a sufficient grid is used for the problem and scheme in question, and that the scale of turbulent fluctuations is larger than the spurious high-frequency oscillations. Below we present a filter model under these assumptions. If the scale of the turbulent fluctuation is in similar scale as the high-frequency oscillations, a different wavelet with a turbulent fluctuation sensor should be added. For example, for a sixth-order central spatial base scheme, we define the 1-D filter numerical flux of the numerical dissipation operator as $H^d_{j-1/2}$.

$$H^d_{j-1/2} = S_{j-1/2}F^*_{j-1/2} + d_j[1 - S_{j-1/2}](h^6 D_-(D_+D_-)^3 U_j, \qquad (5.32)$$

here $S_{j-1/2}$ is a switch computed as described in Section 5.2.4, and $F^*_{j-1/2}$ is the flux function corresponding to the dissipative portion of a shock-capturing scheme (e.g., second order accurate TVD scheme) [82]. The first part of the filter stabilizes the scheme at shock/shear locations. The second part is an eighth-order linear filter which improves nonlinear stability away from shock/shear locations. Analogous eighth-order filters can be used if a sixth-order compact spatial base scheme is used [18, 76]. We switch on the high order part of the filter when we switch off the nonlinear filter. The physical quantity (e.g., local Mach number) can be used to determine the d_j parameter of this high order dissipation term.

To further increase stability properties, it is possible to use the sensor to switch on and off the entropy splitting and adjust the value of the entropy splitting parameter according to flow type and region. For the 1-D shock/turbulence interactions to be presented in the next section, however, we believe a constant

$\beta = 1$ away from the shock waves is sufficient. After the completion of a full time step computation using the sixth-order base scheme (denoting the solution by $\widetilde{U}_j$), we filter this solution by

$$U_j^{n+1} = \widetilde{U}_j + \frac{\Delta t}{h}[H^d_{j+1/2} - H^d_{j-1/2}]. \tag{5.33}$$

Here the filter numerical fluxes $H^d_{j\pm1/2}$ are evaluated at $\widetilde{U}$.

5.5 Spurious Numerical Solutions and Instability Due to Under-Resolved Grids

There has been much discussion on verification and validation processes for establishing the credibility of CFD simulations [68, 7, 75, 21, 52]. Since the early 1990s, many of the aeronautical and mechanical engineering related reference journals mandated that any accepted articles in numerical simulations (without known solutions to compare with) need to perform a minimum of one level of grid refinement and time step reduction. On the other hand, it has become common to regard high order schemes as more accurate, reliable and requiring less grid points. The danger comes when one tries to perform computations with the coarsest grid possible while still hoping to maintain numerical results sufficiently accurate for complex flows and, especially, data-limited problems. On one hand, high order methods when applied to highly coupled multidimensional complex nonlinear problems might have different stability, convergence and reliability behavior than their well studied low order counterparts, especially for nonlinear schemes such as TVD, MUSCL with limiters, ENO, WENO, and spectral elements and discrete Galerkin. See for example references [23, 74, 6, 49, 78, 81, 80, 79]. On the other hand, high order methods involve higher operation counts per grid and systematic grid convergence studies can be time consuming and prohibitively expensive. At the same time it is difficult to fully understand or categorize the different nonlinear behavior of finite discretizations, especially at the limits of under-resolution when different types of numerical (spurious) bifurcation phenomena might occur, depending on the combination of grid spacings, time steps, initial conditions (ICs) and numerical treatments of BCs as well as the nonlinear stability of the scheme in question.

Due to the difficulty in analysis, the effect of under-resolved grids and the nonlinear behavior of available spatial discretizations are scarcely discussed in the literature. Here, an under-resolved numerical simulation, according to Brown & Minion, is one where the grid spacing being used is too coarse to resolve the smallest physically relevant scales of the chosen continuum governing equations that are of interest to the numerical modeler. Before the nineties, it was common in DNS to avoid the use of numerical dissipations. It was standard practice to refine the grid not just to resolve the multiscale physics but also to

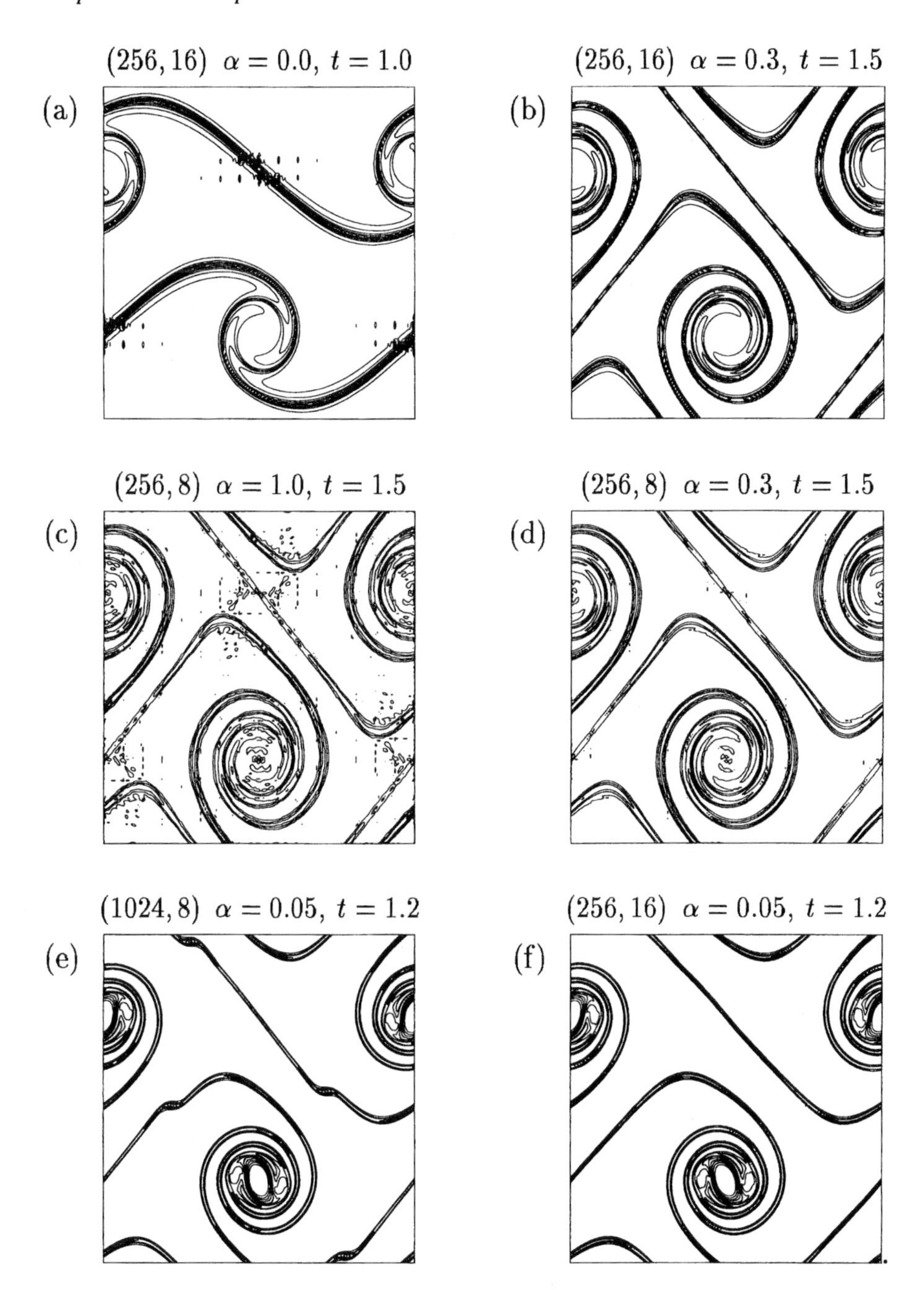

Figure 5.3. Spectral element solutions of doubly periodic shear layer roll-up problem with different (E, N) pairings and filter strengths α: (a-d) thick shear layer case ($\rho = 30$), (e-f) thin shear layer case ($\rho = 100$).

overcome nonlinear instability instead of employing numerical dissipation or filters. In certain cases, the grid is finer than what is needed to resolve the smallest scale. This section illustrates the situation where the necessity of finer grid can be overcome by the use of an appropriate filter and still be able to obtain an accurate and stable solution with a much coarser grid.

Brown and Minion [6, 49] studied the effects of under-resolved grids by considering the shear layer roll-up problem that arises when the Navier-Stokes equations are solved in the unit square with doubly-periodic BCs with ICs given by

$$u = \begin{cases} \tanh(\rho(y - 0.25)) & \text{for y} \leq 0.5 \\ \tanh(\rho(0.75 - y)) & \text{for y} > 0.5 \end{cases}, \tag{5.34}$$

$$v = 0.05 \sin(2\pi x). \tag{5.35}$$

In [6, 49], the behavior of several difference methods was considered. These difference methods include a Godunov projection method, a primitive variable ENO method, an upwind vorticity stream-function method, centered difference methods of both a pressure-Poisson and vorticity stream-function formulation, and a pseudospectral method. They demonstrated that all these methods produce spurious, non-physical vortices. While these extra vortices might appear to be physically reasonable, they disappear when the mesh is refined.

Figure 5.3 shows filter-based spectral element results for the problem (5.34) as computed by Fischer and Mullen [17]. The spectral element method is characterized by the discretization pair (E, N), where E is the number of quadrilateral elements and N is the order of the tensor-product polynomial expansion within each element. This filter presented in [17] is designed to stabilize the $PN2$ spectral element method at high Reynolds numbers. The $PN2$ method, introduced by Maday and Patera [40], is a consistent approximation to the Stokes problem which employs continuous velocity expansions of order N and discontinuous pressure expansions of order $N-2$. The discretizations in Fig. 5.3a - 5.3e consist of a 16 array of elements, while Fig. 5.3f consists of a 32×32 array. Here, α denotes the spectral filter coefficient (not to confused with the Lipschitz exponent or the jump in the characteristics in Sections 5.1 and 5.2), with $\alpha = 0$ corresponding to no filtering. The time step size is $\Delta t = 0.002$ in all cases, corresponding to CFL numbers in the range of 1 to 5. Without filtering, Fischer and Mullen were not able to simulate this problem at any reasonable resolution. Figure 5.3a illustrates the result prior to blow up for the unfiltered case with $(E, N) = (16^2, 16)$, which has a resolution corresponding to an $n \times n$ grid with $n = 256$. Unfiltered results for $N = 8$ ($n = 128$) and $N = 32$ ($n = 512$) are similar. Filtering with $\alpha = 0.3$ yields dramatic improvement for $n = 256$ (Fig. 5.3b) and $n = 128$ (Fig. 5.3d). Though the so-called full projection with filter strength $\alpha = 1$ is stable, the partial filtering of ($\alpha < 1$) gives smoother results and is preferable. The cases in 5.3e and 5.3f correspond to the difficult "thin" shear layer case of [6] and show the benefits of high-order discretizations. Both cases correspond to a resolution of $n^2 = 256^2$. In Fig. 5.3e, this is attained with $(E, N) = (16^2, 16)$, while in Fig 5.3f, $(E, N) = (32^2, 8)$. Although both results are stable (due to the filter), Fig 5.3f reveals the presence of spurious vortices that are absent in the higher-order case.

6. Numerical Examples

This section illustrates the power of entropy splitting, the difference in performance of linear and nonlinear (with sensor controls) filters and the combination of both types of filters with adaptive sensor controls. We use the same notation as in [82, 83, 64]. The artificial compression method (ACM) and wavelet filter schemes using a second-order nonlinear filter with sixth-order spatial central interior scheme for both the inviscid and viscous flux derivatives are denoted by ACM66 and WAV66. See [82, 83, 64] for the forms of these filter schemes. The same scheme without filters is denoted by CEN66. The scheme using the fifth-order WENO for the inviscid flux derivatives and sixth-order central for viscous flux derivatives is denoted by WENO5. Computations using the standard fourth-order Runge-Kutta temporal discretization are indicated by appending the letters "RK4" as in CEN66-RK4. ACM66 and WAV66 use the Roe's average state and the van Leer limiter for the nonlinear numerical dissipation portion of the filter. The wavelet decomposition is applied in density and pressure, and the maximum wavelet coefficient of the two components is used. The nonlinear numerical dissipation is switched on wherever the wavelet analysis gives a Lipschitz exponent [63] less than 0.5. Increasing this number will reduce oscillations, at the price of reduced accuracy (see [63] for other possibilities). Computations using entropy splitting are indicated by appending the letters "ENT" as in WAV66-RK4-ENT. Computations using an eighth-order linear dissipation filter are indicated by appending the letters "D8" as in WAV66-RK4-D8. In order not to introduce additional notation, inviscid flow simulations are designated by the same notation, with the viscous terms not activated.

6.1 A 2-D Vortex Convection Model [82, 83, 63, 65]

The onset of nonlinear instability of long-time numerical integration, the benefit of the entropy splitting and the difference in performance of linear and nonlinear numerical dissipations in improving nonlinear stability for a horizontally convecting vortex (see Fig. 5.4) can be found in in [82, 83, 63, 65]. We summarize the results here.

To show the onset of nonlinear instability, the 2-D perfect gas compressible Euler equations are approximated by CEN66-RK4 with periodic BCs imposed using a 80×79 grid with the time step $\Delta t = 0.01$. Since this is a pure convection problem, the vortex should convect without any distortion if the numerical scheme is highly accurate and non-dissipative. Although CEN66-

Isentropic Vortex Evolution

(Horizontally Convecting Vortex, vortex strength $\hat{\beta}$=5)

Freestream:

$$(u_\infty, v_\infty) = (1,0); \; p_\infty = \rho_\infty = 1$$

IC: Perturbations are added to the freestream (not in entropy)

$$(\delta u, \delta v) = \frac{\hat{\beta}}{2\pi} e^{\frac{1-r^2}{2}} (-(y - y_0), (x - x_0))$$

$$\delta T = -\frac{(\gamma - 1)\hat{\beta}^2}{8\gamma\pi^2} e^{1-r^2}$$

$$r^2 = (x-x_0)^2 + (y-y_0)^2$$

Computational Domain & Grid Size:

$0 \leq x \leq 10$ & $-5 \leq y \leq 5$

80×79 Uniform grid

Periodic BC in x & y

Initial Vortex, Density Contours

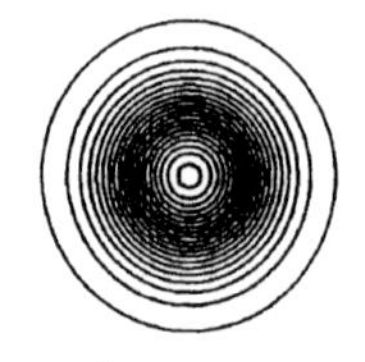

(x_{v_0}, y_{v_0})
center of vortex

Figure 5.4. Vortex convection problem description.

RK4 is linearly stable, the test problem is nonlinear and instability eventually sets in. Almost perfect vortex preservation is observed for up to 5 periods of integrations (5 times after the vortex has convected back to the same position - time = 50). Beyond 5 periods the solution becomes oscillatory, and blows up before the completion of 6 periods. The blow up is associated with an increase in entropy [65]. If we instead use the entropy-split form of the approximation (CEN66-RK4-ENT) with a split parameter $\beta = 1$, almost perfect vortex preservation for up to 40 periods can be obtained. The computation remains stable for up to 67 periods before it breaks down. The time history of the L^2 entropy norm and density contours of the IC and the computed solution after 5, 10, 30, 50 and 67 periods using CEN66-RK4-ENT is shown in Figs. 5.5 and 5.6. The norm is decreasing, although the instabilities break down the solution after 67 periods. Using the second-order nonlinear filter without splitting (ACM66-RK4 or WAV66-RK4), the solution remains stable beyond 67 periods. However, the numerical solution gradually starts to diffuse after 20 periods. If we use the nonlinear filter in conjunction with entropy splitting (ACM66-RK4-ENT or WAV66-RK4-ENT), almost perfect vortex preservation can be obtained for up to 120 periods using a split parameter $\beta = 1$ [83].

The density contours of the solution after 5, 10, 200 and 300 periods for the un-split ($\beta = \infty$) computation using the eighth-order linear dissipation (CEN66-RK4-D8) are shown in Fig. 5.7. The linear dissipation, given by

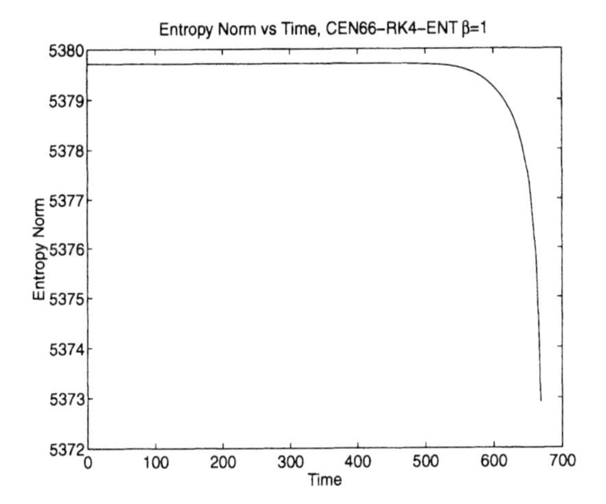

Figure 5.5. Entropy norm history of CEN66-ENT: entropy split parameter $\beta = 1$ and 80×79 grid.

$-dh^7(D_+D_-)^4U_j$ with grid spacing h was added to the sixth-order base scheme to discretize the convection terms. The parameter d is a given constant ($d = 0.002$) and is scaled with the spectral radius of the Jacobian of the flux function, and D_+ and D_- are the forward and backward difference operators, respectively. This numerical dissipation is applied as part of the scheme and not as a post processing filter. The computation can be run for 300 periods without breakdown. However, serious degradation of accuracy occurs after 250 periods. For this particular problem, the CEN66-RK4-D8 out performed the ACM66-RK4-ENT and WAV66-RK4-ENT using $\beta = 1$. Perhaps using a higher than third-order nonlinear filter might improve the performance of the ACM66-RK4-ENT and WAV66-RK4-ENT at the expense of an increase in CPU.

6.2 DNS of 3-D Compressible Turbulent Channel Flow [61]

This numerical example illustrates the power of entropy splitting for DNS computations. To obtain accurate turbulent statistics, very long-time integration and highly accurate methods are required for this DNS computation. The computation employed the SBP-satisfying boundary difference operator [9] with the fourth-order central interior scheme applied to the split form of the inviscid flux derivatives CEN44-RK4-ENT with $\beta = 4$, and a Laplacian viscous formulation. The fluid mechanics of this 3-D wall bounded isothermal compressible turbulent channel flow has been studied in some detail by Coleman et al. [11]. They showed that the only compressibility effect at moderate Mach numbers

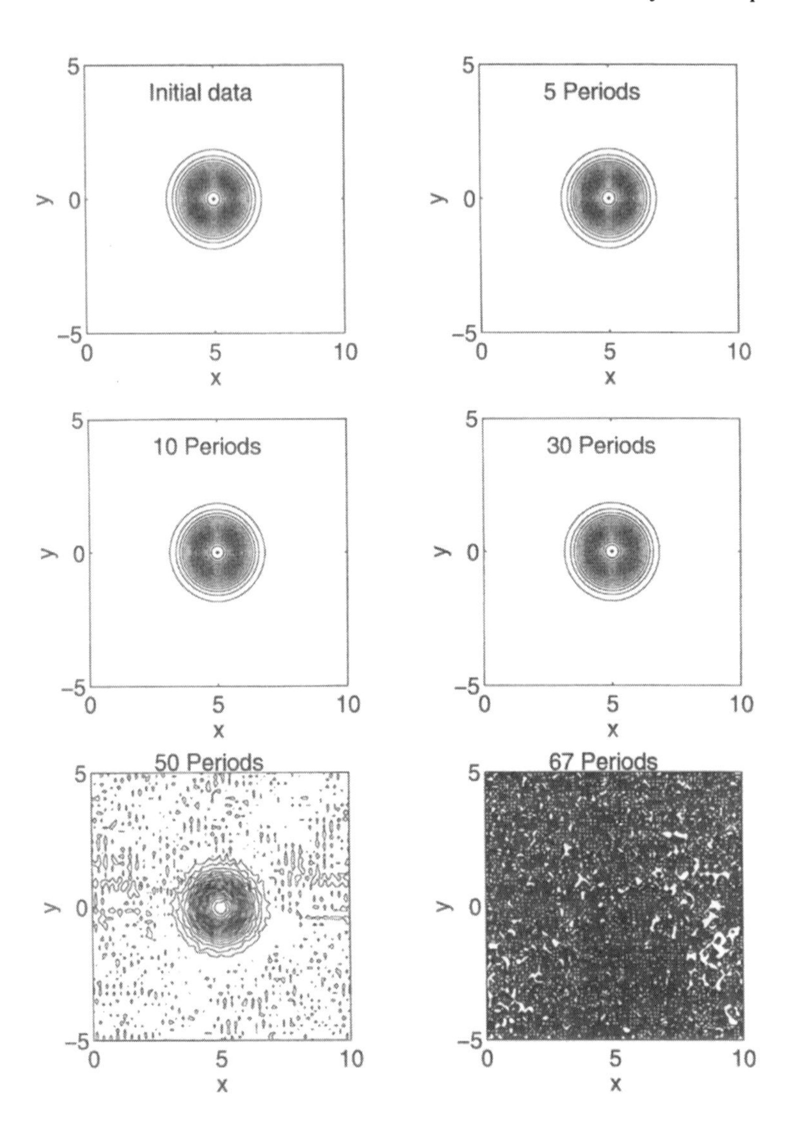

Figure 5.6. Density Contours of CEN66-ENT: entropy split parameter $\beta = 1$ and 80×79 grid.

comes from the variation of fluid properties with temperature. They used a uniform body force term to drive the flow, but recommended the constant pressure gradient approach which was adapted by Sandham et al. [61].

6.2.1 Grid refinement study. A simplified case is taken in which the fluid properties (viscosity and thermal conductivity) are held constant and the

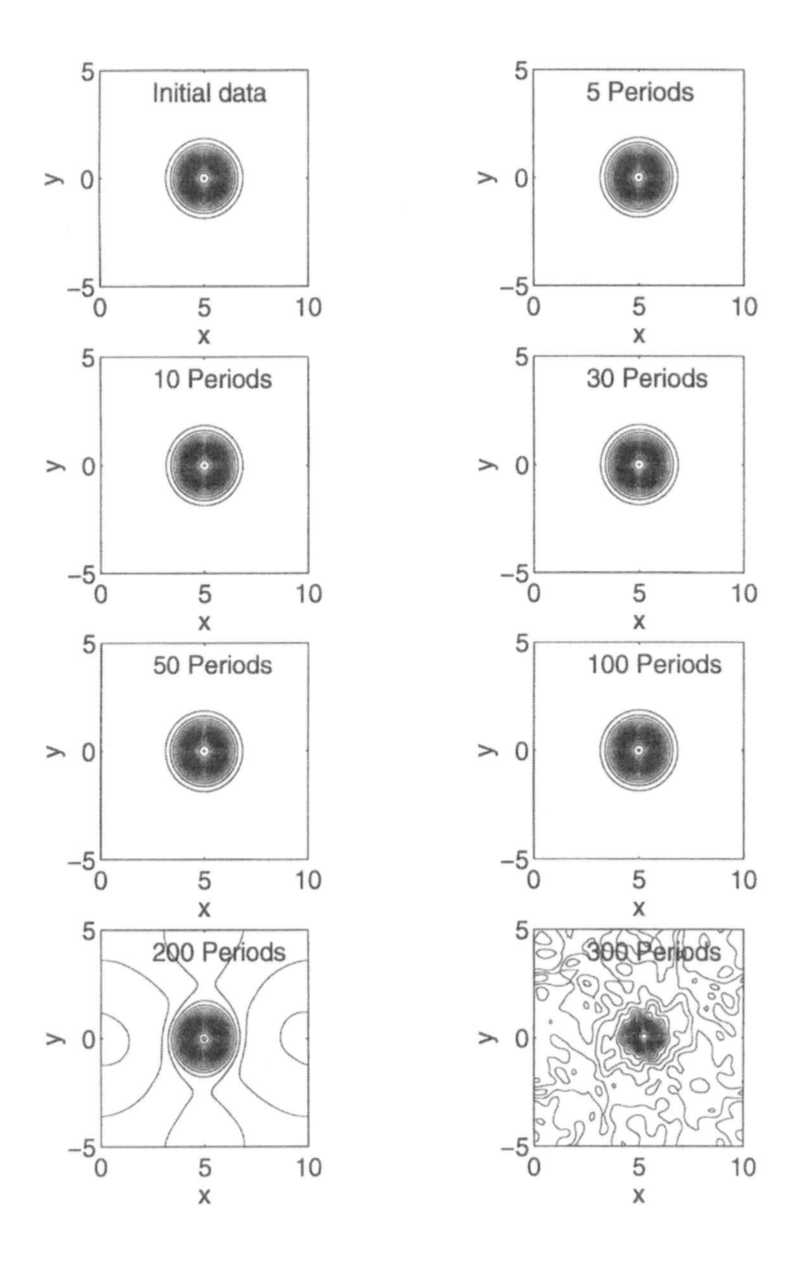

Figure 5.7. Density contours of CEN66-RK4-D8 using 80×79 grid.

computational box size is kept small. The latter is justified as a method of reducing cost as the gross turbulence statistics are relatively insensitive to the computation box size, so long as the domains are still significantly larger than

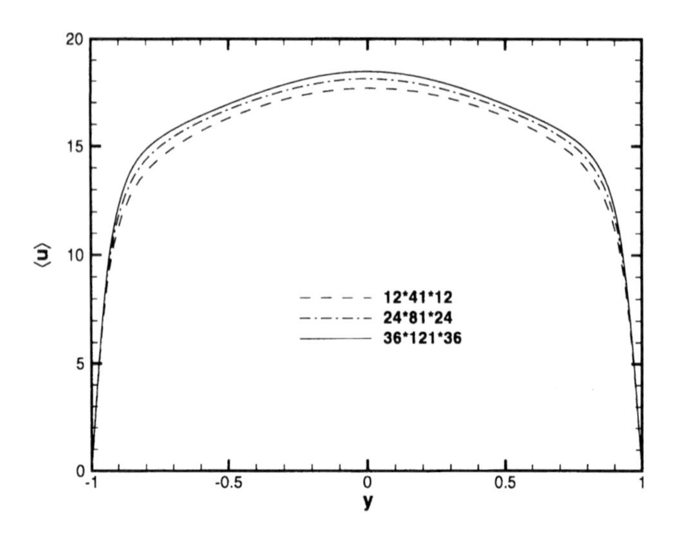

Figure 5.8. Mean velocity profiles

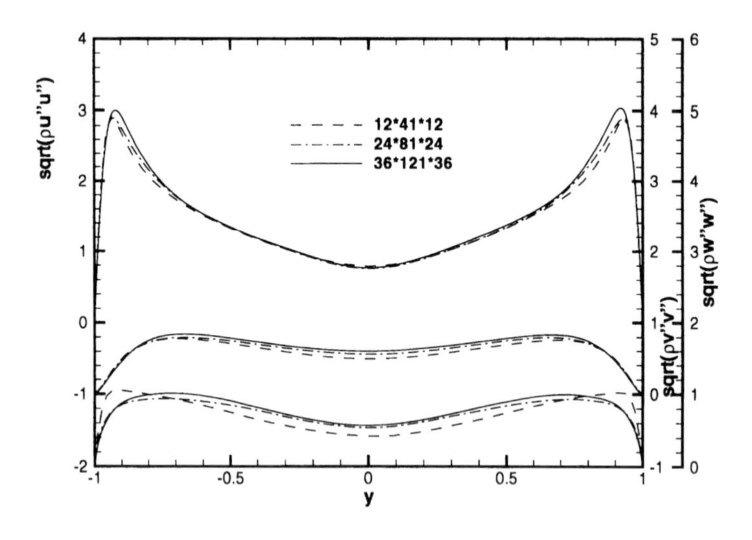

Figure 5.9. Effect of grid refinement on normal stresses. The top curves relate to the left scale, the middle to the right scale and the lowest to the furthest right scale.

the minimal domains on which turbulence can be sustained. A Mach number of 0.1 is chosen, based on friction velocity and sound speed corresponding to the fixed wall temperature. Channel half width h, friction velocity u_τ, wall temperature and bulk (integrated) density are the normalizing quantities for non-dimensionalization with a Reynolds number of 180. Together with the constant property assumption, this choice of Mach number means that results can reasonably be compared to results from previous incompressible flow calculations. The computations were carried out at a fixed CFL=2.0. They were started with artificial ICs and first run to time $t = 50$, by which time dependence on the ICs is lost. Statistics were accumulated over the time interval $t = 200$ to $t = 300$.

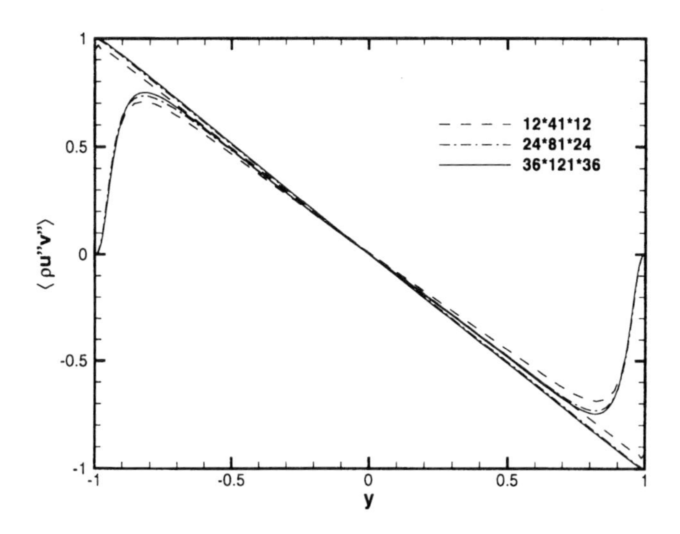

Figure 5.10. Effect of grid refinement on turbulent shear stress (curve falling to zero at the walls). The total stress (straight lines, non-zero at walls) is also shown.

Three grids $12 \times 41 \times 12$, $24 \times 81 \times 24$ and $36 \times 121 \times 36$ were considered. The largest number in each case corresponds to the direction normal to the wall (y). The computational box has non-dimensional length 3 in the (streamwise) x-direction, 1.5 in the (spanwise) z-direction and 2 in the y-direction. The x- and z-directions have periodic BCs with uniform grid spacing. In the $y-$direction, the grid is stretched according to

$$\frac{y}{h} = \frac{\tanh(c_\eta \eta)}{\tanh c_\eta},$$

with η uniformly distributed on [-1,1], $c_\eta = 1.7$. The ratio of grid points in each direction was chosen so that all directions have roughly the same degree of resolution of the relevant turbulence microscales in each direction. Figure 5.8 shows the mean flow velocity, Fig. 5.9 the root mean normal stresses and Fig. 5.10 the stress profiles across the channel. Angle brackets $\langle\rangle$ denote averages over the homogeneous spatial directions and time while in the usual notation double primes denote deviations from mass-weighted (Favre) averages. The convergence is not uniform across the channel but the change from medium to fine grid is smaller than the change from coarse to medium grid. A comparison of the rms quantities with an incompressible flow simulation on the same size computational box (Z. Hu, private communication) is shown in Fig. 5.11. Here we compare the $36 \times 121 \times 36$ fourth-order compressible simulation with a $32 \times 81 \times 32$ fully spectral incompressible simulation using the method described in [59]. As expected, good agreement is found, as expected for this Mach number (0.1 based on friction velocity or 1.8 based on centerline velocity).

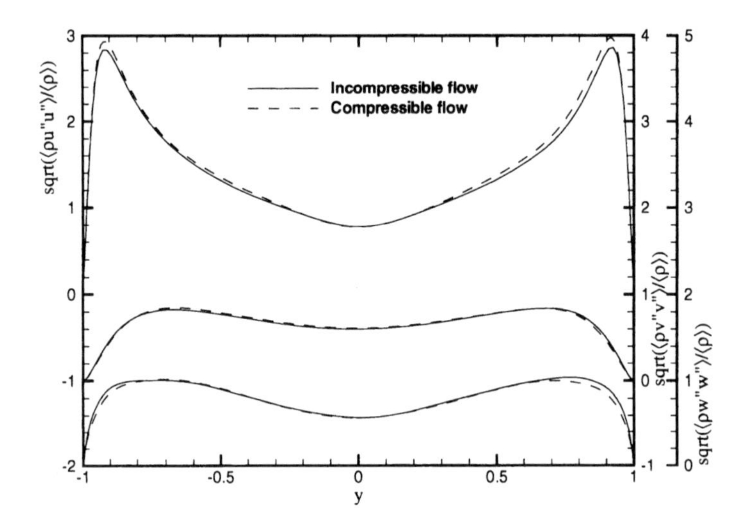

Figure 5.11. Comparison of weighted turbulence quantities on a $36 \times 121 \times 36$ grid with an incompressible flow simulation on a $32 \times 32 \times 81$ grid.

The convergence of various global measures can be found in Table 1 of [61] for the three grids. For the pressure gradient and Reynolds number specified, the velocity gradient at the wall should be 180, the difference away from this being an error of the simulation. Here Re_τ is the Reynolds number based on u_τ, the mean density at the wall $\langle \rho_w \rangle$ and the mean viscosity $\langle \mu_w \rangle$ at the wall. For the finest grid the resolution in wall units (a common check on resolution in DNS) is $\Delta_x^+ = 15$ and $\Delta_z^+ = 7.5$ and approximately 10 points are in the sublayer $y^+ < 10$. The simulations demonstrate a robustness down to very coarse resolutions, comparable with the best incompressible turbulent flow solvers incorporating de-aliasing and skew-symmetric formulation of the convective terms. Without the use of the entropy splitting of the inviscid flux derivatives and without the use of the Laplacian viscous formulation, the CEN44-RK4 (un-split) solutions for the same CFL number, diverge for all three grids before meaningful turbulence statistics can be obtained.

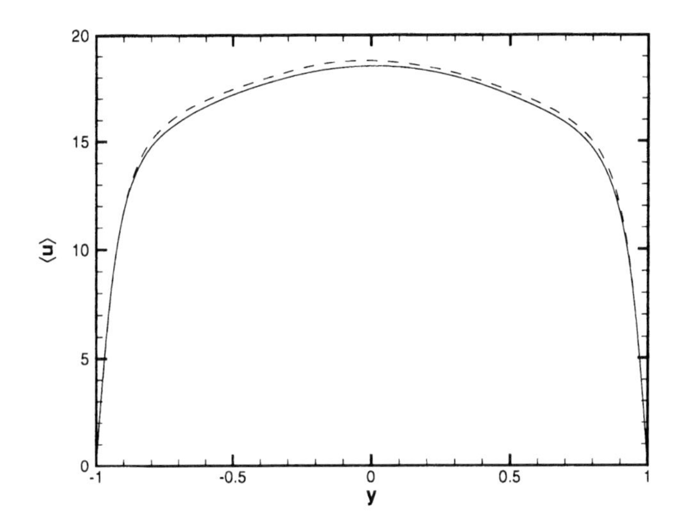

Figure 5.12. Mean velocity profile, comparing current simulation (dashed line) with Coleman *et al.*[11](solid line).

6.2.2 Comparison with Coleman et al.. Coleman *et al.* [11] carried out comparable simulations in their study of the effects of Mach number on turbulence statistics. This section shows the simulation of their case $Re_\tau = 190$ and $M_\tau = 0.095$ with a uniform body force term together with variable fluid properties (power-law temperature dependence of the viscosity with exponent 0.7 and fixed Prandtl number $Pr = 0.7$). With the variable viscosity there is a need to use a larger computational box size than was used in the preceding section, since turbulence structures become larger as the viscosity is reduced (the wall is cold relative to the bulk flow). We chose to use a box of size $6 \times 2 \times 3$, i.e., twice as large in x and z as in the preceding section. This size is still somewhat lower than that of Coleman *et al.*, who used a box of size $4\pi \times 2 \times 4\pi/3$. A computational grid of $60 \times 141 \times 60$ was used, giving $\Delta x^+ = 19$ and $\Delta z^+ = 9.5$. These are comparable to those used by Coleman *et al.* (16.6 and 10.0 respectively). There were 12 points in the sublayer ($y^+ < 10$).

For this simulation a parallel implementation was used, which illustrated the excellent parallel scaling of the method on a Cray T3E-1200E computer (90% efficiency for a 240^3 benchmark on 256 processors and continued good scaling

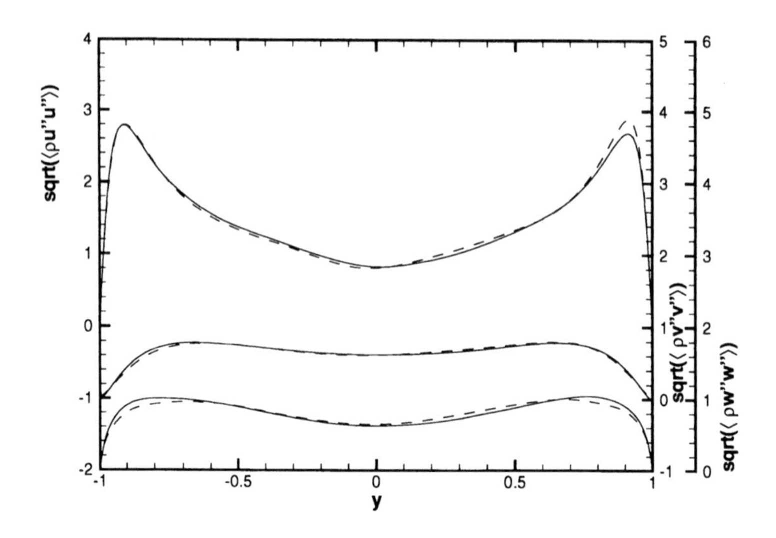

Figure 5.13. Root mean normal turbulent stresses, comparing current simulation (dashed line) with Coleman *et al.*[11](solid line). The top curves relate to the left scale, the middle to the right scale and the lowest to the furthest right scale.

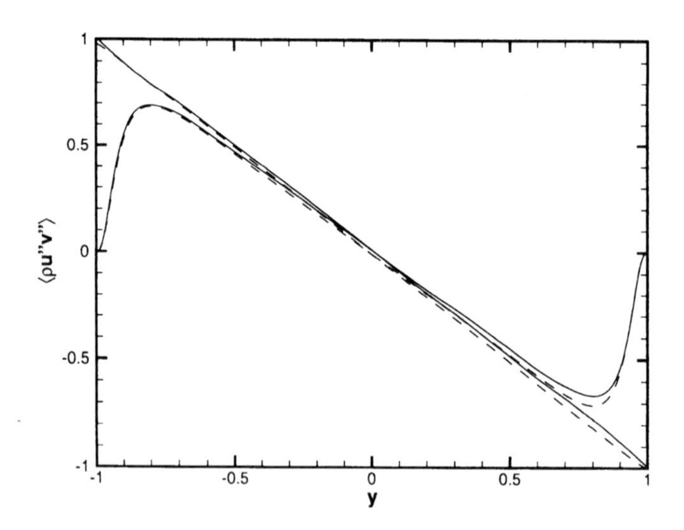

Figure 5.14. Turbulent and total shear stresses, comparing current simulation (dashed line) with Coleman *et al.*[11](solid line).

up to 768 processors, as reported in Ashworth *et al.* [3]). The simulation presented here used 32 processing elements.

Table 5.1. Comparison of centerline velocity U_c, bulk velocity U_b, wall shear stress $\langle \mu \frac{du}{dy} \rangle_w$ and shape factor H with Coleman *et al.*[11]

Simulation	U_c	U_b	$\langle \mu \frac{du}{dy} \rangle_w$	H
Current	18.9	16.3	190.3	1.66
Coleman *et al.*	18.5	15.9	189.5	1.65

Table 1 shows a summary of the output from the simulation. Data from Coleman *et al.* have been re-normalized for comparison with the current simulation results. Figures 5.13 and 5.14 show the shear stress and rms turbulence fluctuations, while Fig. 5.12 shows the mean velocity profile. Overall a good agreement is obtained illustrating the good performance of the method for a resolution comparable to that of a spectral method. Good turbulence kinetic energy budgets have also been obtained [38].

We note that for this well-studied problem with accurate turbulent flow databases for comparison, we can safely conclude that entropy splitting in conjunction with the Laplacian formulation calculations was able to obtain stable and fairly accurate solutions using coarse to moderate grid sizes without added numerical dissipation or filters. Unlike the spectral method, this high order method can be efficiently extended to general geometries [77]. For the same 3-D problem, the finite difference formulation of the WENO5 is more than six times as expensive, yet more diffusive than the present scheme using the same temporal discretization.

The numerical methods are currently being applied to several practical problems. Alam & Sandham [1] have studied shock-free transition to turbulence near the leading edge of an aerofoil, while Lawal & Sandham [35] have used the above method in conjunction with the high order nonlinear filter scheme from Yee *et al.* [82] to study transitional shock boundary layer interaction in flow over a bump. These practical applications have been run without the need for changes to the numerical method and hence are leading to some confidence that the developments presented here are generally applicable for DNS of compressible turbulent flow .

For the performance of ACM66-RK4, WAV66-RK4 and WENO5-RK4 on a spatially or a time-developing mixing layer problem containing shock waves, see [83, 63].

6.3 Computational Aeroacoustics Applications (CAA) [50]

This numerical example illustrates the applicability of the entropy splitting to CAA for low Mach number flows. The numerical prediction of vortex sound has been an important goal in CAA since the noise in turbulent flow is generated by vortices. To verify our numerical approach for CAA, the Kirchhoff vortex is chosen for the numerical test. The Kirchhoff vortex is an elliptical patch of constant vorticity rotating with constant angular frequency in irrotational flow. The acoustic pressure generated by the Kirchhoff vortex is governed by the 2D Helmholtz equation, which can be solved analytically for almost circular Kirchhoff vortices using separation of variables. See [50] and references cited therein for details. The difficulty with this test case is the large gradient of the acoustic pressure adjacent to the Kirchhoff vortex.

The perturbation form of the entropy split 2D Euler equations in conjunction with a fourth-order linear filter operator CEN66-RK4-ENT-D4 was applied to Kirchhoff vortex sound at low Mach number using a high order metric evaluation of the coordinate transformation. SBP operators using the SAT method of implementing the time-dependent physical BC was employed. Due to the large disparity of acoustic and stagnation quantities in low Mach number aeroacoustics, the split Euler equations are formulated in perturbation form to minimize numerical cancellation errors.

A very accurate numerical solution with a relatively coarse grid was obtained using CEN66-RK4-ENT-D4 compared with the un-split (CEN66-RK4-D4), and un-filter cases CEN66-RK4, CEN22-RK4 and CEN44-RK4. The extra CPU due to the use of the split form of the inviscid flux derivatives is more than compensated by the improved accuracy and stability of the numerical simulation, especially near regions of large gradients. For this weakly nonlinear test case that does not require long time integration, the amount of filter needed, although very small, is still important. Higher order filter operators and the formulation of the numerical BCs and filter operators in terms of the entropy variables to satisfy a discrete energy estimate in a nonlinear sense will be considered in the future.

6.4 Multiscale Complex Unsteady Viscous Compressible Flows [64, 66]

Extensive grid convergence studies using WAV66-RK4 and ACM66-RK4 for two complex highly unsteady viscous compressible flows are given in [64, 66]. The first flow is a 2-D complex viscous shock/shear/boundary-layer interaction. This is the same problem and flow conditions studied in Daru & Tenaud [12]. The second flow is a supersonic viscous reacting flow concerning fuel breakup. More accurate solutions were obtained with WAV66-RK4 and ACM66-RK4 than with WENO5-RK4, which is nearly three times as expensive. To illustrate the performance of these nonlinear filter schemes, the first model is considered. The ideal gas compressible full Navier-Stokes equations with no slip BCs at the adiabatic walls are used. The fluid is at rest in a 2-D box $0 \leq x, y \leq 1$. A membrane with an initial shock Mach number of 2.37 located at $x = 1/2$ separates two different states of the gas. The dimensionless initial states are

$$\rho_L = 120, \quad p_L = 120/\gamma; \qquad \rho_R = 1.2, \quad p_R = 1.2/\gamma, \tag{5.1}$$

where ρ_L, p_L are the density and pressure respectively, to the left of $x = 1/2$, and ρ_R, p_R are the same quantities to the right of $x = 1/2$. $\gamma = 1.4$ and the Prandtl number is 0.73. The viscosity is assumed to be constant and independent of temperature, so Sutherland's law is not used. The velocities and the normal derivative of the temperature at the boundaries are set equal to zero. This is done by leaving the value of the density obtained by the one sided difference scheme at the boundary unchanged, and updating the energy at the boundary to make the temperature derivative equal to zero.

At time zero the membrane is removed and wave interaction occurs. An expansion wave and a shock are formed initially. Then, a boundary layer is formed on the lower boundary behind the right going waves. After reflection, the left going shock wave interacts with the newly formed boundary layer, causing a number of vortices and lambda shocks near the boundary layer. Other kinds of layers remain after the shock reflection near the right wall. The complexity of this highly unsteady shock/shear/boundary-layer interactions increases as the Reynolds number increases.

As an illustration, we show here the difficult case of Reynolds number $Re = 1000$. The computations stop at the dimensionless time 1 when the reflected shock wave has almost reached the middle of the domain, $x = 1/2$. The numerical results discussed here are at time 1 with a uniform Cartesian grid spacing as described by Daru and Tenaud. Due to symmetry, only the lower half of the domain is used in the computations; symmetry BCs are enforced at the boundary $y = 1/2$. Figure 5.15 shows the comparison of a second-order MUSCL using a second-order Runge-Kutta method (MUSCL-RK2) with WAV66-RK4, ACM66-RK4 and WENO5-RK4 using a 1000×500 grid. Comparing with the converged solution of WAV66-RK4 and ACM66-RK4 using 3000×1500 (see bottom of figure) and 4000×2000 grids (see [64]), one can conclude that WAV66-RK4 exhibits the most accurate result among the 1000×500 grid computations. We note that, for this Reynolds number, the unsteady problem is extremely stiff, requiring very small time steps and very long-time integrations before reaching the dimensionless time of 1.

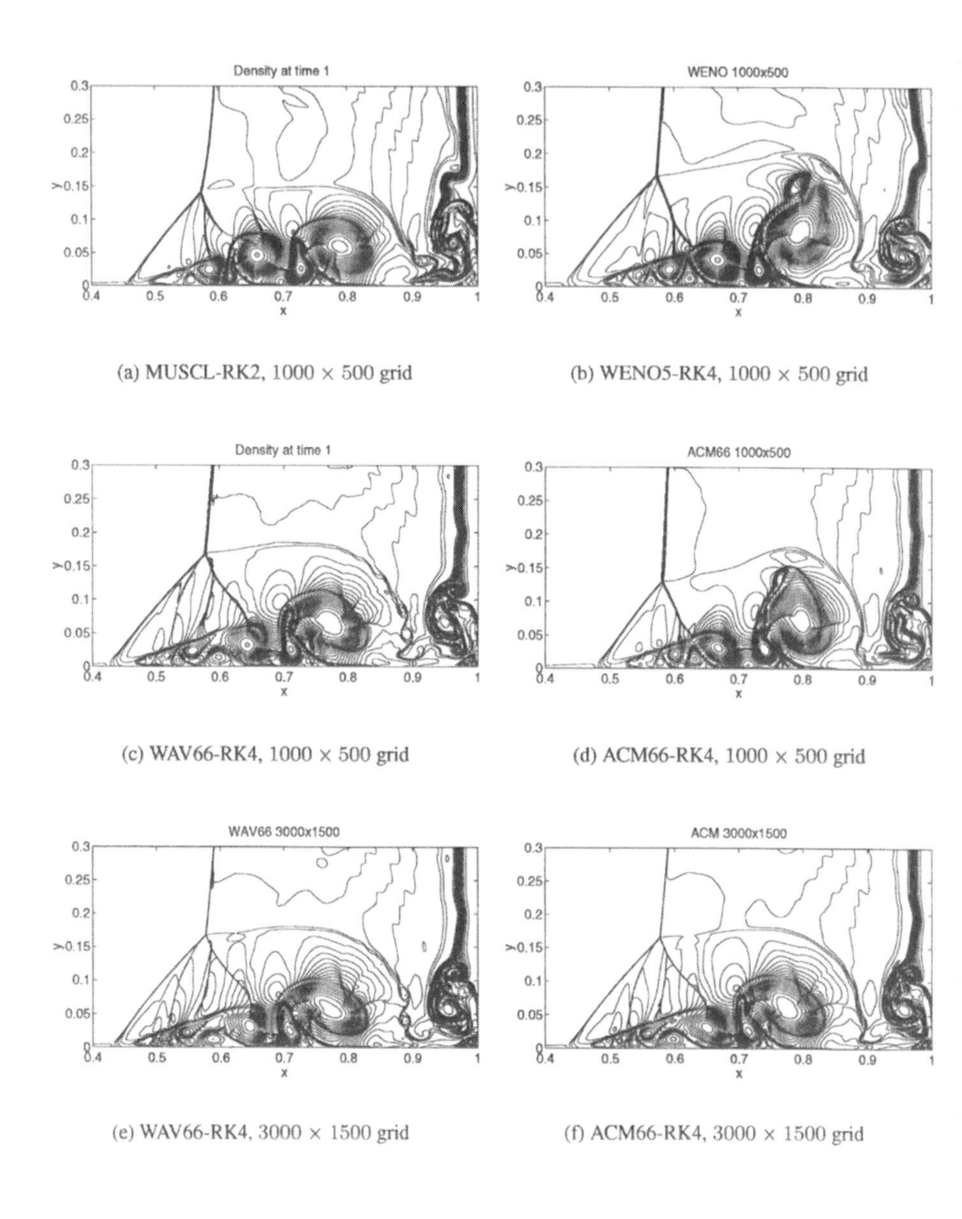

(a) MUSCL-RK2, 1000×500 grid

(b) WENO5-RK4, 1000×500 grid

(c) WAV66-RK4, 1000×500 grid

(d) ACM66-RK4, 1000×500 grid

(e) WAV66-RK4, 3000×1500 grid

(f) ACM66-RK4, 3000×1500 grid

Figure 5.15. Comparison: MUSCL-RK2, WAV66-RK4, WAV66-RK4 and WENO5-RK4 for $Re = 1000$. Density contours using 1000×500 and 3000×1500 grids.

6.5 1-D Shock-Turbulence Interactions Using the Adaptive Numerical Dissipation Model

The dissipative model (5.32) is used to solve a simple, yet difficult, 1-D compressible inviscid shock-turbulence interaction problem with initial data consisting of a shock propagating into an oscillatory density. The initial data are given by

$$(\rho_L,\ u_L,\ p_L) = (3.857143,\ \ 2.629369,\ \ 10.33333) \tag{5.2}$$

to the left of a shock located at $x = -4$, and

$$(\rho_R,\ u_R,\ p_R) = (1 + 0.2\sin(5x),\ \ 0,\ \ 1) \tag{5.3}$$

to the right of the shock where u is the velocity. Fig. 5.16 show the comparison between a second-order MUSCL-RK2 with a sixth-order central scheme and the aforementioned numerical dissipation model using RK4 as the time discretization (WAV66-RK4-D8). The parameter $d = 0.002$ is scaled with the spectral radius of the Jacobian of the flux function. Note that the eighth-order dissipation is a filter, and is different from the CEN66-D8 used in Section 6.3 where the dissipation is part of the scheme. The solution using a second-order uniformly non-oscillatory (UNO) scheme on a 4000 uniform grid is used as the reference solution (solid lines on the first three sub-figures). The bottom of the right figures show the density and Lipschitz exponent distribution for the WAV66-RK4-D8 using 400 grid points. Comparing our result with the most accurate computation found in the literature for this problem, the current approach is highly efficient and accurate using only 800 grid points without grid adaptation or a very high order shock-capturing scheme. For the present computation, the WAV66-RK4-D8 consumed only slightly more CPU than the second-order scheme MUSCL-RK2. With the eighth-order dissipation filter turned off (i.e., only the nonlinear filter remains - WAV66-RK4), the computation is not very stable unless a finer grid and smaller time step is used. Turning on the entropy splitting away from the shocks helps to reduce the amount of the eighth-order dissipation coefficient [65].

7. Concluding Remarks

An integrated approach for the control of the numerical-dissipation/filter in high order schemes for numerical simulations of multiscale complex flow problems is presented. The approach is an attempt to further improve nonlinear stability, accuracy and efficiency of long-time numerical integration of complex shock/turbulence/acoustics interactions and numerical combustion. The required type and amount of numerical-dissipation/filter for these flow problems are not only physical problem dependent, but also vary from one flow region to another. Among other design criteria, the key idea consists of automatic detection of different flow features as distinct sensors to signal the appropriate type and amount of numerical-dissipation/filter for non-dissipative high order schemes. These scheme-independent sensors are capable of distinguishing shocks/shears, turbulent fluctuations and spurious high-frequency

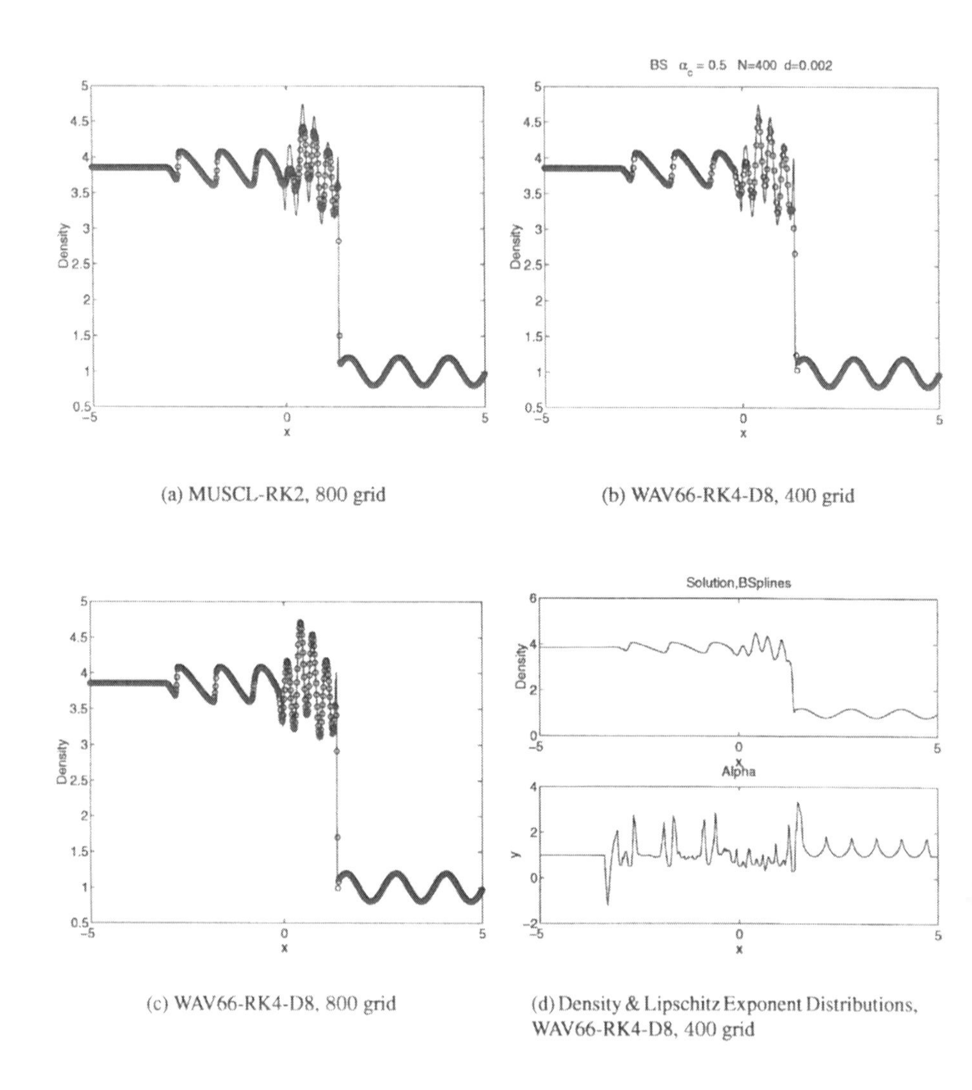

(a) MUSCL-RK2, 800 grid

(b) WAV66-RK4-D8, 400 grid

(c) WAV66-RK4-D8, 800 grid

(d) Density & Lipschitz Exponent Distributions, WAV66-RK4-D8, 400 grid

Figure 5.16. Comparison and Lipschitz exponent distribution of WAV66-RK4-D8, second-order B-spline wavelet.

oscillations. In addition, these sensors are readily available as an improvement over existing grid adaptation indicators. The same shock/shear detector that is designed to switch on the shock/shear numerical dissipation can be used to switch off the entropy splitting form of the inviscid flux derivative in the vicinity the discontinuous regions to further improve nonlinear stability and minimize the use of numerical dissipation. The rest of the sensors in conjunction with the local flow speed and Reynolds number can also be used to adaptively determine the appropriate entropy splitting parameter for each flow type/region.

The minimization of employing very fine grids to overcome the production of spurious numerical solutions and/or instability due to under-resolved grids is also illustrated [79, 17]. Test examples shown are very encouraging. Full implementation of the approach for practical problems is in progress.

Acknowledgment

The authors wish to thank their guest author Paul Fischer for contributing to Section 5.5. They also wish to thank their collaborators Neil Sandham, Jahed Djomehri, Marcel Vinokur and Bernhard Müller for contributing to their earlier work. Special thanks to Marcel Vinokur and Tom Coakley for their critical review of the manuscript. The first author would like to acknowledge the support from the Swedish Natural Science Research Council (NFR).

Appendix

We here give boundary modified numerical dissipation operators, such that the semi bounded property is satisfied for the total operator. The operators are derived by splitting the periodic operator into one boundary part and one interior part,

$$(D_+D_-)_{per} = (D_+D_-)_b + (D_+D_-)_I.$$

The interior operator $(D_+D_-)_I$ has zeros in all rows where the centered operator can not be applied. We define the boundary modified operator as

$$(D_+D_-)^q = (D_+D_-)^p_{per}(D_+D_-)^p_I$$

if $q = 2p$ is even. The zero terms on the boundary of the interior operator, will destroy the periodic terms, so that the boundary modified operator does not have any periodic wrap around terms. If $q = 2p + 1$ is odd, we define

$$(D_+D_-)^q = (D_+D_-)^p_{per}(D_+)_{per}(D_-)_I(D_+D_-)^p_I.$$

The semi boundedness then follows by applying the summation by parts property for the periodic operator, or for the case of q even,

$$\begin{aligned}(u_j, (D_+D_-)^q u_j)_h &= -((D_+D_-)^p_{per}u_j, (D_+D_-)^p_I u_j)_h = \\ &-((D_+D_-)^p_I u_j, (D_+D_-)^p_I u_j)_h \leq 0\end{aligned}$$

where the last equality follows since the interior operator in the scalar product kills all boundary terms in the periodic operator. The case with odd q can be treated similarly.

The order of the operator is reduced on the boundary. To have correct scaling, as numerical dissipation, the $2q$th derivative should be multiplied by h^{2q-1}, thus affecting the accuracy up to order $2q - 1$. However, the boundary terms will be reduced to qth derivatives, and thus have order $q - 1$ on the boundary. For example, the fourth order operator will be third order in the interior, and first order on the boundary. The sixth order operator will be fifth order in the interior, and second order on the boundary, etc. We present below examples for orders 4, 6, and 8.

Fourth derivative

$$\begin{pmatrix} 1 & -2 & 1 & 0 & 0 & 0 & 0 & 0 & 0 & 0 \\ -2 & 5 & -4 & 1 & 0 & 0 & 0 & 0 & 0 & 0 \\ 1 & -4 & 6 & -4 & 1 & 0 & 0 & 0 & 0 & 0 \\ 0 & 1 & -4 & 6 & -4 & 1 & 0 & 0 & 0 & 0 \\ 0 & 0 & 1 & -4 & 6 & -4 & 1 & 0 & 0 & 0 \\ & \ldots & & & & & & & & \end{pmatrix}$$

Sixth derivative

$$\begin{pmatrix} -1 & 3 & -3 & 1 & 0 & 0 & 0 & 0 & 0 & 0 \\ 3 & -10 & 12 & -6 & 1 & 0 & 0 & 0 & 0 & 0 \\ -3 & 12 & -19 & 15 & -6 & 1 & 0 & 0 & 0 & 0 \\ 1 & -6 & 15 & -20 & 15 & -6 & 1 & 0 & 0 & 0 \\ 0 & 1 & -6 & 15 & -20 & 15 & -6 & 1 & 0 & 0 \\ 0 & 0 & 1 & -6 & 15 & -20 & 15 & -6 & 1 & 0 \\ & \ldots & & & & & & & & \end{pmatrix}$$

Eighth derivative

$$\begin{pmatrix} 1 & -4 & 6 & -4 & 1 & 0 & 0 & 0 & 0 & 0 \\ -4 & 17 & -28 & 22 & -8 & 1 & 0 & 0 & 0 & 0 \\ 6 & -28 & 53 & -52 & 28 & -8 & 1 & 0 & 0 & 0 \\ -4 & 22 & -52 & 69 & -56 & 28 & -8 & 1 & 0 & 0 \\ 1 & -8 & 28 & -56 & 70 & -56 & 28 & -8 & 1 & 0 \\ 0 & 1 & -8 & 28 & -56 & 70 & -56 & 28 & -8 & 1 \\ & \ldots & & & & & & & & \end{pmatrix}$$

References

[1] M. Alam and N.D. Sandham, *DNS of Transition Near the Leading Edge of an Aerofoil. Proc Direct and Large Eddy Simulation 4*, B.J.Geurts, ed, to appear Kluwer (2001).

[2] Alpert, P., *Implicit Filtering in Conjunction with Explicit Filtering*, J. Comput. Phys. **44**, 212-219 (1981).

[3] M. Ashworth, D.R. Emerson, N.D. Sandham, Y.F. Yao, and Q. Li, *Parallel DNS Using a Compressible Turbulent Channel Flow Benchmark. Proc. ECCOMAS CFD Conference*, Swansea, Wales, 4-7 Sept. 2001.

[4] A. Arakawa, *Computational Design for Long-Term Numerical Integration of the Equations of Fluid Motion: Two-Dimensional Incompressible Flow. Part I*, J. Comput. Phys., **1**, pp. 119–143 (1966).

[5] G. A. Blaisdell, *Numerical Simulation of Compressible Homogeneous Turbulence*, PhD Thesis, Stanford University, 1991.

[6] Brown, D.L. and Minion, M.L., *Performance of Under-resolved Two-Dimensional Incompressible Flow Simulations*, J. Comput. Phys. **122**, 165-183 (1995).

[7] Brown, D.L., Margolin, L., Sharp, D. and A. White, A., *Predictability of Complex Phenomena, A White Paper*, Los Alamos National Laboratory reprint (1997).

[8] M.H. Carpenter, D. Gottlieb, S. Abarbanel S. and W.-S. Don, *The Theoretical Accuracy of Runge-Kutta Time Discretizations for Initial Value Problem: A Study of the Boundary Error*, SIAM J. Sci. Comput. **16**, 1241-1252 (1995).

[9] M.H. Carpenter, J. Nordstrom and D. Gottlieb, *A Stable and Conservative Interface Treatment of Arbitrary Spatial Accuracy*, ICASE Report 98-12, 1998.

[10] M.H. Carpenter, D. Gottlieb and S. Abarbanel, *Time-Stable Boundary Conditions for Finite-Difference Schemes Solving Hyperbolic Systems: Methodology and Application to High-Order Compact Schemes*, J. Comput. Phys., **111**, 220-236 (1994).

[11] G.N. Coleman, J. Kim, J. and R. Moser, *A Numerical Study of Turbulent Supersonic Isothermal-Wall Channel Flow*, *J. Fluid Mech.* **305**, pp. 159-183 (1995).

[12] V. Daru and C. Tenaud, *Evaluation of TVD High Resolution Schemes for Unsteady Viscous Shocked Flows*, Computers Fluids, **30** 89–113 (2000).

[13] I. Daubechies, *Ten Lectures on Wavelets*, CBMS-NSF Regional Conference Series in Applied Mathematics, No 61, SIAM, 1992.

[14] B. Engquist and B. Sjögreen, *High Order Shock Capturing Methods*, *CFD Reviews*, (Hafez & Oshima, Eds.), John Wiley, New York, 210-233 (1995).

[15] M. Farge, *Wavelet Transforms and their Applications to Turbulence*, Ann. Rev. of Fluid Mech., **24**, 395 (1992).

[16] W. J. Feiereisen, W. C. Reynolds, and J. H. Ferziger, *Numerical Simulation of a Compressible Homogeneous, Turbulent Shear Flow*, Report No. TF-13 (1981), Thermosciences Division, Department of Mechanical Engineering, Stanford University.

[17] P.F. Fischer and J.S. Mullen, *Filter-Based Stabilization of Spectral Element Methods*, Argonne National Lab. Report, August 4, 1999.

[18] D.V. Gaitone and M.R. Visbal, *Further Development of a Navier-Stokes Solution Procedure Based on Higher-Order Formulas*, AIAA Paper 99-0557, Reno, NV, 1999.

[19] P. R. Garabedian, *Partial Differential Equations*, John Wiley & Sons, 1964.

[20] M. Gerritsen and P. Olsson, *Designing an Efficient Solution Strategy for Fluid Flows I.* J. Comput. Phys., **129**, pp. 245–262 (1996).

[21] J. Glimm and D. Sharp, *Prediction and the Quantification of Uncertainty*, Los Alamos National Laboratory LAUR-98-3391 (1998).

[22] D. Gottlieb and J.S. Hesthaven, *Spectral Methods for Hyperbolic Problems*, J. Comput. Applied Math., **128**, 83-131 (2001).

[23] P.M. Gresho, D.K. Gartling, J.R. Torczynski, K.A. Cliffe, K.H. Winters, T.J. Garrett, A. Spence and J.W. Goodrich, *Is the Steady Viscous Incompressible Two-Dimensional Flow Over a Backward-facing Step at Re=800 Stable?*, Intern. J. Numer. Meth. Fluids **17** 501-541 (1993).

[24] B. Gustafsson, H.-O. Kreiss and J. Oliger, J., *Time Dependent Problems and Difference Methods*, John Wiley & Sons, New York (1995).

[25] A. Harten, *On the Symmetric Form of Systems for Conservation Laws with Entropy*, J. Comput. Phys. **49**, 151-164 (1983).

[26] R. D. Henderson, *Adaptive Spectral Element Methods for Turbulence and Transition*, in High-Order Methods for Computational Physics, T. J. Barth and H. Deconinck, editors. Springer 1999.

[27] C.W. Hirt, *Heuristic Stability Theory for Finite-Difference Equations*, J. Comput. Phys., **2**, pp. 339-355 (1968).

[28] Huang, P. G., G.N. Coleman, and P. Bradshaw, *Compressible Turbulent Channel Flows: DNS Results and Modeling. J. Fluid Mech.* **305**, 185-218 (1995).

[29] M. Johansson, *Loss of High Order Spatial Accuracy due to Boundary Error Caused by Runge-Kutta Time Integration*, Technical report 2000-013, Dept. Info. Tech., Uppsala University, May 2000.

[30] H.-O. Kreiss and J. Oliger, *Comparison of accurate methods for the integration of hyperbolic equations*, Tellus, **24**, pp. 199-215 (1972).

[31] H.-O. Kreiss, *Numerical Methods for Solving Time-Dependent Problems for Partial Differential Equations*, Les Presses de l'Université de Montréal, 1978.

[32] J. Kim, P. Moin and R.D. Moser, *Turbulence Statistics in fully-Developed Channel Flow at Low Reynolds Number. J. Fluid Mech.* **177**, 133-166 (1987).

[33] Lafon, A. and Yee, H.C., *Dynamical Approach Study of Spurious Steady-State Numerical Solutions for Nonlinear Differential Equations, Part III: The Effects of Nonlinear Source Terms in Reaction-Convection Equations*, Comp.. Fluid Dyn. **6** 1-36 (1996).

[34] Lafon, A. and Yee, H.C., *Dynamical Approach Study of Spurious Steady-State Numerical Solutions of Nonlinear Differential Equations, Part IV: Stability vs. Numerical Treatment of Nonlinear Source Terms*, Comput. Fluid Dyn. **6** 89-123 (1996).

[35] A.A. Lawal, and N.D. Sandham, *Direct Simulation of Transonic Flow Over a Bump. Proc Direct and Large Eddy Simulation 4*, B.J.Geurts, ed, to appear Kluwer (2001).

[36] LeVeque, R.J. and Yee, H.C., *A Study of Numerical Methods for Hyperbolic Conservation Laws with Stiff Source Terms*, J. Comput. Phys. **86** 187-210 (1990).

[37] Lele. S.A., *Compact Finite Difference Schemes with Spectral-Like Resolution*, J. Comput. Phys., **103**, 16-42 (1992).

[38] Q. Li, *Direct Numerical Simulation of Compressible Turbulent Channel Flow. PhD Thesis, University of Southampton*, in preparation (2001).

[39] R.W. MacCormack, *Numerical Solution of the Interaction of a Shock Wave with a Laminar Boundary Layer*, Proceedings of the 2nd Intern. Conf. on Num. Meths. in Fluid Dynamics, pp. 151-163 (1971).

[40] Y. Maday and A. T. Patera, *Spectral element methods for the Navier-Stokes equations*, State of the Art Surveys in Computational Mechanics, A. K. Noor, ed., ASME, New York, pp. 71–143 (1989).

[41] S.G. Mallat, *A Wavelet Tour of Signal Processing*, Second Edition, Academic Press, San Diego (1999).

[42] S.G. Mallat, *Multifrequency channel Decompositions of Images and Wavelet Models*, IEEE Trans. Acoust. Speech Signal Process., **37**, 2091 (1989).

[43] S.G. Mallat, *Multiresolution Approximations and Wavelet Orthonormal Bases of* $L^2(R)$, Trans. Amer. Math. Soc., **315**, 69 (1989).

[44] S.G. Mallat, *A Theory for Multiresolution Signal Decomposition: The Wavelet Representation*, IEEE Trans. Patt. Anal. Mach. Intell., **11**, 674 (1989).

[45] S.G. Mallat and W. L. Hwang, *Singularity Detection and Processing with Wavelets*, IEEE Transactions on Information Theory, **38**, 617 (1992).

[46] S.G. Mallat and S. Zhong, *Characterization of Signals from Multiscale Edges*, IEEE Transactions on Pattern Analysis and Machine Intelligence, **14** 710-732 (1992).

[47] N. N. Mansour, J. H. Ferziger, and W. C. Reynolds, *Large-Eddy Simulation of a Turbulent Mixing Layer*, Report TF-11, Thermosciences Division, Stanford University, (1978).

[48] K. Mattsson, *Imposing Boundary Conditions with the Injection, the Projection and the Simultaneous Approximation Term Methods*, Proceedings of the First International Conference on CFD, July 10-14, 2000, Kyoto.

[49] M.L. Minion, and D.L. Brown, *Performance of Under-resolved Two-Dimensional Incompressible Flow Simulations II*, J. Comput. Phys. **138**, 734-765 (1997).

[50] B. Müller and H.C. Yee, *Entropy Splitting for High Order Numerical Simulation of Vortex Sound at Low Mach Numbers*, Proceedings of the 5th Internat. Conf. on Spectral and High Order Methods, Uppsala, Sweden, June 11-15, 2001.

[51] J. Nordström and M.H. Carpenter, *Boundary and Interface Conditions for High-Order Finite-Difference Methods Applied to the Euler and Navier-Stokes Equations*, J. Comput. Phys., **148**, 621-645 (1999).

[52] W.L. Oberkampf, K.V. Diegert, K.F. Alvin, and B.M. Rutherford, *Variability, Uncertainty and Error in Computational Simulation*, ASME Proceedings of the 7th. AIAA/ASME Joint Thermophysics and Heat Transfer Conference, HTD-Vol. 357-2 (1998).

[53] P. Olsson, *Summation by Parts, Projections, and Stability. I*, Math. Comp. **64** (1995), pp. 1035–1065.

[54] P. Olsson, *Summation by Parts, Projections, and Stability. II*, Math. Comp. **64**, pp. 1473-1493 (1995).

[55] P. Olsson, *Summation by Parts, Projections, and Stability. III*, RIACS Technical Report 95.06 (1995).

[56] P. Olsson and J. Oliger, *Energy and Maximum Norm Estimates for Nonlinear Conservation Laws*, RIACS Technical Report 94.01, (1994).

[57] V. Perrier, T. Philipovitch, and C. Basdevant, *Wavelet Spectra Compared to Fourier Spectra*, Publication of ENS, Paris, 1999.

[58] R. D. Richtmyer and K. W. Morton, *Difference Methods for Initial-Value Problems*, 2nd ed., John Wiley & Sons, 1967.

[59] N.D. Sandham and R.J.A. Howard, *Direct Simulation of Turbulence Using Massively Parallel Computers. Parallel Computational Fluid Dynamics*, Ed. D.R.Emerson et al., Elsevier (1998).

[60] Sandham, N.D. and Yee, H.C., *Entropy Splitting for High Order Numerical Simulation of Compressible Turbulence*, RIACS Technical Report 00.10, June, 2000, Proceedings of the 1st Intern. Conf. on CFD, July 10-14, 2000, Kyoto, Japan.

[61] N. D. Sandham, Q. Li and H.C. Yee, *Entropy Splitting for High Order Numerical Simulation of Compressible Turbulence*, Revised version of RIACS Technical Report 00.10, June 2000, NASA Ames Research Center;

Proceedings of the First International Conference on CFD, July 10-14, 2000, Kyoto, Japan.

[62] B. Sjögreen, *Numerical Experiments with the Multiresolution Scheme for the Compressible Euler Equations*, J. Comput. Phys., **117**, 251 (1995).

[63] B. Sjögreen and H. C. Yee, *Multiresolution Wavelet Based Adaptive Numerical Dissipation Control for Shock-Turbulence Computations*, RIACS Report 01.01, NASA Ames research center (Oct 2000).

[64] B. Sjögreen and H. C. Yee, *Grid Convergence of High Order Methods for Multiscale Complex Unsteady Viscous Compressible Flows*, RIACS Report 01.06, April, 2001, NASA Ames research center; AIAA 2001-2599, Proceedings of the 15th AIAA CFD Conference, June 11-14, 2001, Anaheim, CA.

[65] B. Sjögreen and H.C. Yee, *On Entropy Splitting, Linear and Nonlinear Numerical Dissipations and Long-Time Integrations*, Proceedings of the 5th Internat. Conf. on Spectral and High Order Methods, Uppsala, Sweden, June 11-15, 2001.

[66] B. Sjögreen and H. C. Yee, *Low Dissipative High Order Numerical Simulations of Supersonic Reactive Flows*, RIACS Report 01-017, NASA Ames Research Center (May 2001); Proceedings of the ECCOMAS Computational Fluid Dynamics Conference 2001, Swansea, Wales, UK, September 4-7, 2001.

[67] B. Sjögreen, *High Order Centered Difference Methods for the Compressible Navier-Stokes Equations*, J. Comput. Phys., **117**, pp. 67–78 (1995).

[68] Special Section of the AIAA Journal, *Credible Computational Fluid Dynamics Simulation*, AIAA J. **36**, 665-759 (1998).

[69] Strand, B., *High-Order Difference Approximations for Hyperbolic Initial Boundary Value Problems*, PhD thesis, Uppsala University, Department of Scientific Computing, 1996.

[70] B. Strand, *Summation by Parts for Finite Difference Approximations for d/dx*, J. Comput. Phys. **110**, pp. 47–67 (1994).

[71] G. Strang, *Wavelet Transforms versus Fourier Transforms*, Bull. Amer. Math. Soc. (N.S.), **28**, 288 (1993).

[72] G. Strang, *Wavelet and Dilation Equations: A Brief Introduction*, SIAM Rev., **31**, 614 (1989).

[73] E. Tadmor, *Skew-Selfadjoint Form for Systems of Conservation Laws*, J. Math. Anal. Appl., **103**, pp. 428–442 (1984).

[74] J.R. Torczynski, *A Grid Refinement Study of Two-Dimensional Transient Flow Over a Backward-facing Step Using a Spectral-Element Method*, FED-Vol. 149, Separated Flows, ASME, J.C. Dutton and L.P. Purtell, Eds. (1993).

[75] T.G. Trucano, *Prediction and Uncertainty in Computational Modeling of Complex Phenomena: A Whitepaper*, Sandia Report, SAND98-2776 (1998).

[76] R. Vichnevetsky, *Numerical Filtering for Partial Differencing Equations*, Numerical Applications Memorandum, Rutgers University, NAM 156 (1974).

[77] M. Vinokur and H.C. Yee, *Extension of Efficient Low Dissipative High Order Schemes for 3-D Curvilinear Moving Grids*, NASA TM 209598, June 2000.

[78] Yee, H.C., Sweby, P.K. and Griffiths, D.F., *Dynamical Approach Study of Spurious Steady-State Numerical Solutions for Nonlinear Differential Equations, Part I: The Dynamics of Time Discretizations and Its Implications for Algorithm Development in Computational Fluid Dynamics*, NASA TM-102820, April 1990; J. Comput. Phys. **97** 249-310 (1991).

[79] H.C. Yee, P.K. Sweby, *Dynamics of Numerics Spurious Behaviors in CFD Computations*, 7th ISCFD Conference, Sept. 15-19, 1997, Beijing, China, RIACS Technical Report 97.06, June (1997).

[80] H.C. Yee and P.K. Sweby, *Some Aspects of Numerical Uncertainties in Time Marching to Steady-State Computations*, AIAA-96-2052, 27th AIAA Fluid Dynamics Conference, June 18-20, 1996, New Orleans, LA., AIAA J., **36** 712-724 (1998).

[81] H.C. Yee, J.R. Torczynski, S.A. Morton, M.R. Visbal and P.K. Sweby, *On Spurious Behavior of CFD Simulations*, AIAA 97-1869, Proceedings of the 13th AIAA Computational Fluid Dynamics Conference, June 29 - July 2, 1997, Snowmass, CO.; also International J. Num. Meth. Fluids, **30**, 675-711 (1999).

[82] H. C. Yee, N. D. Sandham, and M. J. Djomehri, *Low Dissipative High Order Shock-Capturing Methods Using Characteristic-Based Filters*, J. Comput. Phys., **150** 199–238 (1999).

[83] H.C. Yee, M. Vinokur, and M.J. Djomehri, *Entropy Splitting and Numerical Dissipation*, NASA Technical Memorandum 208793, August, 1999, NASA Ames Research Center; J. Comput. Phys., **162**, 33 (2000).

[84] H.C. Yee and B. Sjögreen, *Adaptive Numerical-Dissipation/Filter Controls for High Order Numerical Methods*, Proceedings of the 3rd International Conference on DNS/LES, Arlington, Texas, August 4-9, 2001.

[85] S.T. Zalesak, *High Order "ZIP" Differencing of Convective Terms*, J. Comput. Phys., **40** (1981), pp. 497-508.

[86] T. A. Zang, *On the Rotation and Skew-Symmetric Forms for Incompressible Flow Simulations*, Appl. Numer. Math., **7** (1991), pp. 27–40.

Chapter 6

BUILDING BLOCKS FOR RELIABLE COMPLEX NONLINEAR NUMERICAL SIMULATIONS

H. C. Yee
NASA Ames Research Center, Moffett Field, CA 94035, USA.
yee@nas.nasa.gov

Abstract This chapter describes some of the building blocks to ensure a higher level of confidence in the predictability and reliability (PAR) of numerical simulation of multiscale complex nonlinear problems. The focus is on relating PAR of numerical simulations with complex nonlinear phenomena of numerics. To isolate sources of numerical uncertainties, the possible discrepancy between the chosen partial differential equation (PDE) model and the real physics and/or experimental data is set aside. The discussion is restricted to how well numerical schemes can mimic the solution behavior of the underlying PDE model for finite time steps and grid spacings. The situation is complicated by the fact that the available theory for the understanding of nonlinear behavior of numerics is not at a stage to fully analyze the nonlinear Euler and Navier-Stokes equations. The discussion is based on the knowledge gained for nonlinear model problems with known analytical solutions to identify and explain the possible sources and remedies of numerical uncertainties in practical computations. Examples relevant to turbulent flow computations are included.

Keywords: Dynamics of Numerics, Chaotic Transients, Direct Numerical Simulations, Reliability of Numerical Simulations, Under-Resolved Grids, Spurious Numerical Solutions, Nonlinear Dynamics, Spurious Bifurcations.

1. Introduction

The last two decades have been an era when computation is ahead of analysis and when very large scale practical computations are increasingly used in poorly understood multiscale complex nonlinear physical problems and non-traditional fields. Ensuring a higher level of confidence in the predictability and reliability (PAR) of these numerical simulations could play a major role in furthering the design, understanding, affordability and safety of our next generation air

D. Drikakis and B.J. Geurts (eds.), Turbulent Flow Computation, 199-236.

and space transportation systems, and systems for planetary and atmospheric sciences, and astrobiology research. In particular, it plays a major role in the success of the US Accelerated Strategic Computing Initiative (ASCI) and its five Academic Strategic Alliance Program (ASAP) centers. Stochasticity stands alongside nonlinearity and the presence of multiscale physical processes as a predominant feature of the scope of this research. The need to guarantee PAR becomes acute when computations offer the **ONLY** way of generating this type of data limited simulations, the experimental means being unfeasible for any of a number of possible reasons. Examples of this type of data limited problem are:

- Stability behavior of re-entry vehicles at high speeds and flow conditions beyond the operating ranges of existing wind tunnels

- Flow field in thermo-chemical nonequilibrium around space vehicles traveling at hypersonic velocities through the atmosphere (lack sufficient experimental or analytic validation)

- Aerodynamics of aircraft in time-dependent maneuvers at high angles of attack (free of interference from support structures, wind-tunnel walls etc., and able to treat flight at extreme and unsafe operating conditions)

- Stability issues of unsteady separated flows in the absence of all the unwanted disturbances typical of wind-tunnel experiments (e.g., geometrically imperfect free-stream turbulence)

This chapter describes some of the building blocks to ensure a higher level of confidence in the PAR of numerical simulation of the aforementioned multiscale complex nonlinear problems, especially the related turbulence flow computations. To isolate the source of numerical uncertainties, the possible discrepancy between the chosen model and the real physics and/or experimental data is set aside for the moment. We concentrate only on how well numerical schemes can mimic the solution behavior of the underlying partial different equations (PDEs) for finite time steps and grid spacings. Even with this restriction, the study of PAR encompasses elements and factors far beyond what is discussed here. It is important to have a very clear distinction of numerical uncertainties from each source. These include but are not limited to (a) stability and well-posedness of the governing PDEs, (b) type, order of accuracy, nonlinear stability, and convergence of finite discretizations, (c) limits and barriers of existing finite discretizations for highly nonlinear stiff problems with source terms and forcing, and/or for long time wave propagation phenomena, (d) numerical boundary condition (BC) treatments, (e) finite representation of infinite domains, (f) solution strategies in solving the nonlinear discretized equations, (g) procedures for obtaining the steady-state numerical solutions, (h) grid quality and grid adaptations, (i) multigrids, and (j) domain decomposition (zonal or multicomponent approach) in solving large problems. At present, some of the numerical uncertainties can be explained and minimized by traditional numerical analysis and standard CFD practices. However, such practices, usually based on linearized analysis, might not be sufficient for strongly nonlinear and/or stiff problems. We need a good understanding of the nonlinear behavior of numerical schemes being used as an integral part of code verification, validation and certification.

A major stumbling block in genuinely nonlinear studies is that unlike the linear model equations used for conventional stability and accuracy considerations in time-dependent PDEs, there is no equivalent unique nonlinear model equation for nonlinear hyperbolic and parabolic PDEs for fluid dynamics. On one hand, a numerical method behaving in a certain way for a particular nonlinear differential equation (DE) (PDE or ordinary differential equation (ODE)) might exhibit a different behavior for a different nonlinear DE even though the DEs are of the same type. On the other hand, even for simple nonlinear model DEs with known solutions, the discretized counterparts can be extremely complex, depending on the numerical methods. Except in special cases, there is no general theory at the present time to characterize the various nonlinear behaviors of the underlying discretized counterparts. Herein, the discussion is based on the knowledge gained for nonlinear model problems with known analytical solutions to identify and explain the possible sources and remedies of numerical uncertainties in practical computations.

The term **"discretized counterparts"** is used to mean the finite difference equations resulting from finite discretizations of the underlying DEs. Here **"dynamics"** is used loosely to mean the dynamical behavior of nonlinear dynamical systems (continuum or discrete) and **"numerics"** is used loosely to mean the numerical methods and procedures in solving dynamical systems. We emphasize here that in the study of the dynamics of numerics, unless otherwise stated, we always assume the continuum (governing equations) is nonlinear.

Outline: Section 2 discusses the sources of nonlinearities and the knowledge gained from studying the dynamics of numerics for nonlinear model problems. Sections 3-5 discuss some of the relevant issues and building blocks for a more reliable (and predictability) numerical simulation in more details. Section 6 shows examples of spurious numerics relevant to turbulent flow computations.

2. Sources of Nonlinearities and Knowledge Gained from Nonlinear Model Problems

Two of the building blocks for the PAR of numerical simulations are to identify all the sources of nonlinearities and to isolate the elements and issues of numerical uncertainties due to these nonlinearities.

Sources of Nonlinearities: The sources of nonlinearities that are well known in computational fluid dynamics (CFD) are due to the physics. Examples of nonlinearities due to the physics are convection, diffusion, forcing, turbulence source terms, reacting flows, combustion related problems, or any combination of the above. The less familiar sources of nonlinearities are due to the numerics. There are generally three major sources:

- Nonlinearities due to time discretizations – the discretized counterpart is nonlinear in the time step. Examples of this type are Runge-Kutta methods. If fixed time steps are used, spurious steady-state or spurious asymptotic numerical solutions can occur, depending on the the initial condition (IC). Linear multistep methods (LMMs) (Butcher 1987) are linear in the time step, and they do not exhibit spurious steady states. See Yee & Sweby (1991-1997) and references cited therein for the dynamics of numerics of standard time discretizations.

- Nonlinearities due to spatial discretizations – in this case, the discretized counterpart can be nonlinear in the grid spacing and/or the scheme. Examples of nonlinear schemes are the total variation diminishing (TVD), essentially nonoscillatory (ENO) and weighted ENO (WENO) schemes. The resulting discretized counterparts are nonlinear (in the dependent variables) even though the governing equation is linear. See Yee (1989) and references cited therein for the forms of these schemes.

- Nonlinearities due to complex geometries, boundary interfaces, grid generation, grid refinements and grid adaptations (Yee & Sweby 1995) – each of these procedures can introduce nonlinearities even though the governing equation is linear.

Continuous and Discrete Dynamical Systems: Before analyzing the dynamics of numerics, it is necessary to analyze (or understand) as much as possible the dynamical behavior of the governing equations and/or the physical problems using theories of DEs, dynamical systems of DEs, and also physical guidelines. For stability and well-posedness considerations, whenever it is possible, it is also necessary to condition (not pre-condition) the governing PDEs before the application of the appropriate scheme (Yee & Sjögreen 2001a,b). The discretized counterparts are dynamical systems on their own. They have their own dynamics, and they are different from one numerical method to another in space and time and are different from the underlying governing PDE (Yee & Sweby 1993). The procedures of solving the nonlinear algebraic systems resulting from using implicit methods can interfere with or superpose unwanted behavior on the underlying scheme. Also, the same scheme can exhibit different spurious behavior when used for time-accurate vs. time marching to the steady states. For a combination of initial condition and time step, a super-stable scheme can stabilize unstable physical (analytic) steady states (Yee & Sweby 1993-1996). Super-stable scheme here refers to the region of numerical stability enclosing the physical instability of the true solution of the governing equation. Yee et al. and Yee & Sweby (1991-1997) divide their studies into two categories, steady state and time accurate computations. Within each category they further divide the governing PDEs into homogeneous and nonhomogeneous (i.e., with or without source terms), and rapidly/slowly developing and long time integration problems.

Knowledge Gained from Nonlinear Model Problems: With the aid of elementary examples, Yee et al., Yee & Sweby (1991-1997), Sweby & Yee (1994-1995) and Griffiths et al. (1992a,b) discuss the fundamentals of spurious behavior of commonly used time and spatial discretizations in CFD. Details of these examples can be found in their earlier papers. These examples consist of nonlinear model ODEs and PDEs with known analytical solutions (the most straight forward way of being sure what is "really" happening with the numerics). They illustrate the danger of employing fixed (constant) time steps and grid spacings. They were selected to illustrate the following different nonlinear behavior of numerical methods:

- Occurrence of stable and unstable spurious asymptotes **above** the linearized stability limit of the scheme (for constant time steps)
- Occurrence of stable and unstable spurious steady states **below** the linearized stability limit of the scheme (for constant time steps)
- **Stabilization** of unstable steady states by implicit and semi-implicit methods
- Interplay of initial data and time steps on the occurrence of spurious asymptotes
- Interference with the dynamics of the underlying implicit scheme by procedures in solving the nonlinear algebraic equations (resulting from implicit discretizations of the continuum equations)
- Dynamics of the linearized implicit Euler scheme solving the time-dependent equations to obtain steady states vs. Newton's method solving the steady equation

- Spurious dynamics independently introduced by spatial and time discretizations
- Convergence problems and spurious behavior of high-resolution shock-capturing methods
- Numerically induced & suppressed (spurious) chaos, and numerically induced chaotic transients
- Spurious dynamics generated by grid adaptations

Here "spurious numerical solutions (and asymptotes)" is used to mean numerical solutions (asymptotes) that are solutions (asymptotes) of the discretized counterparts but are not solutions (asymptotes) of the underlying DEs. Asymptotic solutions here include steady-state solutions, periodic solutions, limit cycles, chaos and strange attractors. See Thompson & Stewart (1986) and Hoppensteadt (1993) for the definition of chaos and strange attractors.

3. Minimization of Spurious Steady State via Bifurcation Theory

The use of time-marching approaches to obtain steady-state numerical solutions has been considered the method of choice in computational physics for three decades since the pioneering work of Moretti & Abbett (1966). Moretti and Abbett used this approach to solve the inviscid supersonic flow over a blunt body without resorting to solving the steady form of PDEs of the mixed type. Much success was achieved in computing a variety of weakly and moderately nonlinear fluid flow problems. For highly complex nonlinear problems, the situation is more complicated. The following isolates some of the key elements and issues of numerical uncertainties in time-marching to the steady state. Studies in Yee et al. (1991-1997) indicate that each of the following can affect not only the convergence rate but also spurious numerics other than standard stability and accuracy linearized numerical analysis problems.

- Solving an initial boundary value problem (IBVP) with unknown initial data
- Reliability of residual test
- Methods used to accelerate the convergence process
- Precondition (not condition) the governing PDE (*might introduce additional spurious solutions beyond the solution of the underlying PDE*)
- Precondition the discretized counterparts (*might introduce additional spurious solutions beyond the underlying discretized system*)

- Methods in solving the nonlinear algebraic equations from implicit methods
- Mismatch in implicit schemes
- Nonlinear schemes
- Schemes that are linear vs. nonlinear in time step
- Adaptive time step based on local error control

It is a standard practice in time-marching to the steady-state numerical solutions to use "local time step" (varied from grid point to grid point using the same CFL) for nonuniform grids. However, except in finite element methods, an adaptive time step based on local error control is rarely used. An adaptive time step is built in for standard ODE solver computer packages. It enjoyed much success in controlling accuracy and stability for transient (time-accurate) computations. The issue is to what extent this adaptive local error control confers global properties in long time integration of time-dependent PDEs and whether one can construct a similar error control that has guaranteed and rapid convergence to the correct steady-state numerical solutions in the time-marching approaches for time-dependent PDEs.

One can see that the construction of adaptive time integrators for time-marching to the steady states demands new concepts and guidelines and is distinctively different than for the time-accurate case. Straightforward application of adaptive time integrators for time-accurate computations might be inappropriate and/or extremely inefficient for time-marching to the steady state. For example, an adaptive time step based on local error control for accuracy considerations is irrelevant before a steady state has been reached. Moreover, this type of local error control might hinder the speed of the convergence process with no guarantee of leading to the correct steady state.

The twin requirements of **guaranteed** and **rapid convergence** to the **correct** steady-state numerical solution are most often conflicting, and require a full understanding of the global nonlinear behavior of the numerical scheme as a function of the discretized parameters, grid adaptation parameters, initial data and boundary conditions. We believe tools from bifurcation theory can help to minimize spurious steady-state numerical solutions.

In many fluid problems the solution behavior is well known for certain values of the physical parameters but unknown for other values. For these other values of the parameters, the problem might become very stiff and/or strongly nonlinear, making the available numerical schemes (or the scheme in use) intractable. In this situation, continuation methods in bifurcation theory can become very useful. If possible, one should start with the physical parameter of a known or reliable steady state (e.g., flow behavior is usually known for low angles of attack but not for high angles of attack). One can then use a continuation method such as the improved pseudo arclength continuation method of Keller (1977) (or the recent developments in this area) to solve for the bifurcation curve as a function of the physical parameter. See e.g., Doedel (2000), Shroff & Keller (1993) and Davidson (1997). The equations used are the discretized counterpart of the steady PDEs or the time-dependent PDEs. See Stephen & Shubin (1981) and Shubin et al. (1981) for earlier related work. If time-marching approaches are used, a reliable steady-state numerical solution (as a starting value on the correct branch of the bifurcation curve for a particular value of the physical parameter) is assumed. This starting steady-state numerical solution is assumed to have the proper time step and initial data combination and to have a grid spacing fine enough to resolve the flow feature. The continuation method will produce a continuous spectrum of the numerical solutions as the underlying physical parameter is varied until it arrives at a critical value p_c such that it either experiences a bifurcation point or fails to converge. Since we started on the correct branch of the bifurcation curve, the solution obtained before that p_c should be more reliable than if one starts with the physical parameter in question with unknown initial data and tries to stretch the limitation of the scheme. Note that by starting a reliable solution on the correct branch of the bifurcation curve, the dependence of the numerical solution on the initial data associated with time-marching methods can be avoided before a spurious bifurcation occurs.

Finally, when one is not sure of the numerical solution, the continuation method can be used to double check it. This approach can even reveal the true limitations of the existing scheme. In other words, the approach can reveal the critical physical parameter for which the numerical method breaks down. On the other hand, if one wants to find out the largest possible time step and/or grid spacing that one can use for a particular problem and physical parameter, one can also use continuation methods to trace out the bifurcation curve as a function of the time step and/or grid spacing. In this case, one can start with a small time step and/or grid spacing with the correct steady state and observe the critical discretized parameter as it undergoes instability or spurious bifurcation. Of course, this method for minimizing spurious steady states still can suffer from spurious behavior due to an under-resolved grid because of limited computer resources for complex practical problems. Practical guidelines to avoid under-resolved grids are yet another important building block toward reliable numerical simulations. The efficient treatment of solving the extremely large set of eigenvalue problems to study the type and stability of bifurcation points is yet another challenge. See, e.g., Fortin et al. (1996), Davidson (1997) and Shroff & Keller (1993) for some discussions. Consequently, further development in numerical bifurcation analysis and new concepts in adaptive methods for time-marching to steady state hold a key to the minimization of spurious numerics.

4. Source Term Treatments in Reacting Flows

In the modeling of problems containing finite-rate chemistry or combustion, often, a wide range of space and time scales is present due to the reacting terms, over and above the different scales associated with turbulent flows, leading to additional numerical difficulties. This stems mainly from the fact that the majority of widely used numerical algorithms in reacting flows were originally designed to solve non-reacting fluid flow problems. Fundamental studies on the behavior of these schemes for reacting model problems by the author and collaborators were reported in Yee & Sweby (1997) and references cited therein. In a majority of these studies, theory from dynamical systems was used to gain a better understanding of the nonlinear effects on the performance of these schemes. The main findings are:

- It was shown in LeVeque and Yee (1990) that, for stiff reactions containing shock waves, it is possible to obtain stable solutions that look reasonable and yet are completely wrong, because the discontinuities are in the wrong locations. Stiff reaction waves move at nonphysical wave speeds, often at the rate of one grid cell per time step, regardless of their proper speed. There exist several methods that can overcome this difficulty for a single reaction term. For more than a single reacting term in fully coupled nonlinear systems, more research is needed. One impractical way of minimizing the wrong speed of propagation of discontinuities is to demand orders of magnitude grid size reduction compared with what appears to be a reasonable grid spacing in practice.

- It was shown in Lafon and Yee (1991, 1992) that the numerical phenomenon of incorrect propagation speeds of discontinuities may be linked to the existence of some stable spurious steady-state numerical solutions.

- It was also shown in Lafon and Yee (1991, 1992) that various ways of discretizing the reaction term can affect the stability of and convergence to the spurious numerical steady states and/or the exact steady states. Pointwise evaluation of the source terms appears to be the least stable.

- It was shown in Yee et al. (1991) and Griffiths et al. (1992a,b) that spurious discrete traveling waves can exist, depending on the method of discretizing the source term. When physical diffusion is added, it is not known what type of numerical difficulties will surface.

From the above findings we can safely conclude that understanding the nonlinear behavior of numerical schemes for reacting flows and the effects of finite-rate chemistry on turbulence is in its infancy. However, we believe that knowledge gained from fundamental studies is helpful to improve some of the numerical difficulties that were encountered in the past.

5. Adaptive Numerical Methods

Another important building block for PAR is adaptive numerical methods. This includes adaptive temporal and spatial schemes, grid adaptation as an integral part of the numerical solution process, and, most of all, adaptive numerical dissipation controls. Using tools from dynamical systems, Yee et al. (1991-1997), Yee & Sweby (1993-1997), Griffiths et al. (1992a,b) and Lafon & Yee (1991, 1992) showed that adaptive temporal and adaptive spatial schemes are important in minimizing numerically induced chaos, numerically induced chaotic transients and the false prediction of flow instability by direct numerical simulation (DNS). Their studies further indicate the need in the development of practical adaptive temporal schemes based on error controls to minimize spurious numerics due to the **full discretizations**. In addition, the development of adaptive temporal and spatial schemes based on error controls to minimize numerical artifacts due to the full discretizations is also needed. This is due to the fact that adaptive temporal or adaptive spatial schemes **alone** will not be able to provide an accurate and reliable process to minimize numerical artifacts for time-accurate computations. Guided by the theory of nonlinear dynamics, Yee et al. (1997) and Yee & Sweby (1997) presented practical examples which illustrated the danger of using nonadaptive temporal and spatial schemes for studying flow instability.

On the subject of adaptive numerical dissipation controls, it is well known that reliable, accurate and efficient direct numerical simulation (DNS) of turbulence in the presence of shock waves represents a significant challenge for numerical methods. Standard TVD, ENO, WENO and discontinuous Galerkin types of shock-capturing methods for the Euler equations are now routinely used in high speed blast wave simulations with virtually non-oscillatory, crisp resolution of discontinuities (see reference section). For the unaveraged unsteady compressible Navier-Stokes equations, it was observed that these schemes are still too dissipative for turbulence and transition predictions. On the other hand, hybrid schemes, where spectral and/or higher-order compact (Padé) schemes are switched to higher-order ENO schemes when shock waves are detected, have their deficiencies. One shortcoming of this type of hybridization is that the numerical solution might experience a non-smooth transition at the switch to a different type of scheme. For 2-D and 3-D complex shock wave and shear surface interactions, the switch mechanism can become non-trivial and frequent activation of shock-capturing schemes is possible.

The recent work of Yee et al. (1999, 2000), Sjögreen & Yee (2000, 2001), and Yee & Sjögreen (2001a,b) indicates that appropriate adaptive numerical dissipation control is essential to control nonlinear instability in general, and for long time integration, in particular. An integrated design approach on the construction of adaptive numerical dissipation controls can be found in Yee & Sjögreen (2001a,b).

6. Spurious Numerics Relevant to Turbulent Flow Computations

This section illustrates four numerical examples that exhibit spurious numerics relevant to turbulent flow computations. The first example discusses spurious vortices related to under-resolved grids and/or lack of appropriate numerical dissipation/filter controls. The second example discusses spurious behavior of super-stable implicit time integrators. The last two examples discuss spurious behavior near the onset of turbulence and/or the onset of instability of the steady state solution. If care is not taken, spurious bifurcation of the discretized counterpart and/or a numerically induced chaotic transient can be mistaken for the onset of physical turbulence of the governing equation. These examples can serve to illustrate the connection between the spurious numerical phenomena observed in simple nonlinear models and CFD computations.

6.1 Spurious Vortices in Under-Resolved Incompressible Thin Shear Layer Flow Simulations

Brown & Minion (1995) performed a thorough study of a second-order Godunov-projection method and a fourth-order central difference method for the 2-D incompressible Navier-Stokes equations, varying the resolution of the computational mesh with the rest of the physical and discretization parameters fixed. This is a good example of isolating the cause of spurious behavior. The physical problem is a doubly periodic double shear layer. The shear layers are perturbed slightly at the initial time, which causes the shear layer to roll up in time into large vortical structures. For a chosen shear layer width that is considered to be thin and a fixed perturbation size, they compared the solution for four different grid sizes (64×64, 128×128, 256×256, 512×512) with a reference solution using a grid size of 1024×1024. For the 256×256 grid, a spurious vortex was formed midway between the periodically repeating main vortex on each shear layer. The 128×128 solution showed three spurious vortices along the shear layer. The spurious vortex disappeared with a 512×512 mesh. They also disabled the flux limiters (a strictly upwind Fromm's method), and found the behavior to be similar. A subsequent study (Minion & Brown 1997) using five different formulations and six different commonly used schemes in CFD found similar behavior. They concluded that the spurious vortex is the artifact of an under-resolved grid and the behavior is caused by a nonlinear effect. Linking this behavior with nonlinear dynamics, we interpret their observation as follows. For the particular grid size and time step combination, stable spurious equilibrium points were introduced by the numerics into a portion of the flow field while the major portion of the flow field was predicted correctly. In other words, the spurious vortices are the solution of the discretized counterpart for that particular range of grid size and time step. The number of stable spurious vortices is a function of the grid size. As the grid spacing decreases, the spurious equilibria gradually become unstable and the numerical solution mimics the true solution.

Instead of merely increasing the grid size, there are situations where under-resolved grids can be overcome by proper control of numerical dissipation/filters. It was shown in Fischer & Mullen that high-order spectral element methods (Maday & Patera 1989), coupled with filter-based dissipation, can remove what is believed to be an underresolution-induced spurious vortex numerical solution. See Fischer & Mullen or Yee & Sjögreen (2001a,b) for a discussion. Fischer & Mullen or Yee & Sjögreen (2001a,b) illustrate the added benefit of adaptive numerical dissipation/filter controls for high order or high resolution shock-capturing schemes.

6.2 Stabilizing Unstable Steady States with Implicit Time Integrators (Poliashenko & Yee 1995, unpublished)

This is a joint work with Maxim Poliashenko in 1995. This unpublished work was presented at the 10th International Conference in Finite Element Methods, January 5-8, 1998, Tucson, Arizona, and also has been presented at various invited lectures during the last four years. We consider a 2-D lid driven cavity (LDC) problem. The PDE governing equations are the ideal viscous incompressible Navier-Stokes equations of the form

$$\begin{aligned} &\mathbf{u}_t + (\mathbf{u}^T \nabla)\mathbf{u} = -\nabla p + \tfrac{1}{Re}\Delta \mathbf{u}, \\ &div\, \mathbf{u} = 0, \end{aligned} \tag{6.1}$$

with boundary conditions in the domain (x, y)

$$\mathbf{u}(y = a) = 1, \tag{6.2}$$

$$\mathbf{u}(y = 0) = \mathbf{u}(x = 0) = \mathbf{u}(x = 1) = 0. \tag{6.3}$$

Here $\mathbf{u}$ is a 2-D velocity vector, p is pressure and Re is the Reynolds number, a dimensionless parameter of the problem that describes the relationship between kinematic and viscous forces in the fluid. Here a is the cavity aspect ratio. The velocity vector $\mathbf{u}$, pressure p and the time t are normalized with Re being proportional to the velocity of the lid and inversely proportional to the viscosity of the fluid.

Several numerical time integrators indicated that while the steady state is unique and stable for small Reynolds numbers, the flow becomes time-periodic as Re is increased to a few thousand. Poliashenko and Aidun (1995) applied their direct method for computations of co-dimension one bifurcations to show that the steady state of the LDC problem indeed loses its stability via Hopf bifurcation (Thompson & Stewart 1986) as the Reynolds number increases, giving rise to a time-periodic solution. This bifurcation can be subcritical or supercritical depending on the aspect ratio a. With a given spatial resolution, they found that the Hopf bifurcation point is supercritical for $a = 0.8$ at $Re = 5220$, and for $a = 1.0$ with $Re = 7760$, and subcritical for $a = 1.5$ at $Re = 7220$, and for $a = 2.0$ with $Re = 5120$.

For the current numerical experiment, a 47×47 mildly clustered finite element mesh is used to spatially discretize the incompressible Navier-Stokes equations. The flow solver is the finite element code FIDAP. Nine-node quadrilateral elements with biquadratic interpolation functions for velocity components are used. The bilinear pressure interpolation functions are projected onto the four Gauss points inside each element. In order to reduce the number of nodal unknowns, a penalty approach to remove the pressure is used. The elements are 5 times thinner at the side walls and the bottom and 7 times thinner at the moving lid boundary than at the cavity center. After the spatial discretization, and the use of the weighted residual Galerkin method and the penalty formulation for the pressure, we obtain

$$\mathbf{M}\frac{d\mathbf{U}}{dt} + \mathbf{K}(\mathbf{U})\mathbf{U} = \mathbf{F}. \tag{6.4}$$

Here $\mathbf{U}$ is the global vector of system unknowns of size $2 * N$ where N is the total number of non-boundary nodes. $\mathbf{M}$ is a block diagonal mass matrix. The nonlinear matrix $\mathbf{K}$ represents contributions from the convective and diffusive terms. $\mathbf{F}$ is a generalized force vector which includes contribution from body forces.

The dynamics of two implicit predictor-corrector time integrators are studied. The first is a first-order implicit Euler

$$\mathbf{M}\frac{\mathbf{U}^{n+1} - \mathbf{U}^{p}}{h} + \mathbf{K}(\mathbf{U}^{n+1})\mathbf{U}^{n+1} = \mathbf{F}^{n+1}, \tag{6.5}$$

with the explicit Euler scheme as predictor

$$\mathbf{M}\frac{\mathbf{U}^{p} - \mathbf{U}^{n}}{h} + \mathbf{K}(\mathbf{U}^{n})\mathbf{U}^{n} = \mathbf{F}^{n}. \tag{6.6}$$

The second is the Trapezoidal rule

$$\mathbf{M}\frac{\mathbf{U}^{n+1}-\mathbf{U}^{p}}{h_n}+\frac{1}{2}[\mathbf{K}(\mathbf{U}^{n+1})\mathbf{U}^{n+1}+\mathbf{K}(\mathbf{U}^{p})\mathbf{U}^{p}]=\frac{1}{2}[\mathbf{F}^{n+1}+\mathbf{F}^{p}], \quad (6.7)$$

with the Adams formula as predictor

$$\mathbf{U}^{p}=\mathbf{U}^{n}+\frac{h_n}{2}[(2+\frac{h_n}{h_{n-1}})\dot{\mathbf{U}}^{n}-\frac{h_n}{h_{n-1}}\dot{\mathbf{U}}^{n-1}], \quad (6.8)$$

where h is a constant time step, h_n is a variable time step and $\dot{\mathbf{U}}^n$ is an acceleration vector, approximated by

$$\dot{\mathbf{U}}^{n}=\frac{2}{h_{n-1}}(\mathbf{U}^{n}-\mathbf{U}^{n-1})-\dot{\mathbf{U}}^{n-1}. \quad (6.9)$$

The local time truncation error $\Delta\mathbf{U}^{n+1}$ is computed as follows:

$$\Delta\mathbf{U}^{n+1}=\frac{\mathbf{U}^{n+1}-(\mathbf{U}^{n+1})^{p}}{3(1+\frac{h_{n-1}}{h_n})}+O(h_n^4). \quad (6.10)$$

After the norm $||\Delta\mathbf{U}^{n+1}||$ is evaluated, the next time step is computed according to the formula

$$h_{n+1}=h_n(\frac{\epsilon}{||\Delta\mathbf{U}^{n+1}||})^{1/3}, \quad (6.11)$$

where ϵ is a truncation error tolerance. We also set an upper limit, h_{max}, that restricts growth of the time step.

We first study the dynamics of these time integrators using a fixed time step (h_k= constant). Standard Newton-Raphson and quasi-Newton iterative methods are used to solve the nonlinear algebraic equations for this system of ODEs with the predicted solution as an initial guess. With a fixed time step of $h = 0.001$, both time integrators produce a periodic solution. However, as the time step is increased up to $h = 0.01$, the first time integrator is attracted to the steady state solution. This phenomenon of spurious stabilization of an unstable steady state is very typical for implicit LMM schemes. This spurious steady state remains stable as h increases further.

The dynamics of the second integrator is different. The solution remains time periodic up to $h = 0.75$. For h between 0.75 and 0.8, the solution appears quasiperiodic, indicating the occurrence of the secondary Hopf bifurcation. With h increased to 1.0, the quasiperiodic oscillations become increasingly disturbed until the solution appears very irregular for $h > 1.0$, which is indicative of numerically induced chaos. With further increases in h, more complex bifurcations occur with the computed solution becoming regular again at $h = 1.3$

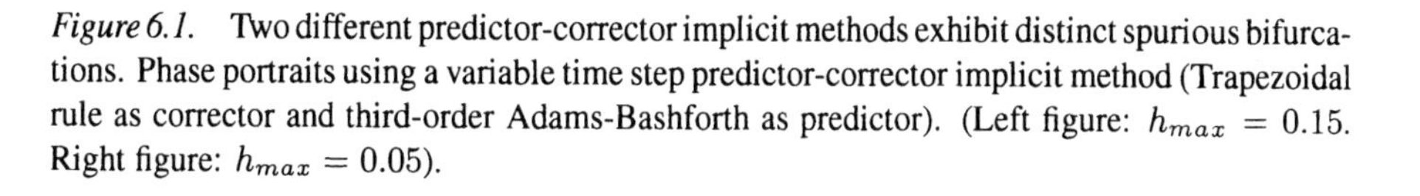

Figure 6.1. Two different predictor-corrector implicit methods exhibit distinct spurious bifurcations. Phase portraits using a variable time step predictor-corrector implicit method (Trapezoidal rule as corrector and third-order Adams-Bashforth as predictor). (Left figure: $h_{max} = 0.15$. Right figure: $h_{max} = 0.05$).

and then returning to chaotic behavior. With fully developed chaotic behavior and the solution being non-smooth, more and more singular eruptions occur. This makes it difficult for the Newton-Raphson procedure to converge. At some point around $h = 1.5$, the Newton-Raphson method fails to converge, implying that there is a nonlinear instability of the chosen time integrator at this time step.

Similar behavior of the numerical solution is observed if the quasi-Newton procedure is applied instead of the standard Newton-Raphson method in solving the underlying nonlinear algebraic system. However, in this case quasiperiodic solutions tend to become irregular with smaller time steps and the transition to chaos occurs earlier.

In Poliashenko & Aidun (1995) the variable time step control version of the Adams-Bashforth/Trapezoidal predictor-corrector scheme with truncation error tolerance of 0.005 and maximum time step of 0.15 for variable time step control and the quasi-Newton nonlinear solver are used. They found that after $Re >$

6200, a weak modulation of the oscillation envelope occurs and the solution becomes quasiperiodic with the indication of a secondary Hopf bifurcation. As Re approaches 6500, they observed a strong resonance between the two independent frequencies which transformed the solution from quasiperiodic to strictly periodic. This phenomenon is known in dynamical systems as "phase-lock" on the torus. This resulted in the birth of limit cycles. This limit cycle is observed in a fairly wide range of Re up to 6700. At $Re = 6700$, the numerical solution exhibits more complex bifurcations and transitions to weakly turbulent, or chaotic motion.

As we decrease $h_{max} < 0.12$, the limit cycle shown in Fig. 6.1 is replaced by a stable 2-D torus which remains qualitatively unchanged as h is further decreased and appears to be close to the "true" solution of the ODEs. The latter example demonstrates that a variable time step integrator with local error control, although more reliable, does not guarantee no spurious numerics.

6.3 Chaotic Transients Near the Onset of Turbulence in Direct Numerical Simulations of Channel Flow (Keefe 1988, Yee & Sweby 1997)

In addition to the inherent chaotic and chaotic transient behavior in some physical systems, numerics can independently introduce and suppress chaos as well as chaotic transients. Loosely speaking, a chaotic transient behaves like a chaotic solution (Grebogi et al. 1983). A chaotic transient can occur in a continuum or a discrete dynamical system. One of the major characteristics of a numerically induced chaotic transient is that if one does not integrate the discretized equations long enough, the numerical solution has all the characteristics of a chaotic solution. The required number of integration steps might be far beyond those found in standard CFD simulation practice before the numerical solution can get out of the chaotic transient mode. Furthermore, standard numerical methods, depending on the initial data, usually experience drastic reductions in step size and convergence rate near a bifurcation point of the continuum in addition to the bifurcation points due solely to the discretized parameters. See Yee & Sweby (1992, 1995, 1996, 1997) for a discussion. Consequently, a possible numerically induced chaotic transient is especially worrisome in direct numerical simulations of the transition from laminar to turbulent flows. Except for special situations, it is extremely difficult to bracket closely the physical transition point by mere DNS of the Navier-Stokes equations. Even away from the transition point, this type of numerical simulation is already very CPU intensive and the convergence rate is usually rather slow. Due to limited computer resources, the numerical simulation can result in chaotic transients indistinguishable from sustained turbulence, yielding a spurious picture of the flow for a given Reynolds number. Consequently, it casts some doubt on the reliability of numerically predicted transition points and chaotic flows. It also influences the true connection between chaos and turbulence. See also Moore et al. (1990).

Assuming a known physical bifurcation or transition point, Fig. 6.2 illustrates the schematic of four possible spurious bifurcations due to constant time steps and constant grid spacings. This section and the next (Section 6.4) illustrate the occurrence of these scenarios. Section 6.4 discusses the stability of the steady state (as a function of the Reynolds number) of a 2-D backward facing step problem using direct simulations. The present section is the computation by Laurence Keefe performed in the late 1980s. In 1996 we made use of the knowledge from continuum and discrete dynamical systems theory to interpret his result. We identified some of the aforementioned numerical uncertainties in his computations. The result is reported in Yee & Sweby 1997.

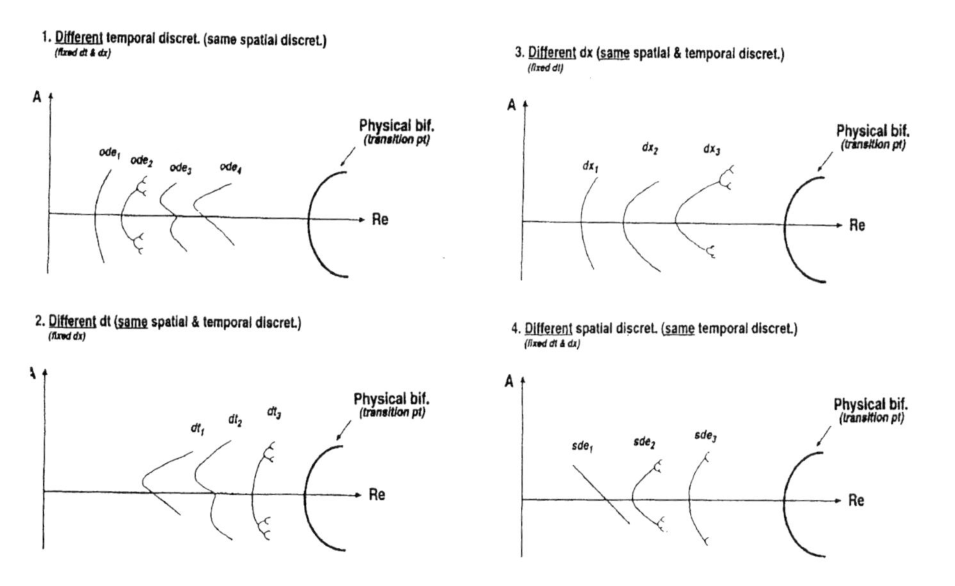

Figure 6.2. Schematic of possible spurious bifurcation for constant time steps and grid spacings. (1) Different temporal discretizations ode_1, ode_2, ode_3 and ode_4 (same spatial discretization and the same constant dt and dx). (2) Different constant time steps dt_1, dt_2, dt_3 and dt_4 (same temporal and spatial discretizations, and the same constant dx). (3) Different constant grid spacings dx_1, dx_2, dx_3 and dx_4 (same spatial and temporal discretizations, and the same constant dt). (4) Different spatial discretizations sde_1, sde_2 and sde_3 (same temporal discretization and the same constant dt).

The physical problem that Keefe considered is depicted in Fig. 6.3, where a flow is confined between planes at $y = \pm 1$ and is driven in the x-direction by a mean pressure gradient dp/dx. The flow is characterized by a Reynolds number $Re = U_\infty L/\nu$, where U_∞ is the mean centerline velocity, L is the channel half-height, and ν is the kinematic viscosity. Within the channel the flow satisfies the incompressible Navier-Stokes equations and no-slip boundary conditions are applied at the walls. In the particular calculations shown here these equations have been manipulated into velocity-vorticity form, where one integrates equations for the wall-normal velocity v and normal vorticity η, and recovers the other two velocity components from the incompressibility condition and the definition of η.

$$\frac{\partial}{\partial t}\Delta^2 v = h_v + \frac{1}{Re}\Delta^4 v, \tag{6.12}$$

$$\frac{\partial}{\partial t}\eta = h_g + \frac{1}{Re}\Delta^2 \eta, \tag{6.13}$$

$$f + \frac{\partial v}{\partial y} = 0, \tag{6.14}$$

where

$$f = \frac{\partial u}{\partial x} + \frac{\partial w}{\partial z}, \qquad \eta = \frac{\partial u}{\partial z} - \frac{\partial w}{\partial x}, \tag{6.15}$$

$$h_v = -\frac{\partial}{\partial y}\left(\frac{\partial H_1}{\partial x} + \frac{\partial H_3}{\partial z}\right) + \left(\frac{\partial^2}{\partial x^2} + \frac{\partial^2}{\partial z^2}\right) H_2, \tag{6.16}$$

$$h_g = \frac{\partial H_1}{\partial z} - \frac{\partial H_3}{\partial x}. \tag{6.17}$$

Here the H_i contain the nonlinear terms in the primitive form of the Navier-Stokes equations and the mean pressure gradient.

The velocity increases extremely rapidly normal to the wall, and turbulent channel flows are essentially homogeneous in planes parallel to the wall. The first requires a concentration of grid points near the wall, and the second suggests use of a doubly periodic domain in planes parallel to the wall. A spectral representation of the velocity field (u, v, w)

$$\vec{u} = \sum_l \sum_m \sum_n \vec{A}_{lmn}(t) T_l(y) e^{im\alpha x + in\beta z}, \tag{6.18}$$

where the $T_l(y)$ are Chebyshev polynomials used for the spatial discretization. The numerical problem then becomes dependent on α and β in addition to Re. For the time discretization, mixed explicit-implicit methods are used.

The nonlinear terms in the equations are advanced using second-order Adams-Bashforth or a low storage, third-order Runge-Kutta scheme (Spalart et al. 1991), while the viscous terms are advanced by Crank-Nicholson.

One of the central problems in studies of wall bounded shear flows is the determination of when a steady laminar flow becomes unstable and transitions to turbulence. In dynamical systems terms, the Navier-Stokes equations always have a fixed point solution for low enough Reynolds numbers, but for each flow geometry the Reynolds number at which this fixed point bifurcates needs to be determined. In channel flow the fixed point solution (a parabolic velocity profile across the channel, $u(y) = (1 - y^2))$ becomes linearly unstable at $Re = 5,772$ (Orszag 1971). However, since turbulence appears in experiments at much lower Reynolds numbers, it was conjectured that this bifurcation must be sub-critical. Subsequent numerical solution of the nonlinear stability equations (Herbert 1976, Ehrenstein & Koch 1991) demonstrated this to be true, showing that limit cycle solutions with amplitude ϵ branch back to lower Reynolds numbers before subsequently passing through a turning point and curving back toward higher Reynolds numbers. Thus for Reynolds numbers just above the turning point the flow equations have at least four solutions: the fixed point; two unstable limit cycles; and a chaotic solution (experimentally observed turbulence). Determining the location of the turning point in $(\alpha, \beta, \epsilon, Re)$ space is known as the minimum-critical-Reynolds-number problem, and its solution is by no means complete.

One way to investigate the turning point problem is to perform DNS of channel flow for conditions believed to be near this critical condition. Beginning with a known turbulent initial condition from higher Reynolds number, one integrates in time at the target Reynolds number to determine whether the flow decays back to the fixed point or sustains itself as turbulence. Although this may not be the most efficient way to bracket the turning point, it has the advantage that the peculiar dynamics of the flow near the turning point, whether in decay or sustained turbulence, are observable. This yields information about the path along which flows become turbulent at these low Reynolds numbers.

Unfortunately the flow dynamics are very peculiar near the turning point, and extremely long chaotic transients are observed in the computations that make a fine determination of that point all but impossible by this method. This can be seen in Fig. 6.3, where a time history of the turbulent energy in a channel flow (energy above that in the laminar flow) is plotted for a Reynolds number of 2,191. To understand the time scale of the phenomenon some experimental facts need to be recalled. In typical experimental investigations of channel flow, the infinite transverse and streamwise extent of the ideal flow are approximated by studying flow in high aspect ratio (10-40) rectangular ducts that typically are 50-100 duct

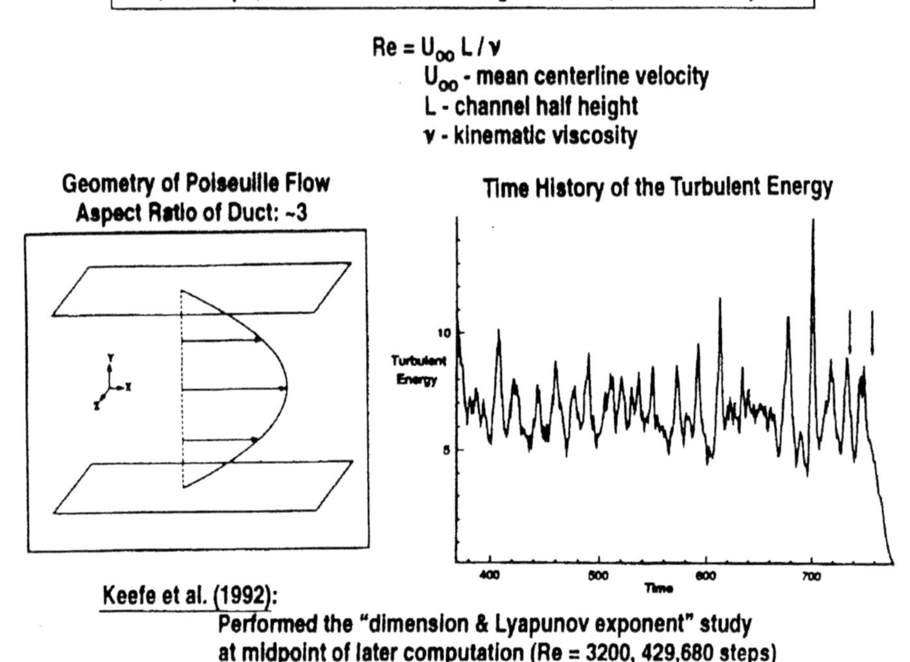

Figure 6.3. 3-D channel flow computation by Keefe, 1988.

heights long. If times are non-dimensionalized by the centerline mean velocity U_∞ and the duct half height L, then statistics on turbulence are gathered by averaging hot-wire data over intervals $\Delta t U_\infty/L \sim 200$. In the simulations and figure the time scale is based on the friction velocity u_τ and L, where typically 15-20 $u_\tau \sim U_\infty$. Thus averaging over intervals $\Delta t u_\tau/L \sim 10$ should and does yield stable flow statistics that compare well with experiments. The near-wall velocity profile, cross- channel turbulence intensities, and Reynolds and shear stress distribution for the $\Delta t u_\tau/L \sim 10$ interval near the end of the transient, delineated by the arrows in Fig. 6.3, indicate the good comparison. In each case they correspond well to available experimental data. Yet look at the time scale of the transient; it spans $\Delta t u_\tau/L \sim 300$, thirty times longer than the time needed to obtain stable statistics that would convince most experimentalists that they are viewing a fully developed turbulent channel flow. This is further complicated by the wide variation of the transient length, depending upon both the grid resolution (number of modes in the spectral representation) and the linearly stable time step of the integration. In fact, for fixed (α, β, Re) it is possible to obtain sustained turbulence for one time step, but see it rapidly decay to laminar flow for another, lower value of the step.

Extended chaotic transients near bifurcation points are not an unknown phenomenon; the "meta-chaos" of the Lorenz system is but one of many known examples. However, the practicalities of numerical computation in fluid dynamics usually interfere with one's ability to discern whether transient, or sustained turbulence, is being calculated. The computations required to obtain the transient plot in Fig. 6.3 needed 40 hours of single processor time on a Cray XMP, some ten years ago. Such a small amount of expended time was only possible because the spatial resolution of the calculation was relatively coarse (32 x 33 x 32), in keeping with the large scales of the phenomena expected at these flow conditions. Higher resolution calculations (192 x 129 x 160) (Kim et al. 1987) at greater Reynolds numbers typically have taken hundreds of hours ($\sim$ 250) to barely obtain the $\Delta t u_\tau / L = 10$ averaging interval that is so inadequate for detecting transients. Because such calculations are so time consuming, one typically chooses an integration time step that is a substantial fraction of the linear stability limit of the algorithm, so as to maximize the calculated "flow time" for expended CPU time. However, it is clear from these transient results that this practice has some dangers when close to critical points of the underlying continuous dynamical system. Thus it appears that just as pseudo-time integration to obtain steady solutions can result in spurious results, genuine time integration can result in chaotic transients indistinguishable from sustained turbulence, also yielding a spurious picture of the flow for a given Reynolds number.

6.4 Temporal & Spatial Refinement Studies of 2-D Incompressible Flow over a Backward-Facing Step

The 2-D incompressible flow over a backward-facing step has been addressed by many authors using a wide variety of numerical methods. Figure 6.4 shows the flow geometry. Fluid with constant density ρ and viscosity μ enters the upstream channel of height h with a prescribed velocity profile (usually parabolic). After traveling a distance l, the fluid passes over a backward-facing step of height s and enters the downstream channel of height $H = h + s$. After traveling a distance L downstream of the step, the fluid exits the region of interest. For Reynolds numbers considered here, the flow separates at the corner and forms a recirculating region behind the step. Additional recirculating regions form on the upper and subsequently the lower walls of the downstream channel as the Reynolds number is increased.

Backward Facing Step

(2-D Incompressible Flow Simulations)

Early 90's Controversy: Transition point Reynolds #

- **Reports of sustained unsteady flow for Reynolds # in the range of (250, 2500)**
- **Formulations**

  vortex method, unsteady Eqns. in stream function form,
  unsteady Eqns. & the associated linear-stability problem,
  unsteady Eqns. in primitive variable form
- **Numerical Methods**

  All of the existing schemes in the literature

Gresho et al. (1993): Provided an answer to the above controversy
(the steady solution at Re=800 is stable)

- **Kaiktsis et al. (1991) - transition to turbulent flow has occurred at Re=800**
- **Torczynski (1993) - the result of Kaiktsis et al. (1991) is an artifact of inadequate spatial resolution**
- **Torczynski's conclusion was confirmed by a subsequent study of Kaiktsis et al. (1996) & Fortin et al. (1996)**

Figure 6.4. Schematic of the backward facing step problem and background.

Results of sustained unsteady flow from various numerical simulations have been reported for Reynolds numbers (Re) ranging from 250 up to 2500. The formulations included the vortex method, unsteady equations in stream function form, steady equations and the associated linear-stability problem, and the unsteady equations in primitive variable form. The numerical methods used cover almost all of the existing schemes in the literature. The majority of the numerical results are summarized in Gresho et al. (1993). The work of Gresho et al. was an answer to a controversy concerning the stability of the stationary solution at $Re = 800$. It was concluded by Kaiktsis et al. (1991) that transition to turbulent flow has occurred at $Re = 800$. Kaiktsis et al. examined the long-time temporal behavior of the flow and found that the flow is steady at

$Re = 500$, time-periodic at $Re = 700$, and chaotic at $Re = 800$. Gresho et al. did a detailed grid refinement study using four different numerical methods and concluded that the backward-facing step at $Re = 800$ is a stable steady flow.

In addition to the study of Gresho et al., an extensive grid refinement study of this flow using a spectral element method was conducted in Torczynski (1993). The simulated geometry and the numerical method corresponds to that of Kaiktsis et al. (1991). Flow was examined at Reynolds numbers of 500 and 800. His systematic grid refinement study was performed by varying both the element size and the order of the polynomial representation within the elements. For both Reynolds number values with the transient computations stopped at $t = 800$, it was observed that low-resolution grid cases exhibit chaotic-like temporal behavior whereas high-resolution grid cases evolve toward asymptotically steady flow by a monotonic decay of the transient. The resolution required to obtain asymptotically steady behavior is seen to increase with Reynolds number. These results suggest that the reported transition to sustained chaotic flow (Kaiktsis et al., 1991) at Reynolds numbers around 700 is an artifact of inadequate spatial resolution. Torczynski's conclusion was further confirmed by a subsequent study of Kaiktsis et al. (1996) and Fortin et al. (1996). Fortin et al. employed tools from dynamical systems theory to search for the Hopf bifurcation point (transition point). They showed that the flow remains steady at least up to $Re = 1600$.

Grid Refinement Study of Torczynski (1993): In Torczynski (1993), the $Re = \rho\overline{u}2h/\mu$ is based on upstream conditions. The variable $\overline{u}$ is the spatial average of the horizontal velocity u over h. The geometry is specified to match that of Kaiktsis et al. (1991). The upstream channel height h and step height s have values of $h = 1$ and $s = 0.94231$, yielding a downstream channel height of $H = 1.94231$. The corner of the step is at $(x, y) = (1, 0)$. The channel extends a distance $L = 1$ upstream from the step and a distance $L = 34$ downstream from the step to preclude undue influence of the finite channel length on the flow at $Re = 800$. The following conditions are applied on the boundaries of the computational domain: $u = v = 0$ on the upper and lower channel walls, $-p + \mu\partial u/\partial n = 0$ and $\partial v/\partial n = 0$ on the outflow boundary, and $u = [\tanh(t/16)]u_B(y) + [1 - \tanh(t/16)]u_P(y)$ and $v = 0$ on the inflow boundary and the step surface. Here, $u_B(y) = max[0, 3y(1-y)]$ is the correct boundary condition for flow over a backward-facing step and $u_P(y) = 3(1-y)(s+y)/(1+s)^3$ is the Poiseuille flow observed infinitely far downstream whenever steady flow is asymptotically obtained. The initial velocity field is set equal to $u = u_P(y)$ and $v = 0$ throughout the domain. Here v is the vertical velocity and p is the pressure. Thus, the above combination of boundary and initial conditions initially allows flow through the step surface so that the simulations can be initialized using an exact divergence-free solution of the Navier-Stokes equations. Furthermore, since the inflow boundary condition is varied smoothly in time from Poiseuille flow to flow over a backward-facing step, the flow experiences an order-unity transient that is probably strong enough to excite sustained unsteady behavior, if that is the appropriate asymptotic state for the numerical solution.

The simulations were performed using the commercial code NEKTON v2.8, which employs a time-accurate spectral-element method with the Uzawa formulation (NEKTON, 1991). Let D be the dimensionality. Each element has N^D velocity nodes located at Gauss-Lobatto Legendre collocation points, some of which are on the element boundaries, and $(N-2)^D$ pressure nodes located at Gauss Legendre collocation points, all of which are internal. Within each element, the velocity components and the pressure are represented by sums of D-dimensional products of Lagrangian-interpolant polynomials based on nodal values. This representation results in continuous velocity components but discontinuous pressure at element boundaries. Henceforth, the quantity N is referred to as the element order, even though the order of the polynomials used to represent the velocity is $N-1$. NEKTON employs mixed explicit and implicit temporal discretizations. To avoid solving a nonlinear nonsymmetric system of equations at each time step, the convective term is advanced explicitly in time using a third-order Adams-Bashforth scheme. All other terms are treated implicitly (implicit Euler for the pressure and for the viscous terms).

Three spectral-element grids of differing resolution, denoted L (low), M (medium), and H (high), are employed. Figure 6.6 shows the computational domain and the grid distribution of the three spectral element grids in which the **distribution of nodes within each spectral element is not shown**. The L grid with $N=9$ is identical to the grid of Kaiktsis et al. (1991). Four general classes of behavior are observed for the numerical solutions. First, "steady monotonic" denotes evolution of the numerical solution toward an asymptotically steady state. Second, "steady oscillatory" denotes evolution toward an asymptotically steady state with a decaying oscillation superimposed on the monotonic decay. Third, "unsteady chaotic" denotes irregular transient behavior of the numerical solution that shows no indication of evolving toward steady behavior. Fourth, "diverge" denotes a numerical solution terminated by a floating-point exception. In Fig. 6.6, the first character denotes the grid resolution L, M or H, the first digit indicates the Reynolds number 500 or 800 and the last two digits indicate the order of the spectral element being used. For example, $L807$ means $Re = 800$ using the L grid with $N = 7$.

The extensive grid refinement study of Torczynski resulted in grid-independent steady-state numerical solutions for both $Re = 500$ and $Re = 800$. As the

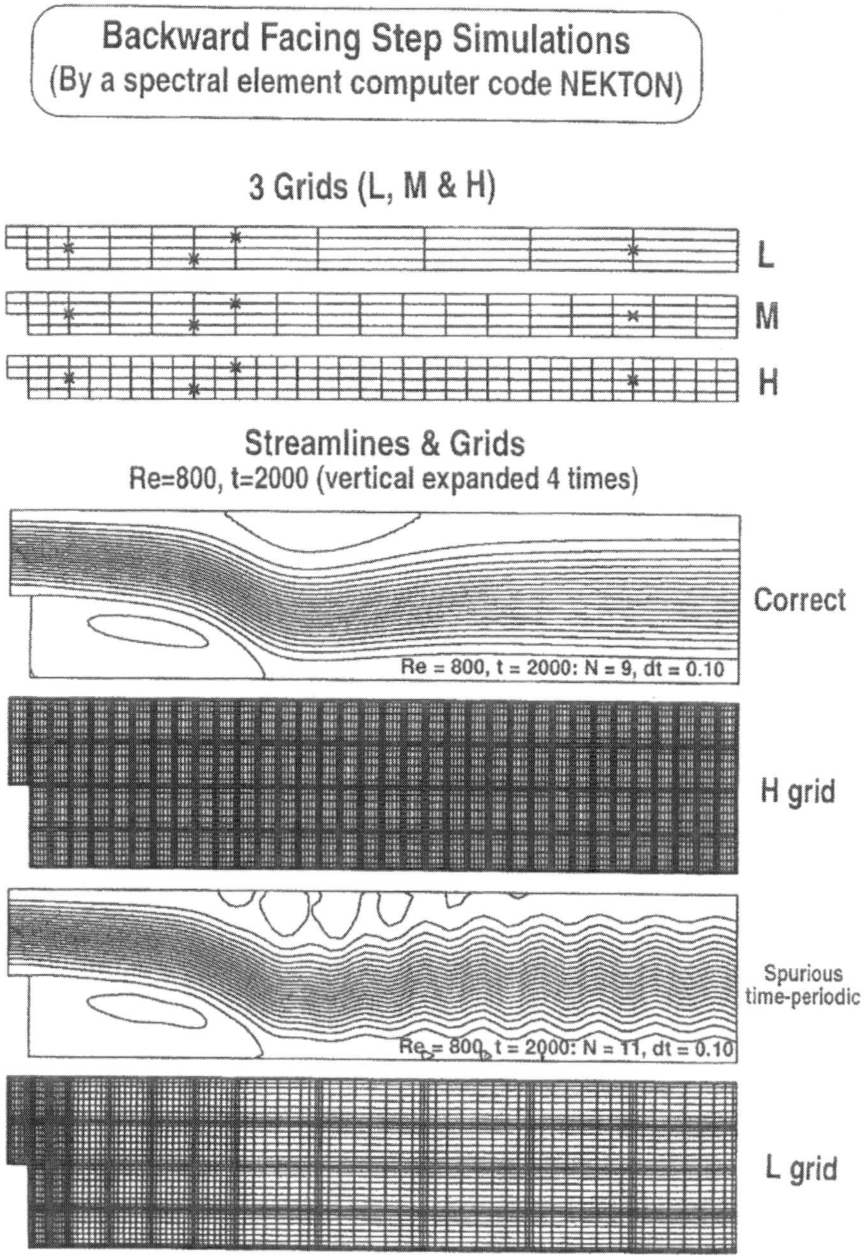

Figure 6.5. Three different grids, and streamlines of $H809$ and $L811$ and their corresponding grids by the spectral element method.

grid resolution is reduced below the level required to obtain grid independent solutions, chaotic-like temporal behavior occurred. The degree of grid resolution required to obtain a grid-independent solution was observed to increase as the Reynolds number is increased. Figure 6.5 shows the streamlines for for $H809$ (steady solution) and $L811$ (spurious time-periodic solution) and the corresponding grids with the distribution of the nodes of the spectral elements shown.

Temporal Refinement Studies Using Knowledge from Dynamical Systems Theory: All of Torczynski's numerical solutions integrate to $t = 800$. With the knowledge of possible nonlinear behavior of numerical schemes such as long time transients before a steady state is reached, numerically induced chaotic transients, numerically induced or suppressed chaos, existence of spurious steady states and asymptotes, and the intimate relationship among initial data, time step and grid spacing observed in discrete dynamical systems theory, Yee et al. (1997) examined the Torczynski cases in more detail.

In the Yee et al. (1997) study, in addition to grid refinement, temporal refinements are made on all of the under-resolved grid cases to determine if these cases sustain the same temporal behavior at a much later time or evolve into a different type of spurious behavior. At $t = 800$, cases $L506, L507, L508, L509, L811$, $M807$ and $M808$ either exhibit "unsteady chaotic" or "steady oscillatory" behavior. We integrate these cases to $t = 2000$ to determine if a change in solution behavior occurs. From the phenomena observed in Keefe's 3-D channel flow computation and others, $t = 2000$ might not be long enough for a long time transient or long chaotic transient to die out. There is also the potential of evolving into a different type of spurious or divergent behavior at a much later time. However, for this study it appears that $t = 2000$ is sufficient. For $Re = 500$, we also recomputed some of these cases with a sequence of Δt that bracketed the benchmark study of Torczynski. The Δt values are $0.02, 0.05, 0.10, 0.125, 0.2, 0.3, 0.4$, and 0.5 for $Re = 500$. The CFL number for all of these cases is above 1 for $\Delta t > 0.10$. The reason for the investigation of $\Delta t = 0.3, 0.4$ and 0.5 is to find out, after the transients have died out, if the solution converges to the correct steady state for Δt that are a few times larger than 0.10.

For $Re = 800$, we integrate $L811$ and $M808$ with $\Delta t = 0.10$ and $M807$ with $\Delta t = 0.02, 0.05$ and 0.10 to $t = 2000$. Aside from integrating to $t = 2000$, five different initial data were examined for cases $M807$, $M809$ and $M811$ for $\Delta t = 0.10$ to determine the influence of the initial data and the grid resolution on the final numerical solution. The five initial data are:

(a) Uniform: $u, v = 0$
(b) Shear layer: $u = u_B(y) = max[0, 3y(1-y)], v = 0$

(c) Solution from solving the steady Stokes equation (with no convection terms)

(d) Torczynski (1993): $u = u_P(y) = 3(1-y)(s+y)/(1+s)^3, v = 0$

(e) Channel flow both upstream & downstream of step: Same as (d) except the boundary conditions

The boundary conditions for (a), (b), (c) and (e) were parabolic inflow and no-slip at walls, whereas the boundary conditions for (d) were those of Torczynski (1993): $u = [\tanh(t/16)]u_B(y) + [1 - \tanh(t/16)]u_P(y)$ and $v = 0$. The CPU time required to run the above cases ranged from less than a day to several days on a Sparc Center 2000 using one processor.

The chaotic-like behavior evolves into a time-periodic solution beyond $t = 800$ for $L506$ and $L507$, whereas the chaotic-like behavior evolves into a time-periodic solution beyond $t = 800$ for $L811$ and a divergent solution for $M807$. The "steady oscillatory" case $L508$ slowly evolves to the correct steady state with an amplitude of oscillation of 10^{-5}. The oscillation is not detectable within the plotting accuracy. The "steady oscillatory" time evolution of $M808$ is similar to that of $L508$. The numerical solutions with "steady oscillatory" and "steady monotonic" behavior at early stages of the time integration are almost identical at later stages of the time integration. They all converge to the correct steady state. The initial data study at $Re = 800$ with $\Delta t = 0.10$ is summarized in Table 5.5 of Yee et al. (1997). It illustrates the intimate relationship between initial data and grid resolution.

Figure 6.6 shows the vertical velocity time histories at $(x, y) = (30, 0)$ advanced to a time of $t = 2000$ for $M807$ with $\Delta t = 0.02, 0.05$ and 0.10, and $L811$ for $\Delta t = 0.10$. Case $M807$ diverges at $t = 1909.2$ for $\Delta t = 0.02$, at $t = 972.4$ for $\Delta t = 0.05$, and at $t = 827.77$ for $\Delta t = 0.10$. The time histories for these three time steps appear to show chaotic-like behavior if one stops the computations at $t = 800$. The bottom plot of Fig. 6.6 shows the vertical velocity time histories advanced to a time of $t = 2000$ for $L811$ with $\Delta t = 0.10$. It shows the definite time-periodic spurious solution pattern. On the other hand, the time history for this case appears to show an aperiodic-like pattern if one stops the computation at $t = 800$. Note that the $L809$ grid case was used by Kaiktsis et al. (1991) and they concluded that "2-D transition" has already occurred at $Re = 800$.

In summary, without the temporal refinement study (longer time integration), the $L506$, $L507$, $L811$ and $M807$ cases can be mistaken to be chaotic-like (or aperiodic-like) flow. Although the time history up to $t = 800$ appears chaotic-like, one cannot conclude it is chaotic without longer transient computations. One can conclude that with transient computations that are 2.5 times longer than Torczynski's original computations, what appeared to be aperiodic-like or

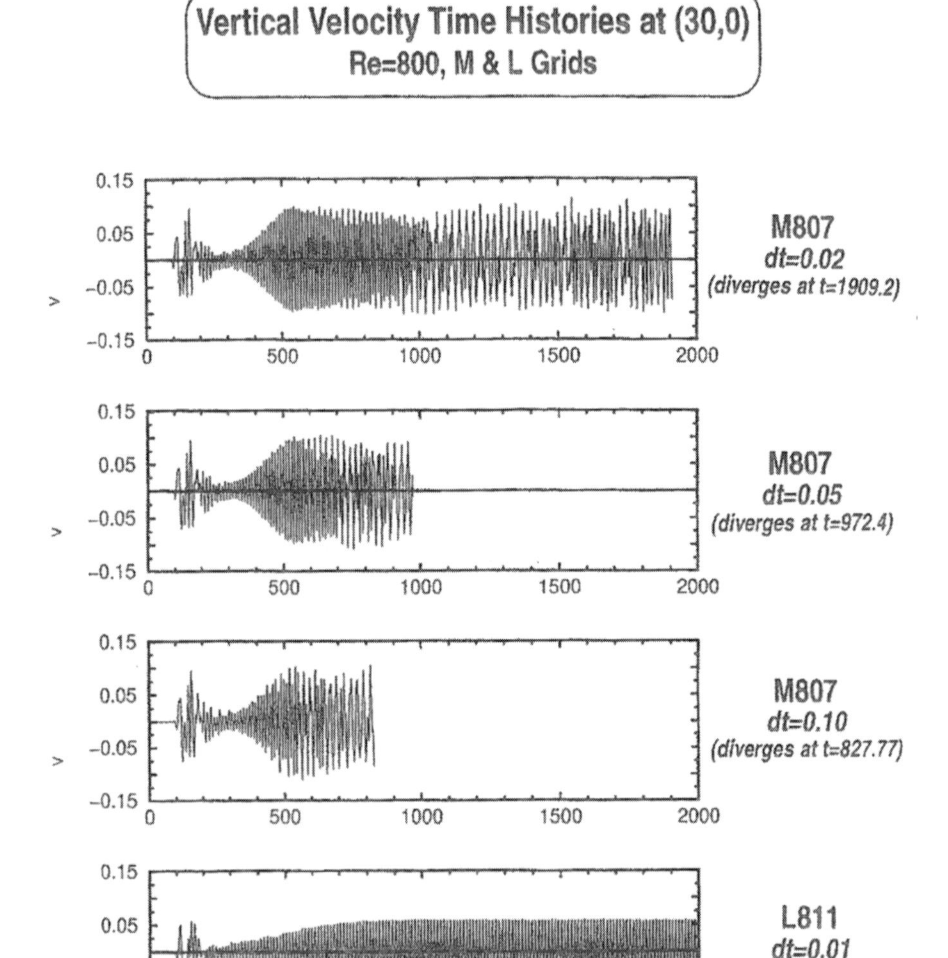

Figure 6.6. Vertical velocity time histories for the $M807$ with time steps $0.02, 0.05, 0.10$, and $L811$ with time step 0.01 for $t = 2000$.

chaotic-like behavior at earlier times evolved toward either a time-periodic or divergent solution at later times. These temporal behaviors appear to be long time aperiodic-like transients or numerically induced chaotic-like transients. For $Re = 800$, five different initial data were examined to determine if the flow exhibits strong dependence on initial data and grid resolution. Results showed that the numerical solutions are sensitive to these five initial data. Note that the results presented pertain to the characteristic of the studied scheme and the direct simulations. However, if one is certain that $Re = 800$ is a stable steady flow, a non-time-accurate method such as time-marching to obtaining the steady-state numerical solution would be a more efficient numerical procedure.

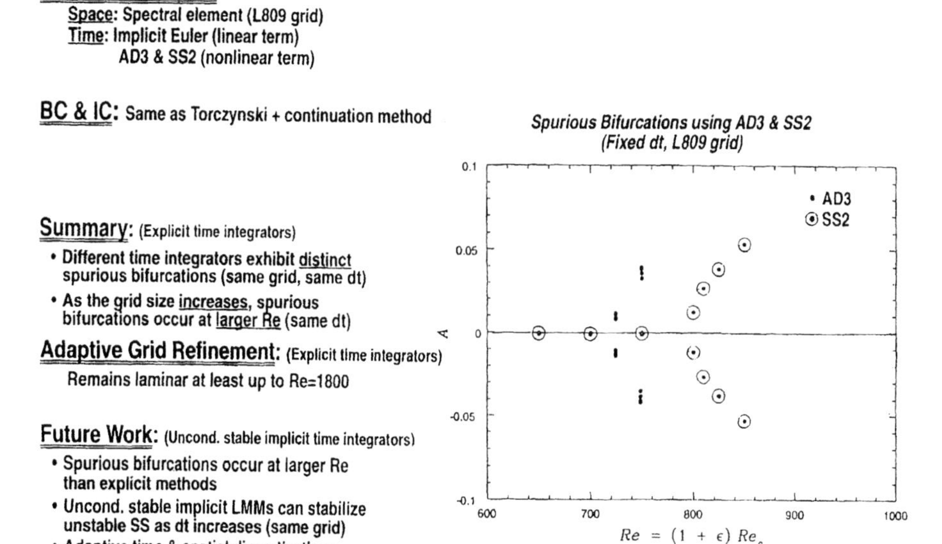

Figure 6.7. Different explicit time integrators (for the nonlinear terms of the Navier-Stokes equations) exhibit distinct spurious bifurcation using the same spatial discretization $L809$ and a time step of 0.10.

Spurious Bifurcation by Different Time Integrators (Henderson & Yee 1998, unpublished): This is a joint work with Ronald Henderson in 1998. The unpublished work was presented at the 10th International Conference in Finite Element Methods, January 5-8, 1998, Tucson, Arizona, and also has been presented at various invited lectures during the last four years. Our joint work illustrates the situation where solving the nonlinear terms of the Navier-Stokes equations by two different explicit time integrators (same implicit time integrator for the linear terms) results in spurious bifurcation. This spurious bifurcation is shown in Fig. 6.7 as a function of the Reynolds number. These computations use the implicit Euler time integrator for the linear terms. Also the same spatial discretization $L809$ is used with a fixed time step of $t = 0.10$. The two explicit time integrators are the third-order Adams-Bashforth (AB3) and a second-order explicit stiffly stable method (SS2) (Henderson 1999). The AB3 method experiences a spurious bifurcation near $Re = 720$, whereas the SS2 method experiences a spurious bifurcation at a larger Reynolds number near $Re = 800$. The method and the scaling for this figure can be found in Henderson (1999). Finding the exact location of these spurious bifurcation points requires more complicated computation which is not performed here. In addition, the exact representation of this bifurcation plot is rather complicated to explain, and is not important for the current discussion. They are not the main illustration for this study. When an adaptive version of the spectral element method (Henderson, 1999) is used, the problem remains laminar up to $Re = 1800$. Future work which is indicated on Fig. 6.7 is planned.

Minimization of Spurious Bifurcation by a Suitable Filter (Fischer 2001, unpublished): Recently, Fischer (2001) computed the same $L811$ spectral element grid using a time integrator based on the operator integration-factor splitting (OIFS) developed by Maday, Patera and Rønquist (1990). This scheme decouples the convective step from the Stokes update, thereby allowing CFL numbers in excess of unity. At the end of each step, Fischer applies a filter to the velocity that effectively scales the Nth-order Legendre modes within each element by $(1-\alpha)$, where, typically, $0.05 \leq \alpha \leq 0.30$ (Fischer & Mullen 2001). Because the filter is applied on each step, its strength is a function of Δt as well as α. The spurious behavior observed by Kaiktsis et al. (1991) is cured by the filter and a stable steady-state numerical solution is obtained without further grid refinement. Figure 6.8 illustrates the velocity time histories at $(30, 0)$ by the filtered and un-filtered spectral element methods with $\Delta t = 0.10$.

In summary, Sections 6.3 and 6.4 illustrate all of the possible scenarios of spurious bifurcations indicated on the schematic diagram of Fig. 6.2. The last scenario, discussed briefly at the beginning of Section 6.4, is quite common and is not shown here. See Gresho et al. (1993) and references cited therein for some examples.

VI. Concluding Remarks

Some building blocks to ensure a higher level of confidence in PAR of numerical simulations have been discussed. The discussion concentrates only on how well numerical schemes can mimic the solution behavior of the underlying PDEs. The possible discrepancy between the chosen model and the real physics and/or experimental data is set aside. These building blocks are based largely on the author's view, background and integrated experience in computational physics, numerical analysis and the dynamics of numerics. They also represent the end result of the various studies with the author's collaborators indicated

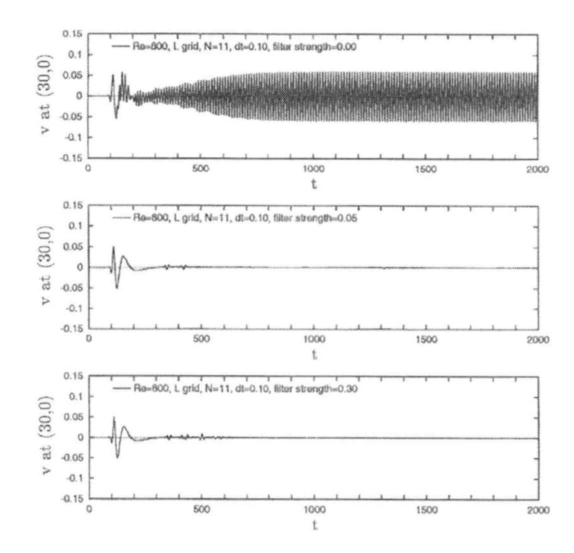

Figure 6.8. Comparison of the filtered with the un-filtered solutions of $L811$ with time step $dt = 0.10$. Top: no filter; Middle: filter (filter strength 0.05); Bottom: filter (filter strength 0.3).

in the acknowledgment Section. Among many other important building blocks for the PAR of numerical simulations, the author believes the following five building blocks are essential. The first building block is to analyze as much as possible the dynamical behavior of the governing equation. For stability and well-posedness considerations, whenever it is possible, it is also necessary to condition (not pre-condition) the governing PDEs before the application of the appropriate scheme (Yee & Sjögreen 2001a,b). The second building block is to understand the nonlinear behavior, limits and barriers, and to isolate spurious behavior of existing numerical schemes. A third building block is to include nonlinear dynamics and bifurcation theories as an integral part of the numerical process whenever it is possible. A fourth building block is to construct appropriate adaptive spatial and temporal discretizations that are suitable for the underlying governing equation. The last building block is to construct appropriate adaptive numerical dissipation/filter controls for long time integrations, complex high speed turbulent and combustion simulations (Sjögreen & Yee 2001, Yee & Sjögreen 2001).

The need for the study of dynamics of numerics is prompted by the fact that the type of problem studied using CFD has changed dramatically over the past two decades. CFD is also undergoing an important transition, and it is increasingly used in nontraditional areas. But even within its field, many algorithms widely used in practical CFD applications were originally designed for much simpler problems, such as perfect or ideal gas flows. As can be seen in the literature, a straightforward application of these numerical methods to high speed flows, nonequilibrium flows, advanced turbulence modeling or combustion related problems can lead to wrong results, slow convergence, or even nonconvergent solutions. The need for new algorithms and/or modification and improvement to existing numerical methods in order to deal with emerging disciplines is evident. We believe the nonlinear dynamical approach for CFD can contribute to the success of this goal.

We have revealed some of the causes of spurious phenomena due to the numerics in an attempt to improve the understanding of the effects of numerical uncertainties in CFD. We have shown that guidelines developed using linearization methods are not always valid for nonlinear problems. We have gained an improved understanding of long time behavior of nonlinear problems and nonlinear stability, convergence, and reliability of time-marching approaches. We have learned that numerics can introduce and suppress chaos and can also introduce chaotic transients. The danger of relying on DNS to bracket closely the the onset of turbulence and chaos is evident.

We illustrated with practical CFD examples that exhibit properties and qualitative behavior similar to those of elementary examples in which the full dynamical behavior of the numerics can be analyzed. The observed spurious behavior related to under-resolved grid cases is particularly relevant to DNS and large Eddy simulation (LES). Spatial resolutions in DNS and LES are largely dictated by the computer power. These studies serve to point out the various possible dangers of misinterpreting numerical simulations of realistic complex flows that are constrained by available computing power.

As can be seen, recent advances in dynamics of numerics show the advantage of adaptive step size error control for long time integration of nonlinear ODEs. Although much research is needed to construct suitable yet practical similar adaptive methods for PDEs, these early developments lead our way to future theories. There remains the challenge of constructing practical spatial and temporal adaptive methods for time-accurate computations, and constructing adaptive step size control methods that are suitable yet practical for time marching to the steady state for aerospace CFD applications. Another even more challenging area is the quest for an adaptive numerical scheme that leads to guaranteed and rapid convergence to the correct steady-state numerical solutions.

Acknowledgments

The author wishes to thank her collaborators Peter Sweby, Laurence Keefe, John Torczynski, Ronald Henderson, Maxim Poliashenko, David Griffiths, Andre Lafon and Andrew Stuart, for contributing to her earlier work. The contributions of Section 6.2 by Laurence Keefe, Section 6.3 by Maxim Poliashenko, and Section 6.4 by Paul Fischer, John Torczynski and Ronald Henderson are gratefully acknowledged. Special thanks to Dennis Jespersen, Unmeel Mehta, Murray Tobak and Marcel Vinokur for their suggestions and critical review of the manuscript.

References

[1] Brown, D.L. and Minion, M.L. (1995), "Performance of Under-resolved Two-Dimensional Incompressible Flow Simulations," **J. Comput. Phys.**, Vol. 122, pp. 165-183.

[2] Butcher, J.C. (1987), *Numerical Analysis of Ordinary Differential Equations*, John Wiley & Son, Chichester.

[3] Davidson, B. (1997), Large Scale Continuation and Numerical Bifurcation for PDE's, **SIAM J. of Numer. Analy.**, Vol. 34, No. 5, pp. 2008-2027.

[4] Doedel, E. (2000), AUTO: Software for Continuation and Bifurcation Problems in Ordinary Differential Equations, Concordia University, Montreal, Canada and Cal. Tech., Pasadena, Calif.

[5] Ehrenstein, U. and Koch, W. (1991), "Nonlinear bifurcation study of plane Poiseuille flow," **J. Fluid Mech.**, Vol. 228, pp. 111-148.

[6] Fischer, P.F. and Mullen, J.S., (2001) "Filter-Based Stabilization of Spectral Element Methods", Comptes Rendus de l'Académie des sciences Paris, t. 332, Série I - Analyse numérique, 265-270 (2001).

[7] Fischer, P.F., (2001) Private communication.

[8] Fortin, A., Jardak, M., Gervais, J.J. and Pierre, R. (1996), "Localization of Hopf Bifurcations in Fluid Flow Problems," **Intern. J. Numer. Meth. Fluids**.

[9] Gresho, P.M., Gartling, D.K., Torczynski, J.R., Cliffe, K.A., Winters, K.H., Garrett, T.J., Spencer, A. and Goodrich, J.W. (1993), "Is the Steady Viscous Incompressible Two-Dimensional Flow Over a Backward-Facing Step at Re=800 Stable?" **Intern. J. Numer. Meth. Fluids**, Vol. 17, pp. 501-541.

[10] Grebogi, C., Ott, E. and Yorke, J.A. (1983), "Crises, Sudden Changes in Chaotic Attractors, and Transient Chaos," Physica 7D, pp. 181-200.

[11] Griffiths, D.F., Sweby, P.K. and Yee, H.C. (1992a), "On Spurious Asymptotes Numerical Solutions of Explicit Runge-Kutta Schemes," **IMA J. Numer. Anal.**, Vol. 12, pp. 319-338.

[12] Griffiths, D.F., Stuart, A.M. and Yee, H.C. (1992b), "Numerical Wave Propagation in Hyperbolic Problems with Nonlinear Source Terms," **SIAM J. of Numer. Analy.**, Vol. 29, pp. 1244-1260.

[13] Henderson, R. D., (1999) "Adaptive Spectral Element Methods for Turbulence and Transition, in High-Order Methods for Computational Physics", T. J. Barth and H. Deconinck, editors. Springer.

[14] Herbert, Th. (1976), *Lecture Notes in Physics 59, Springer-Verlag*, p235.

[15] Hoppensteadt, F.C. (1993), *Analysis and Simulation of Chaotic Systems*, Springer-Verlag, New York.

[16] Kaiktsis, L., Karniadakis, G.E. and Orszag, S.A. (1991), "Onset of Three-Dimensionality, Equilibria, and Early Transition in Flow Over a Backward-Facing Step," **J. Fluid Mech.**, Vol. 231, pp. 501-528.

[17] Kaiktsis, L., Karniadakis, G.E. and Orszag, S.A. (1996), "Unsteadiness and Convective Instabilities in Two-Dimensional Flow Over a Backward-Facing Step," **J. Fluid Mech.**, Vol. 321, pp. 157-187.

[18] Keefe, L., Moin, P. and Kim, J. (1992), "The Dimension of Attractors Underlying Periodic Turbulent Poiseuille Flow," J. Fluid Mech., Vol. 242, pp. 1-29.

[19] Keefe, L. (1988-1996), unpublished; private communication.

[20] Kim, J., Moin, P. and Moser, R. (1987), "Turbulence statistics in fully developed channel flow at low Reynolds number," **J. Fluid Mech.**, Vol. 177, pp. 133-166.

[21] Keller, H.B. (1977), "Numerical Solution of Bifurcation and Nonlinear Eigenvalue Problems," *Applications of Bifurcation Theory*, P.H. Rabinowitz, ed., Academic Press, pp. 359-384.

[22] Lafon, A. and Yee, H.C. (1991), "Dynamical Approach Study of Spurious Steady-State Numerical Solutions for Nonlinear Differential Equations, Part III: The Effects of Nonlinear Source Terms in Reaction-Convection Equations," NASA Technical Memorandum 103877, July 1991; **International J. Comput. Fluid Dyn.**, Vol. 6, pp. 1-36, 1996.

[23] Lafon, A. and Yee, H.C. (1992), "Dynamical Approach Study of Spurious Steady-State Numerical Solutions of Nonlinear Differential Equations, Part IV: Stability vs. Numerical Treatment of Nonlinear Source Terms," ONERA-CERT Technical Report DERAT 45/5005.38, also, **International J. Comput. Fluid Dyn.**, Vol. 6, pp. 89-123, 1996.

[24] LeVeque, R.J. and Yee, H.C. (1990), "A Study of Numerical Methods for Hyperbolic Conservation Laws with Stiff Source Terms," **J. Comput. Phys.**, Vol. 86, pp. 187-210.

[25] Maday, Y. and Patera, A.T., (1989) "Spectral Element Methods for the Navier-Stokes Equations", State of the Art Survey in Computational Mechanics, A.K. Noor, ed., ASME, New York, pp. 71-143.

[26] Maday, Y., Patera, T. and Rønquist, E.M., (1990) "An operator-integration-factor splitting method for time-dependent problems: Application to incompressible fluid flow. newblock *J.Sci.Comput.*, Vol. 5, pp. 263–292.

[27] Minion, M.L. and Brown, D.L., (1997) "Performance of Under-resolved Two-Dimensional Incompressible Flow Simulations II", **J. Comput. Phys. 138**, 734-765.

[28] Moin, P., and Kim, J. (1982), "Numerical investigation of turbulent channel flow," J. Fluid Mech., Vol. 118, pp. 341-378.

[29] Moore, D.R., Weiss, N.O. and Wilkins, J.M. (1990), "The Reliability of Numerical Experiments: Transitions to Chaos in Thermosolutal Convection," **Nonlinearity**, Vol. 3, pp. 997-1014.

[30] Moretti, G. and Abbett, M., (1966), "A Time-Dependent Computational Method for Blunt Body Flows," **AIAA Journal**, Vol. 4, pp. 2136-2141.

[31] NEKTON User's Guide, Version 2.8, 1991, Nektonics Inc., Cambridge, MA.

[32] Orszag, S. (1971), "Accurate solution of the Orr-Sommerfeld stability equation," **J. Fluid Mech.**, Vol. 50, pp. 689-703.

[33] Poliashenko, M. and Aidun, C.K. (1995), "Computational Dynamics of Ordinary Differential Equations," **Intern. J. Bifurcation and Chaos**, Vol. 5, pp. 159-174.

[34] Schreiber, R. and Keller, H.B. (1983), "Spurious Solution in Driven Cavity Calculations," **J. Comput. Phys.**, Vol. 49, pp. 310-333.

[35] Shroff, G.M. and Keller, H.B. (1993), Stabilisation of unstable procedures: The RPM, **SIAM J. of Numer. Analy.**, Vol. 30, No. 4, pp. 1099-1120.

[36] Shubin, G.R., Stephens, A.B. and Glaz, H.M. (1981), "Steady Shock Tracking and Newton's Method Applied to One-Dimensional Duct Flow," **J. Comput. Phys.**, Vol. 39, pp. 364-374.

[37] Sjögreen, B. and Yee, H.C., (2000) "Multiresolution Wavelet Based Adaptive Numerical Dissipation Control for Shock-Turbulence Computations", RIACS Report 01.01, NASA Ames research center (Oct 2000).

[38] Sjögreen, B. and Yee, H.C., (2001) "On Entropy Splitting, Linear and Nonlinear Numerical Dissipations and Long-Time Integrations, Proceedings of the 5th Internat. Conf. on Spectral and High Order Methods, Uppsala, Sweden, June 11-15, 2001.

[39] Spalart, P.R., Moser, R.D. and Rogers, M.M. (1991), "Spectral methods for the Navier-Stokes equations with one infinite and two periodic directions," **J. Comput. Phys.**, Vol. 96, p297.

[40] Stephens, A.B. and Shubin, G.R. (1981) "Multiple Solutions and Bifurcation of Finite Difference Approximations to Some Steady Problems of Fluid Dynamics," **SIAM J. Sci. Statist Comput.**, Vol. 2, pp. 404-415.

[41] Sweby, P.K. and Yee, H.C. (1994), "On the Dynamics of Some Grid Adaptation Schemes," Proceedings of the 4th International Conference on Numerical Grid Generation in CFD and Related Fields, University College of Swansea, UK, also RIACS Technical Report 94.02, Feb. 1994.

[42] Sweby, P.K., Lafon, A. and Yee, H.C. (1995), "On the Dynamics of Computing a Chemically Relaxed Nonequilibrium Flow," presented at the ICFD Conference on Numerical Methods for Fluid Dynamics, April 3-6, 1995, Oxford, UK.

[43] Thompson, J.M.T. and Stewart, H.B. (1986), *Nonlinear Dynamics and Chaos*, John Wiley, New York.

[44] Torczynski, J.R. (1993), "A Grid Refinement Study of Two-Dimensional Transient Flow Over a Backward-Facing Step Using a Spectral-Element Method," FED-Vol. 149, Separated Flows, ASME 1993, J.C. Dutton and L.P. Purtell, editors.

[45] Yee, H.C., Sweby, P.K. and Griffiths, D.F. (1991), "Dynamical Approach Study of Spurious Steady-State Numerical Solutions for Nonlinear Differential Equations, Part I: The Dynamics of Time Discretizations and Its Implications for Algorithm Development in Computational Fluid Dynamics," NASA TM-102820, April 1990; **J. Comput. Phys.**, Vol. 97, pp. 249-310.

[46] Yee, H.C. and Sweby, P.K. (1992), "Dynamical Approach Study of Spurious Steady-State Numerical Solutions for Nonlinear Differential Equations, Part II: Global Asymptotic Behavior of Time Discretizations," RNR-92-008, March 1992, NASA Ames Research Center; also **International J. Comput. Fluid Dyn.**, Vol. 4, pp. 219-283, 1995.

[47] Yee, H.C. and Sweby, P.K. (1993), "Global Asymptotic Behavior of Iterative Implicit Schemes," RIACS Technical Report 93.11, December 1993, NASA Ames Research Center, also **Intern. J. Bifurcation & Chaos**, Vol. 4, pp. 1579-1611.

[48] Yee, H.C. and Sweby, P.K. (1995), "On Super-Stable Implicit Methods and Time-Marching Approaches," RIACS Technical Report 95.12, NASA Ames Research Center, July 1995; also, Proceedings of the Conference on Numerical Methods for Euler and Navier-Stokes Equations, Sept. 14-16, 1995, University of Montreal, Canada; **International J. Comput. Fluid Dyn.** Vol. 8, pp. 265-286, 1997.

[49] Yee, H.C. and Sweby, P.K. (1996), "Some Aspects of Numerical Uncertainties in Time Marching to Steady-State Computations," AIAA-96-2052, 27th AIAA Fluid Dynamics Conference, June 18-20, 1996, New Orleans, LA., AIAA J., Vol. 36, No. 5, pp. 712-724, 1998

[50] Yee, H.C., Torczynski, J.R., Morton, S.A., Visbal, M.R. and Sweby, P.K. (1997), "On Spurious Behavior of CFD Simulations," AIAA 97-1869, Proceedings of the 13th AIAA Computational Fluid Dynamics Conference, June 29 - July 2, 1997, Snowmass, CO.; also **International J. Num. Meth. Fluids**, Vol. 30, pp. 675-711, 1999.

[51] Yee, H.C. and Sweby, P.K. (1997), "Dynamics of Numerics & Spurious Behaviors in CFD Computations," **Keynote paper, 7th ISCFD Conference, Sept. 15-19, 1997, Beijing, China**, RIACS Technical Report 97.06, June 1997.

[52] Yee, H.C., Sandham, N.D. and Djomehri, M.J., (1999) ' 'Low Dissipative High Order Shock-Capturing Methods Using Characteristic-Based Filters", **J. Comput. Phys., 150** 199–238.

[53] Yee, H.C., Vinokur, M. and Djomehri, M.J., (1999) "Entropy Splitting and Numerical Dissipation," NASA Technical Memorandum 208793, August, 1999, NASA Ames Research Center; **J. Comput. Phys., 162**, 33 (2000).

[54] Yee, H.C. and Sjögreen, B., (2001a) "Adaptive Numerical-Dissipation/Filter Controls for High Order Numerical Methods," Proceedings of the 3rd International Conference on DNS/LES, Arlington, Texas, August 4-9, 2001.

[55] Yee, H.C. and Sjögreen, B., (2001b) "Designing Adaptive Low Dissipative High Order Schemes for Long-Time Integrations," *Turbulent Flow Computation, (Eds. D. Drikakis & B. Geurts), Kluwer Academic Publisher; also RIACS Technical Report, Dec. 2001.*

Chapter 7

ALPHA-MODELING STRATEGY FOR LES OF TURBULENT MIXING

Bernard J. Geurts

Faculty of Mathematical Sciences, University of Twente, P.O. Box 217, 7500 AE Enschede, The Netherlands ‡

b.j.geurts@math.utwente.nl

Darryl D. Holm

Theoretical Division and Center for Nonlinear Studies, Los Alamos National Laboratory, MS B284 Los Alamos, NM 87545, USA

dholm@lanl.gov

Abstract The α-modeling strategy is followed to derive a new subgrid parameterization of the turbulent stress tensor in large-eddy simulation (LES). The LES-α modeling yields an explicitly filtered subgrid parameterization which contains the filtered nonlinear gradient model as well as a model which represents Leray-regularization. The LES-α model is compared with similarity and eddy-viscosity models that also use the dynamic procedure. Numerical simulations of a turbulent mixing layer are performed using both a second order, and a fourth order accurate finite volume discretization. The Leray model emerges as the most accurate, robust and computationally efficient among the three LES-α subgrid parameterizations for the turbulent mixing layer. The evolution of the resolved kinetic energy is analyzed and the various subgrid-model contributions to it are identified. By comparing LES-α at different subgrid resolutions, an impression of finite volume discretization error dynamics is obtained.

Keywords: large-eddy simulation, dispersion, dissipation, numerical error dynamics

1. Introduction

Accurate modeling and simulation of turbulent flow is a topic of intense ongoing research. The approaches to this problem area can be distinguished,

‡Also: Department of Engineering, Queen Mary, University of London, Mile End, London E1 4NS, UK

D. Drikakis and B.J. Geurts (eds.), Turbulent Flow Computation, 237–278.

e.g., by the amount of detail that is intended to be included in the physical and numerical description. Simulation strategies that aim to calculate the full, unsteady solution of the governing Navier-Stokes equations are known as direct numerical simulations (DNS). The DNS approach does not involve any modeling or approximation except its numerical nature and in principle it can provide solutions that possess all dynamically relevant flow features [1, 2]. In turbulent flow, these features range from large, geometry dependent scales to very much smaller dissipative length-scales. While accurate in principle, the DNS approach is severely restricted by limitations in spatial and temporal resolution, even with modern computational capabilities, because of the tendency of fluid flow to cascade its energy to smaller and smaller scales.

This situation summons alternative, restricted simulation approaches to the turbulent flow problem that are aimed at capturing the primary features of the flow above a certain length-scale only. A prominent example of this is the large-eddy simulation (LES) strategy [3]. Rather than aiming for a precise and complete numerical treatment of all features that play a role in the evolution of the flow, an element of turbulence modeling is involved in LES [4]. In the filtering approach to LES, this modeling element is introduced by applying a spatially localized filter operation to the Navier-Stokes equations [5]. This introduces a smoothing of the flow features and a corresponding reduction in the flow complexity [6]. One commonly adopts spatial convolution filters which effectively remove the small-scale flow features that fall below an externally introduced length-scale Δ, referred to as the filter-width. This smoothing can significantly reduce the requirements on the resolution and, thus, allow LES to be performed for much more realistic situations than DNS, e.g., at higher Reynolds number, within the same computational capabilities [7]. This constitutes the main virtue of LES.

The LES approach is conceptually different from the Reynolds Averaged Navier-Stokes (or, RANS) approach, which is based on statistical arguments and exact ensemble averages that raise the classic turbulence closure problem. When the spatially localized smoothing operation in LES is applied to the nonlinear convective terms in the Navier-Stokes equations, this also gives rise to a closure problem that needs to be resolved. Thus, the LES approach must face its own turbulence closure problem: How to model the effects of the filtered-out scales in terms of the remaining resolved fields?

In the absence of a comprehensive theory of turbulence, empirical knowledge about subgrid-scale modeling is essential but still incomplete. Since in LES only the dynamical effects of the smaller scales need to be represented, the modeling is supposed to be simpler and more straightforward, compared to the setting encountered in statistical modeling such as in RANS. To guide the construction of suitable models we advocate the use of constraints based on rigorous properties of the LES modeling problem such as realizability conditions

[8] and algebraic identities [5, 6]. A thoughtful overview of these constraints is given in [9].

In this paper, we follow the α-modeling approach to the LES closure problem. The α-modeling approach is based on the Lagrangian-averaged Navier-Stokes$-\alpha$ equations (LANS$-\alpha$, or NS$-\alpha$) described below. The LANS$-\alpha$ approach eliminates some of the heuristic elements that would otherwise be involved in the modeling. The original LANS$-\alpha$ theory also involves an elliptic operator inversion in defining its stress tensor. When we apply filtering in defining the LANS$-\alpha$ stress tensor, instead of the operator inversion in the original theory, we call it LES-α .

Background and references for LANS$-\alpha$, or NS$-\alpha$ equations. The inviscid LANS$-\alpha$ equations (called Euler$-\alpha$, in the absence of viscosity) were introduced through a variational formulation in [10], [11] as a generalization to $3D$ of the integrable inviscid $1D$ Camassa–Holm equation discovered in [12]. A connection between turbulence and the solutions of the viscous $3D$ Camassa–Holm, or Navier–Stokes–alpha (NS$-\alpha$) equations was identified, when viscosity was introduced in [13]–[15]. Specifically, the steady analytical solution of the NS$-\alpha$ equations was found to compare successfully with experimental and numerical data for mean velocity and Reynolds stresses for turbulent flows in pipes and channels over a wide range of Reynolds numbers. These comparisons suggested the NS$-\alpha$ equations could be used as a closure model for the mean effects of subgrid excitations. Numerical tests further substantiating this intuition were performed and reported in [16].

An alternative more "physical" derivation for the inviscid NS$-\alpha$ equations (Euler$-\alpha$), was introduced in [17] (see also [14]). This alternative derivation was based on substituting in Hamilton's principle the decomposition of the Lagrangian fluid-parcel trajectory into its mean and fluctuating components at linear order in the fluctuation amplitude. This was followed by making the Taylor hypothesis for frozen-in turbulence and averaging at constant Lagrangian coordinate, before taking variations. Hence, the descriptive name Lagrangian-averaged Navier-Stokes$-\alpha$ equations (LANS$-\alpha$) was given for the viscous version of this model. A variant of this approach was also elaborated in [18] but this resulted in a second-grade fluid model, instead of the viscous LANS$-\alpha$ equations, because the choice of dissipation made in [18] differed from the Navier-Stokes dissipation chosen in [13]–[17]. The geometry and analysis of the inviscid Euler$-\alpha$ equations was presented in [19], [20]. The analysis of global existence and well-posedness for the viscous LANS$-\alpha$ was given for periodic domains in [21] and was modified for bounded domains in [22]. For more information and a guide to the previous literature specifically about the NS$-\alpha$ model, see paper [23]. The latter paper also discusses connections to standard concepts and scaling laws in turbulence modeling, including relationships of the NS$-\alpha$ model to large eddy simulation (LES) models that are pursued farther in the present paper. Related results interpreting the NS$-\alpha$ model as an extension of scale similarity LES models of turbulence are also reported in [24]. A numerical comparison of LANS$-\alpha$ model results with LES models for the late stages of decaying homogeneous turbulence is discussed in [25]. Vortex interactions in the early stages of $3D$ turbulence decay are studied numerically with LANS$-\alpha$ and compared with both DNS and the Smagorinsky eddy viscosity approach in [26].

Three contributions in the present approach. Stated most simply, the LANS$-\alpha$ approach may be interpreted as a closure model for the turbulent stress tensor that is derived from Kelvin's circulation theorem, using a smoothed transport velocity, as discussed in [23], [24], [27]. A new development within this approach is introduced here that gives rise to an explicitly filtered similarity-type model [28] for the turbulent stress tensor, composed of three different contributions. The first contribution is a filtered version of the nonlinear gradient model. The unfiltered version of this model is also known as the 'Clark' model [29, 30], the 'gradient' model [31] or the 'tensor-diffusivity' model [32]. The second contribution, when combined with the filtered nonlinear gradient model, represents the so-called 'Leray regularization' of Navier-Stokes dynamics [33]. Finally, a new third contribution emerges from the derivation which completes the full LES-α model and endows it with its own Kelvin's circulation theorem.

To investigate the physical and numerical properties of the resulting three-part LANS$-\alpha$ subgrid parameterization, we consider a turbulent mixing layer [34]. This flow is well documented in literature and provides a realistic canonical flow problem [35] suitable for testing and comparison with predictions arising from more traditional subgrid model developments [7]. In particular, we consider similarity, and eddy-viscosity modeling, combined with the dynamic procedure based on Germano's identity [5, 6, 36], to compare with LES-α . In addition to the full LES-α model, in our comparisons we also consider the two models that are contained in it, i.e., the filtered nonlinear gradient model and the Leray model. We will refer to all three as LES-α models. For all these models, the explicit filtering stage is essential. Without this filtering operation in the definition of the models, a finite time instability is observed to arise in the simulations. The basis for this instability can be traced back to the presence of anti-diffusion in the nonlinear gradient contribution. We sketch an analysis of the one-dimensional Burgers equation, following [31], to illustrate this instability and show, through simulation, that increasing the subgrid resolution further enhances this instability. Analyzing the resolved kinetic energy dynamics reveals that this instability is associated with an excessive contribution to back-scatter.

The 'nonlinearly dispersive' filtered models that arise in LES-α are reminiscent of similarity LES models [24]. The LES-α model separates the resolved kinetic energy (RKE) of the flow into the sum of two contributions: namely, the energies due to motions at scales that are either greater, or less than an externally determined length-scale (α). The two contributions are modeled by

$$RKE = RKE^{(>)} + RKE^{(<)} \tag{7.1}$$

As we shall describe later in reviewing the LES-α strategy, the kinetic energy $RKE^{(<)}$ of turbulent motions at scales less than α is modeled by a term proportional to the *rate of dissipation* of the kinetic energy $RKE^{(>)}$ at scales greater than α. (The time-scale in the proportionality constant is the viscous diffusion time $\alpha^2/2\nu$.)

A key aspect of the LES-α dynamics is the exchange, or conversion, of kinetic energy between $RKE^{(>)}$ and $RKE^{(<)}$. We focus on the contributions to the dynamics of the resolved kinetic energy $RKE^{(>)}$ at scales greater than α that arise from the different terms in the LES-α models. The filtered nonlinear gradient model, the Leray model and the full LES-α model all contribute to the reduction of the $RKE^{(>)}$ in the laminar stages of the flow. This corresponds to forward scatter of $RKE^{(>)}$ into $RKE^{(<)}$ outweighing backward scatter. In the developing turbulent flow regime, the resolved kinetic energy $RKE^{(>)}$ of the full LES-α model may decrease too slowly compared to DNS, and for some settings of the (numerical) parameters can even become reactive in nature, thereby back-scattering too much kinetic energy from $RKE^{(<)}$ into $RKE^{(>)}$. In contrast, the contribution of the Leray model to the $RKE^{(>)}$ dynamics remains forward in nature and appears to settle around some negative, nonzero value in the turbulent regime. All the LES-α models show contributions to both forward and backward scatter of RKE. It is observed that in the full LES-α model two of the three terms almost cancel in the evolution of resolved kinetic energy $RKE^{(>)}$. This cancellation nearly reduces the full LES-α model to the filtered nonlinear gradient model.

The mixing layer simulations indicate that the Leray subgrid model provides more accurate predictions compared to both the filtered nonlinear gradient and the full LES-α model. This is based on comparisons that include mean flow quantities, fluctuating flow properties and the energy spectrum. In addition, the Leray LES-α model appears more robust with respect to changes in numerical parameters. Predictions based on this model compare quite favorably with those obtained using dynamic (mixed) models and filtered DNS results. The Leray model combines this feature with a strongly reduced computational cost and is favored for this reason, as well. In addition, a number of classic mathematical properties (e.g., existence and uniqueness of strong solutions) can be proven rigorously for fluid flows that are modeled with Leray's regularization. These can be used to guide further developments of this model such as extensions to more complex flows at higher Reynolds number. This is a topic of current research and will be published elsewhere [37].

Apart from the problem of modeling the subgrid-scale stresses, any actual realization of LES is inherently endowed with (strongly) interacting errors arising from the required use of marginal numerical resolution [38, 39, 40, 41]. The accuracy of the predictions depends on the numerical method and subgrid resolution one uses. We consider in some detail numerical contamination of a 'nonlinear gradient fluid' and a 'Leray fluid,' which are defined as the hypothetical fluids governed by the corresponding subgrid model. In this analysis we are consequently not concerned with how accurately the modeled equations represent filtered DNS results. Rather, we focus on the numerical contamination of the predictions. For this purpose we compare two finite volume spatial discretization methods, one at second order, and the other at fourth order accuracy.

The subgrid modeling and the spatial discretization of the equations give rise to a computational dynamical system whose properties are intended to simulate those of the filtered Navier-Stokes equations. The success of this simulation depends of course on the properties of the model, as well as of the spatial discretization method and the subgrid resolution. The model properties are particularly important in view of the marginal subgrid resolution used in present-day LES. We consider the role of the numerical method at various resolutions and various ratios of the filter-width Δ compared to the grid-spacing h. Let Δ be a fixed constant. In cases of large ratios $\Delta/h \gg 1$ one approximates the grid-independent LES solution corresponding to the given value of Δ, and the accuracy of its predictions will be limited by the quality of the assumed subgrid model. At the other extreme, one may assume Δ/h to be rather small and numerical effects can constitute a large source of error. Through a systematic variation of the ratio Δ/h at constant Δ we can identify the contributions of the numerical method at coarse resolutions. This will give an impression of how the computational dynamical system is affected by variations in the resolution and the numerical method.

The organization of this chapter is as follows. In section 2 we introduce the large-eddy simulation problem and identify the closure problem and some of its properties. The treatment of this closure problem using the α-framework is sketched, together with more conventional subgrid parameterization that involves the introduction of similarity, and eddy-viscosity modeling. Finally, we analyze the instabilities associated with the use of the unfiltered nonlinear gradient model. In section 3 we introduce the numerical methods used and consider the simulation of a turbulent mixing layer. Some direct and large-eddy simulation results will be shown. In section 4 we focus on the LES-α models and consider the dynamics of the resolved kinetic energy in each of the three cases. This comparison provides a framework for understanding how the different LES-α subgrid models function. We proceed with an assessment of the numerical error dynamics at relatively coarse subgrid resolutions. A summary and concluding remarks for the chapter are given in section 5.

2. Large-eddy simulation and α-modeling

This section sketches the traditional approach to large-eddy simulation, which arises from direct spatial filtering of the Navier-Stokes equations (section 2.1). The algebraic and analytic properties of the LES modeling problem will be discussed first. The LES closure problem will then be considered in the α-framework of turbulent flow, derived via Kelvin's circulation theorem for a smoothed, spatially filtered transport velocity (section 2.2). The closure of the filtered fluid flow problem achieved this way will be compared with the more traditional methods of similarity, and eddy-viscosity modeling for LES. The latter is introduced in section 2.3 together with the dynamic procedure based on Germano's identity. We also sketch a stability analysis of the one-dimensional filtered Burgers equation involving the nonlinear gradient subgrid model that illustrates the instabilities associated with this model (section 2.4).

2.1 Spatially filtered fluid dynamics

We consider the incompressible flow problem in d spatial dimensions. The Cartesian velocity fields u_i ($i = 1, \ldots, d$) and the normalized pressure field p constitute the complete solution. The velocity field is considered to be solenoidal and the evolution of the solution is described by the Navier-Stokes equations. These are conservation laws for mass and momentum, respectively, that can be written in the absence of forcing as

$$\partial_j u_j = 0 \tag{7.2}$$

$$\partial_t u_i + \partial_j(u_i u_j) + \partial_i p - \frac{1}{Re}\partial_{jj} u_i = 0 \tag{7.3}$$

where ∂_t and ∂_j denote, respectively, partial differential operators in time t and Cartesian coordinate x_j, $j = 1, \ldots, d$. The quantity $Re = u_r l_r / \nu_r$ is the Reynolds number based on reference velocity (u_r), reference length (l_r) and reference kinematic viscosity (ν_r), which were selected to non-dimensionalize the governing equations. Repeated indices are summed over their range, except where otherwise noted.

Equations (7.2 - 7.3) model incompressible flow in all its spatial and temporal details. In deriving approximate equations that are specialized to capture the generic large-scale flow features only, one applies a spatial filter operation $L : u \to \overline{u}$ to (7.2 - 7.3). For simplicity, we restrict to linear convolution filters:

$$\overline{u}(\mathbf{x}, t) = L(u)(\mathbf{x}, t) = \int_{-\infty}^{\infty} G(\mathbf{x} - \boldsymbol{\xi})\, u(\boldsymbol{\xi}, t)\, d\boldsymbol{\xi} = \Big(G * u\Big)(\mathbf{x}, t) \tag{7.4}$$

in which the filter-kernel G is normalized, i.e., $L(c) = c$ for any constant solution $u = c$. We assume that the filter-kernel G is localized as a function of $\mathbf{x} - \boldsymbol{\xi}$ and a filter-width Δ can be assigned to it. Typical filters which are commonly considered in LES are the top-hat, the Gaussian and the spectral cut-off filter. Here, we restrict ourselves to the top-hat filter which has a filter-kernel given by

$$G(\mathbf{z}) = \begin{cases} \Delta^{-3} & \text{if } |z_i| < \Delta_i/2 \\ 0 & \text{otherwise} \end{cases} \tag{7.5}$$

where Δ_i denotes the filter-width in the x_i direction and the total filter-width Δ is specified by

$$\Delta^3 = \Delta_1 \Delta_2 \Delta_3 \tag{7.6}$$

Apart from the filter-kernel in physical space, the Fourier-transform of $G(\mathbf{z})$, denoted by $H(\mathbf{k})$, is important, e.g., for the interpretation of the effect of the filter-operation on signals which are composed of various length-scales. The Fourier-transform of the top-hat filter is given by (no sum in $\Delta_i k_i$)

$$H(\mathbf{k}) = \prod_{i=1}^{3} \frac{\sin(\Delta_i k_i/2)}{\Delta_i k_i/2} \tag{7.7}$$

If we consider a general Fourier-representation of a solution $u(\mathbf{x}, t)$,

$$u(\mathbf{x}, t) = \sum_{\mathbf{k}} c_{\mathbf{k}}(t) e^{i\mathbf{k}\cdot\mathbf{x}} \tag{7.8}$$

the filtered solution can directly be written as

$$\overline{u}(\mathbf{x}, t) = \sum_{\mathbf{k}} \Big(H(\mathbf{k}) c_{\mathbf{k}}(t) \Big) e^{i\mathbf{k}\cdot\mathbf{x}} \tag{7.9}$$

We notice that each Fourier-coefficient $c_{\mathbf{k}}(t)$ is attenuated by a factor $H(\mathbf{k})$. The normalization condition of the filter-operation implies $H(\mathbf{0}) = 1$. For small values of $|\Delta_i k_i|$ one infers from a Taylor expansion

$$H(\mathbf{k}) = 1 - (1/24)\Big((k_1\Delta_1)^2 + (k_2\Delta_2)^2 + (k_3\Delta_3)^2\Big) + \ldots \tag{7.10}$$

which shows the small attenuation of flow features which are considerably larger than the filter-width Δ, i.e., $|\Delta_i k_i| \ll 1$ for $i = 1,\ 2,\ 3$. As $|\mathbf{k}|$ increases $H(\mathbf{k})$ becomes smaller while oscillating as a function of $\Delta_i k_i$. Consequently, the coefficients $c_{\mathbf{k}}(t)$ are strongly reduced as $|\Delta_i k_i| \gg 1$ and the small scale features in the solution are effectively taken out by the filter operation. Similarly, the Gaussian filter can be shown to have the same expansion for small $|\Delta_i k_i|$ and reduces to zero monotonically as $|\Delta_i k_i|$ becomes large.

The filter operation L is a convolution integral. Hence, it is a linear operation that commutates with partial derivatives [42, 43]. This property facilitates the application of the filter to the governing equations (7.2 - 7.3). A straightforward application of such filters leads to

$$\partial_j \overline{u}_j = 0 \tag{7.11}$$

$$\partial_t \overline{u}_i + \partial_j(\overline{u}_i \overline{u}_j) + \partial_i \overline{p} - \frac{1}{Re}\partial_{jj}\overline{u}_i = -\partial_j \tau_{ij} \tag{7.12}$$

where we introduced the turbulent stress tensor

$$\tau_{ij} = \overline{u_i u_j} - \overline{u}_i \overline{u}_j \tag{7.13}$$

We observe that the filtered solution $\{\overline{u}_i,\ \overline{p}\}$ represents an incompressible flow ($\partial_j \overline{u}_j = 0$). The same differential operator as in (7.3) acts on $\{\overline{u}_i,\ \overline{p}\}$ and due to the filtering a non-zero right-hand side has arisen which contains the divergence of the turbulent stress tensor τ_{ij}. This latter term is the so-called subgrid term, and expressing it in terms of the filtered velocity and its derivatives constitutes the closure problem in large-eddy simulation.

The LES modeling problem as expressed above has a number of important, rigorous properties which may serve as guidelines for specifying appropriate subgrid-models for τ_{ij}. In particular, we will briefly review realizability conditions, algebraic identities and transformation properties. Adhering to these basic features of τ_{ij} limits some of the heuristic elements in the subgrid-modeling.

Realizability It is well known that the Reynolds stress $\overline{u_i' u_j'}$ in RANS is positive semi-definite [44, 45] and the following inequalities hold [46]

$$\tau_{ii} \geq 0 \quad \text{for} \quad i \in \{1,2,3\} \quad \text{(no sum)} \tag{7.14}$$

$$|\tau_{ij}| \leq \sqrt{\tau_{ii}\tau_{jj}} \quad \text{for} \quad i,j \in \{1,2,3\} \quad \text{(no sum)} \tag{7.15}$$

$$\det(\tau_{ij}) \geq 0 \tag{7.16}$$

If the filtering approach is followed, in general $\tau_{ij} \neq \overline{u_i' u_j'}$ and, therefore, it is relevant to know the conditions under which τ_{ij} is positive semi-definite. Following Vreman *et al.* [8], it can be proved that τ_{ij} in LES is positive semi-definite if and only if the filter kernel $G(\mathbf{x}, \boldsymbol{\xi})$ is positive for all $\mathbf{x}$ and $\boldsymbol{\xi}$. If we assume $G \geq 0$, the expression

$$(f,g) = \int_\Omega G(\mathbf{x}, \boldsymbol{\xi}) f(\boldsymbol{\xi}) g(\boldsymbol{\xi}) d\boldsymbol{\xi} \tag{7.17}$$

defines an inner product and we can rewrite the turbulent stress tensor as:

$$\tau_{ij}(\mathbf{x}) = \int_\Omega G(\mathbf{x}, \boldsymbol{\xi})(u_i(\boldsymbol{\xi}) - \overline{u}_i(\mathbf{x}))(u_j(\boldsymbol{\xi}) - \overline{u}_j(\mathbf{x}))d\boldsymbol{\xi} = (v_i^{\mathbf{x}}, v_j^{\mathbf{x}}) \tag{7.18}$$

with $v_i^{\mathbf{x}}(\boldsymbol{\xi}) \equiv u_i(\boldsymbol{\xi}) - \overline{u}_i(\mathbf{x})$. In this way the tensor τ_{ij} forms a 3×3 Grammian matrix of inner products. Such a matrix is always positive semi-definite and consequently τ_{ij} satisfies the realizability conditions. The reverse statement can likewise be established, showing that the condition $G \geq 0$ is both necessary and sufficient.

One prefers the turbulent stress tensor τ_{ij} in LES to be realizable for a number of reasons. For example, if τ_{ij} is realizable, the generalized turbulent kinetic energy $k = \tau_{ii}/2$ is a positive quantity. This quantity is required to be positive in subgrid models which involve the k-equation [47]. Several further benefits of realizability and positive filters can be identified [8]; here we restrict to adding that the kinetic energy of $\overline{u}$ is bounded by that of u for positive filter-kernels:

$$\frac{1}{2}\int_{\Omega} |\overline{u}|^2 d\mathbf{x} \leq \frac{1}{2}\int_{\Omega} |u|^2 d\mathbf{x} \tag{7.19}$$

Requiring realizability places some restrictions on subgrid models. For example, if $G \geq 0$ models for τ should be realizable. Consider, e.g., an eddy-viscosity model m_{ij} given by

$$m_{ij} = -\nu_e \sigma_{ij} + \frac{2}{3} k \delta_{ij} \tag{7.20}$$

In order for this model to be realizable, a lower bound for k in terms of the eddy-viscosity ν_e arises, i.e., $k \geq \frac{1}{2}\sqrt{3}\sigma\nu_e$ where $\sigma = \frac{1}{2}\sigma_{ij}\sigma_{ij}$ and σ_{ij} is the rate of strain tensor given by $\sigma_{ij} = \partial_i u_j + \partial_j u_i$.

Algebraic identities The introduction of the product operator $S(u_i, u_j) = u_i u_j$ allows to write the turbulent stress tensor as [6]:

$$\tau_{ij}^L = \overline{u_i u_j} - \overline{u}_i \overline{u}_j = L(S(u_i, u_j)) - S(L(u_i), L(u_j)) = [L, S](u_i, u_j) \tag{7.21}$$

in terms of the central commutator $[L, S]$ of the filter L and the product operator S. This commutator shares a number of properties with the Poisson-bracket in classical mechanics. Leibniz' rule of Poisson-brackets is in the context of LES known as Germano's identity [5]

$$[\mathcal{L}_1 \mathcal{L}_2, S] = [\mathcal{L}_1, S]\mathcal{L}_2 + \mathcal{L}_1[\mathcal{L}_2, S] \tag{7.22}$$

This can also be written as

$$\tau^{\mathcal{L}_1 \mathcal{L}_2} = \tau^{\mathcal{L}_1} \mathcal{L}_2 + \mathcal{L}_1 \tau^{\mathcal{L}_2} \tag{7.23}$$

and expresses the relation between the turbulent stress tensor corresponding to different filter-levels. In these identities, $\mathcal{L}_1$ and $\mathcal{L}_2$ denote any two filter operators and $\tau^K = [K, S]$. The first term on the right-hand side of (7.23) is interpreted as the 'resolved' term which in an LES can be evaluated without further approximation. The other two terms require modeling of τ at the corresponding filter-levels.

Similarly, Jacobi's identity holds for S, $\mathcal{L}_1$ and $\mathcal{L}_2$:

$$[\mathcal{L}_1, [\mathcal{L}_2, S]] + [\mathcal{L}_2, [S, \mathcal{L}_1]] = -[S, [\mathcal{L}_1, \mathcal{L}_2]] \tag{7.24}$$

The expressions in (7.22) and (7.24) provide relations between the turbulent stress tensor corresponding to different filters and these can be used to dynamically model τ^L. The success of models incorporating Germano's identity (7.22) is by now well established in applications for many different flows. In the traditional formulation one selects $\mathcal{L}_1 = \mathcal{H}$ and $\mathcal{L}_2 = L$ where $\mathcal{H}$ is the so called test-filter. In this case one can specify Germano's identity [5] as

$$\tau^{\mathcal{H}L}(u) = \tau^{\mathcal{H}}(L(u)) + \mathcal{H}\left(\tau^L(u)\right) \tag{7.25}$$

The first term on the right hand side involves the operator $\tau^{\mathcal{H}}$ acting on the resolved LES field $L(u)$ and during an LES this is known explicitly. The remaining terms need to be replaced by a model. In the dynamic modeling [36] the next step is to assume a base-model m^K corresponding to filter-level K and optimize any coefficients in it, e.g., in a least squares sense [48]. The operator formulation can be extended to include approximate inversion defined by $\mathcal{L}^{-1}(L(x^k)) = x^k$ for $0 \le k \le N$ [49]. Dynamic inverse models have been applied in mixing layers [50].

Transformation properties The turbulent stress tensor τ_{ij} can be shown to be invariant with respect to Galilean transformations. This property also holds for the divergence, i.e., $\partial_j \tau_{ij}$, referred to as the subgrid scale force. Hence, the filtered Navier-Stokes dynamics is Galilean invariant. Suitable subgrid models should at least maintain the Galilean invariance of the divergence of the model, i.e., $\partial_j m_{ij}$. In fact, most subgrid models are represented by tensors which are Galilean invariant. Examples of non-symmetric tensor models have been reported in [24]; in that case however, the divergence of the stress tensor model was verified to be Galilean invariant.

Likewise, it is of interest to consider a transformation of the subgrid scale stress tensor to a frame of reference rotating with a uniform angular velocity. The full subgrid scale stress tensor transforms in such a way that the subgrid scale force is the same in an inertial and in a rotating frame. Horiuti [51] has recently analyzed several subgrid scale models and showed that some of them do not satisfy this condition. This is an example of how transformation properties of the exact turbulent stress tensor can be used to guide propositions for subgrid modeling.

After closing the filtered equations (7.11-7.12) by a subgrid model m_{ij} for the turbulent stress tensor τ_{ij} we arrive at the modeled filtered dynamics. The model needs to be expressed in terms of operations on the filtered solution only. In the absence of forcing these equations can be written as

$$\begin{aligned} \partial_j v_j &= 0 \\ \partial_t v_i + \partial_j(v_i v_j) + \partial_i P - \frac{1}{Re}\partial_{jj} v_i &= -\partial_j m_{ij} \end{aligned} \tag{7.26}$$

whose solution is denoted as $\{v_i,\ P\}$. Ideally, if m_{ij} and the numerical treatment were correct and had no undesirable effects on the dynamics, one might expect $v_i = \overline{u}_i$. In view of possible sensitive dependence of an actual solution, e.g., on the initial condition, one should not expect instantaneous and pointwise equality of v_i and $\overline{u}_i$ but rather one should expect statistical properties of the filtered and modeled solution to be equal. Assessing the extent to which the properties of $\{v_i, P\}$ and $\{\overline{u}_i, \overline{p}\}$ are correlated allows an evaluation of the quality of the subgrid model, the dynamic effects arising from the numerical method and the interactions between modeling and numerics. In what follows, we will use the notation $\{v_i, P\}$ to distinguish the solution of the subgrid model from the filtered solution $\{\overline{u}_i, \overline{p}\}$.

2.2 Subgrid model derived from Kelvin's theorem

The LES-α modeling scheme we shall use here is based on the well-known viscous Camassa-Holm equations, or LANS$-\alpha$ model. This modeling strategy imposes a "cost" in resolved kinetic energy (RKE) for creation of smaller and smaller excitations below a certain, externally specified length scale, denoted by α. This cost in converting $RKE^{(>)}$ to $RKE^{(<)}$ implies a nonlinear modification of the Navier-Stokes equations which is reactive, or dispersive, in nature instead of being diffusive, as is more common in present-day LES modeling. The modification appears in the nonlinear convection term and can be rewritten in terms of a subgrid model for the turbulent stress tensor. In the LANS$-\alpha$ model, the processes of nonlinear conversion of $RKE^{(>)}$ to $RKE^{(<)}$ and sweeping of the smaller scales by the larger ones are still included in the modeled dynamics. We will sketch the LES-α approach in this subsection and extract the subgrid models used in this study. For more details and applications of this approach, see [13]–[17], [23]–[26].

It is well known that the Navier-Stokes equations satisfy Kelvin's circulation theorem, i.e.,

$$\frac{d}{dt}\oint_{\gamma(\mathbf{u})} u_j\, dx_j = \oint_{\gamma(\mathbf{u})} \frac{1}{Re}\,\partial_{kk} u_j\, dx_j \tag{7.27}$$

Here $\gamma(\mathbf{u})$ represents a fluid loop that moves with the Eulerian fluid velocity $\mathbf{u}(\mathbf{x}, t)$. The basic equations in the LES-α modeling may be introduced by modifying the velocity field by which the fluid loop is transported. The governing LES-α equations will provide the smoothed solution $\{v_j, P\}$ and we specify the equations for $\mathbf{v}$ through the Kelvin-filtered circulation theorem. Namely, we integrate an approximately 'defiltered' velocity $\mathbf{w}$ around a loop $\gamma(\mathbf{v})$ that moves with the regularized spatially filtered fluid velocity $\mathbf{v}$, cf. [23], [24], [27]

$$\frac{d}{dt}\oint_{\gamma(\mathbf{v})} w_j\, dx_j = \oint_{\gamma(\mathbf{v})} \frac{1}{Re}\,\partial_{kk} w_j\, dx_j \tag{7.28}$$

Hence, the basic transport properties of the LES-α model arise from filtering the 'loop-velocity' to obtain $\mathbf{v}$, then approximately defiltering $\mathbf{v}$ to obtain the velocity $\mathbf{w}$ in the Kelvin integrand. As we shall show, this approach will yield the model stress tensor m_{ij} needed to complete the filtering approach outlined in section 2.1. Direct calculation of the time derivative in this modified circulation theorem yields the Kelvin-filtered Navier-Stokes equations,

$$\partial_t w_i + v_j \partial_j w_i + w_k \partial_i v_k + \partial_i \hat{P} - \frac{1}{Re}\partial_{jj} w_i = 0\,, \;\; \partial_j v_j = 0 \tag{7.29}$$

where we introduce the scalar function $\hat{P}$ in removing the loop integral. The relation between the 'defiltered' velocity components w_i and the LES-α velocity components v_i of the Kelvin loop needs to be specified separately. The Helmholtz defiltering operation was introduced in [10], [11] for this purpose:

$$w_i = v_i - \alpha^2 \partial_{jj} v_i = (1 - \alpha^2 \partial_{jj}) v_i = \mathcal{H}_\alpha(v_i) \tag{7.30}$$

where $\mathcal{H}_\alpha$ denotes the Helmholtz operator. We recall that all explicit filter operations L with a non-zero second moment, have a Taylor expansion whose leading order terms are of the same form as (7.30). Consequently, we infer that the leading order relation between α and Δ follows as $\alpha^2 = \Delta^2/24$ for the top-hat and the Gaussian filter. We will use this as the definition of α in the sequel.

The LES-α equations can be rearranged into a form similar to the basic LES equations (7.26), by splitting off a subgrid model for the turbulent stress tensor. For the Helmholtz defiltering, we obtain from (7.29):

$$\partial_t v_i + \partial_j(v_i v_j) + \partial_i P + \partial_j m_{ij}^{\alpha} - \frac{1}{Re}\partial_{jj} v_i = 0\,, \;\; \partial_j v_j = 0 \tag{7.31}$$

after absorbing gradient terms into the redefined pressure P. Thus, we arrive at the following parameterization for the turbulent stress tensor

$$\mathcal{H}_\alpha(m_{ij}^{\alpha}) = \alpha^2 \Big(\partial_k v_i\, \partial_k v_j + \partial_k v_i\, \partial_j v_k - \partial_i v_k\, \partial_j v_k \Big) \tag{7.32}$$

In the evaluation of the LES-α dynamics in the above formulation, an inversion of the Helmholtz operator $\mathcal{H}_\alpha$ is required. The 'exponential' (or 'Yukawa') filter [52] is the exact, explicit filter which inverts $\mathcal{H}_\alpha$. Thus, an inversion of $\mathcal{H}_\alpha$ corresponds to applying the exponential filter to the right-hand side of (7.32) in order to find m_{ij}^α. However, since the Taylor expansion of the exponential filter is identical at quadratic order to that of the top-hat and the Gaussian filters, we will approximate the inverse of $\mathcal{H}_\alpha$ by an application of the explicit top-hat filter, for reasons of computational efficiency. Moreover, in actual simulations the numerical realization of the exponential filter is only approximate and can just as well be replaced by the numerical top-hat filter. This issue of (approximately) inverting the Helmholtz operator will be studied separately and published elsewhere [37].

The full LES-α subgrid model m_{ij}^α has three distinct contributions. The first term on the right-hand side is readily recognized as the nonlinear gradient model which we will denote by A_{ij}. This term is closely related to the similarity model proposed by Bardina [28], as will be shown in the next subsection. The second term will be denoted by B_{ij} and combined with the first term, corresponds to the Leray regularization of the convective terms in the Navier-Stokes equations. This regularization arises if the familiar contribution $u_j \partial_j u_i$ in the Navier-Stokes equations is replaced by $\overline{v}_j \partial_j v_i$ in the smoothed description. The third term will be denoted by C_{ij}. Further details of the derivation and mathematical properties of the LES-α model will be published elsewhere [37]. We can explicitly write the stress tensor for the LES-α model as

$$\begin{aligned} m_{ij}^\alpha &= \frac{\Delta^2}{24}\left(\overline{\partial_k v_i\, \partial_k v_j} + \overline{\partial_k v_i\, \partial_j v_k} - \overline{\partial_i v_k\, \partial_j v_k}\right) \\ &= \overline{A_{ij}} + \overline{B_{ij}} - \overline{C_{ij}} \end{aligned} \qquad (7.33)$$

The explicit filter, represented by the overbar in this expression, is realized by the numerical top-hat filter in this study. It does not necessarily have to coincide with the LES-filter. While the LES-filter specifies the relation between the Navier-Stokes solution u_i and the LES-α solution v_i, the explicit LES-α filter is used to approximate $\mathcal{H}_\alpha^{-1}$. We will consider the effects associated with variations in the filter-width $\tilde{\Delta}$ of the explicit LES-α filter with filter-width $\tilde{\Delta}/\Delta = \kappa$. Typical values that will be considered are $\kappa = 1$ and $\kappa = 2$.

In the next subsection we will describe some familiar subgrid models used in LES which are based on the similarity and eddy-viscosity concepts.

2.3 Similarity modeling and eddy-viscosity regularization

We distinguish two main contributions in present-day traditional subgrid modeling of the turbulent stress tensor, i.e., dissipative and similarity subgrid models. In this subsection we briefly describe these two basic approaches, as well as subgrid models that consist of combinations of an eddy-viscosity and a similarity part, so-called mixed models. The relative importance of the two components in such mixed models is obtained by using the dynamic procedure which is based upon Germano's identity (7.25). This mixed approach effectively regularizes and stabilizes similarity models.

As a result of the filtering, flow features of length-scales (much) smaller than the filter-width Δ are considerably attenuated. This implies that the natural molecular dissipation arising from the viscous fluxes, is strongly reduced, compared to the unfiltered flow-problem. In order to compensate for this, dissipative subgrid-models have been introduced to model the turbulent stress tensor. The prime example of such eddy-viscosity models is the Smagorinsky model [2, 53]:

$$m_{ij}^{S} = -(C_S\Delta)^2|\sigma(\mathbf{v})|\sigma_{ij}(\mathbf{v}) \quad \text{with} \quad |\sigma(\mathbf{v})|^2 = \frac{1}{2}\sigma_{ij}(\mathbf{v})\sigma_{ij}(\mathbf{v}) \qquad (7.34)$$

where σ_{ij} is the strain rate, introduced above ($\sigma_{ij} = \partial_i v_j + \partial_j v_i$). This model adds only little computational overhead. The major short-coming of the Smagorinsky model is its excessive dissipation in laminar regions with mean shear, because σ_{ij} is large in such regions [36]. Furthermore, the correlation between the Smagorinsky model and the actual turbulent stress is quite low (reported to be ≈ 0.3 in several flows).

In trying to compensate for these short-comings of the Smagorinsky model, a second main branch of subgrid models emerges from the similarity concept [28]. Using the commutator notation, the turbulent stress tensor can be expressed as $\tau_{ij}(\mathbf{u}) = [L, S](u_i, u_j)$. In terms of this short-hand notation, the basic similarity model can be written as

$$m_{ij}^{B} = [L, S](v_i, v_j) \qquad (7.35)$$

i.e., directly following the definition of the turbulent stress tensor, but expressed in terms of the available modeled LES velocity field. Generalizations of this similarity model arise by replacing v_i in (7.35) by an approximately defiltered field $\widehat{v}_i = \mathcal{L}(v_i)$ where $\mathcal{L}(L(u)) \approx u$, i.e., $\mathcal{L}$ approximates the 'inverse' of the filter L [49]. In detail, a generalized similarity model arises from $m^G = [L, S]\left(\mathcal{L}^{-1}(v)\right)$ using the approximate inversion. This approach is also known as the deconvolution model [54] and is reminiscent to the subgrid estimation model [55]. The correlation with τ_{ij} is much better with correlation coefficients reported in the range 0.6 to 0.9 in several flows. The low level of dissipation associated with these models renders them quite sensitive to the spatial resolution. At relatively coarse resolutions, the low level of dissipation can give rise to instability of the simulations. Moreover, these models add significantly to the required computational effort. At suitable resolution, however, the predictions arising from generalized similarity models are quite accurate.

An interesting subgrid model which follows the similarity approach to some degree and avoids the costly additional filter-operations is the nonlinear gradient model, mentioned earlier. This model can be derived from the Bardina scale-similarity model by using Taylor expansions of the filtered velocity. One may arrive at $\tau_{ij} = \frac{1}{12}\sum_k \Delta_k^2(\partial_k\overline{u}_i)(\partial_k\overline{u}_j) + \mathcal{O}(\Delta^4)$. The first term on the right-hand side is referred to as the 'nonlinear gradient model' or tensor-diffusivity model:

$$m_{ij}^{TD} = \frac{1}{24}\sum_k \Delta_k^2(\partial_k v_i)(\partial_k v_j) \tag{7.36}$$

Since this model is part of all three different LES-α models identified in the previous subsection, we will analyze the dynamics and the instabilities arising from this model in some more detail in the next subsection.

The three subgrid models, i.e., (7.34), (7.35) and (7.36) constitute well-known examples in LES-literature, which represent basic dissipative and reactive, or dispersive, properties of subgrid models for the turbulent stress tensor. These basic similarity and eddy-viscosity models can be combined in mixed models using the dynamic procedure, which provides a way of combining the two basic components of a mixed model without introducing additional external ad hoc parameters.

We consider simple mixed models based on eddy-viscosity and similarity. In these models the eddy-viscosity component reflects local turbulence activities and the local value of the eddy-viscosity adapts itself to the instantaneous flow. The dynamic procedure starts from Germano's identity (7.25). A common way to write Germano's identity is:

$$T_{ij} - \widehat{\tau_{ij}} = R_{ij} \tag{7.37}$$

where

$$T_{ij} = \widehat{\overline{u_i u_j}} - \widehat{\overline{u}}_i\widehat{\overline{u}}_j \tag{7.38}$$

$$R_{ij} = \widehat{(\overline{u}_i\overline{u}_j)} - \widehat{\overline{u}}_i\widehat{\overline{u}}_j \tag{7.39}$$

Here, in addition to the basic LES-filter $\overline{(\cdot)}$ of width $\overline{\Delta}$ a so-called 'test'-filter $\widehat{(\cdot)}$ of width $\widehat{\Delta}$ is introduced. Usually, this test-filter is wider than the LES-filter and the combined filter $\widehat{\overline{(\cdot)}}$ has a width that follows from $\widehat{\overline{\Delta}}^2 = \widehat{\Delta}^2 + \overline{\Delta}^2$. This relation is exact for the composition of two Gaussian filters and can be shown to be 'optimal' for other filters such as the top-hat filter [7]. The only external parameter that needs to be specified in the dynamic procedure is the ratio $\widehat{\Delta}/\overline{\Delta}$ which is commonly set equal to two. The terms at the left-hand side of the Germano identity (7.37) are the turbulent stress tensor on the 'combined' filter level (T_{ij}) and the turbulent stress tensor, filtered with the test-filter ($\widehat{\tau_{ij}}$), respectively. Finally, R_{ij} represents the resolved stress tensor which can be explicitly calculated using the modeled LES fields.

The general procedure for obtaining 'locally' optimal model parameters in a mixed formulation starts by assuming a basic model m_{ij} to approximate the turbulent stress tensor τ_{ij}, and a corresponding model M_{ij} for T_{ij}. We consider m_{ij} to be of 'mixed' type, i.e.,

$$m_{ij} = a_{ij} + cb_{ij} \tag{7.40}$$

where a_{ij} and b_{ij} are basic models. These basic models involve operations on $\mathbf{v}$ only; $a_{ij} = a_{ij}(\mathbf{v})$, $b_{ij} = b_{ij}(\mathbf{v})$. Furthermore, in standard mixed models, c is a scalar coefficient-field which is to be determined. The model M_{ij} is represented as:

$$M_{ij} = A_{ij} + CB_{ij} \tag{7.41}$$

where $A_{ij} = a_{ij}(\widehat{\mathbf{v}})$, $B_{ij} = b_{ij}(\widehat{\mathbf{v}})$. It is essential in this formulation that the coefficient C corresponding to the composed filter-level is well approximated by the coefficient c; i.e., we assume $C \approx c$. Insertion in Germano's identity yields $M_{ij} + \widehat{m_{ij}} = R_{ij}$, or in more detail,

$$\left(A_{ij} + \widehat{a_{ij}}\right) + c\left[B_{ij} + \widehat{b_{ij}}\right] = R_{ij} \tag{7.42}$$

where we have used the approximation $\widehat{cb_{ij}} \approx c\widehat{b_{ij}}$. Introducing the short-hand notation $\mathcal{A}_{ij} = A_{ij} + \widehat{a_{ij}}$, $\mathcal{B}_{ij} = B_{ij} + \widehat{b_{ij}}$, the coefficient c is required to obey $c\mathcal{B}_{ij} = R_{ij} - \mathcal{A}_{ij}$. This relation should hold for all tensor-components, which of course is not possible for a scalar coefficient field c. To resolve this situation we introduce an averaging operator $\langle f \rangle$ and define the 'Germano-residual' by

$$\varepsilon(c) = \langle \frac{1}{2}\{(R_{ij} - \mathcal{A}_{ij}) - c\mathcal{B}_{ij}\}^2 \rangle \tag{7.43}$$

From this we obtain an optimality condition for c from $\varepsilon'(c) = 0$ and we can solve the local coefficient as

$$c = \frac{\langle (R_{ij} - \mathcal{A}_{ij})\mathcal{B}_{ij} \rangle}{\langle \mathcal{B}_{ij}\mathcal{B}_{ij} \rangle} \tag{7.44}$$

where we assumed $\langle cfg \rangle \approx c\langle fg \rangle$. The averaging operator $\langle f \rangle$ is usually defined in terms of an integration over homogeneous directions of the flow-domain. In the case of the mixing layer, considered here, the averaging over the homogeneous streamwise and spanwise direction results in a dynamic coefficient c which is a function of the normal coordinate x_2 and time t. In more complex flow-domains, averaging over homogeneous directions may no longer be possible. Taking a running-average over time t is then a viable alternative, as was recently established, e.g., for flow in a spatially developing mixing layer [35].

As an example we consider the Smagorinsky model as the base model. The corresponding models on the two filter-levels can be written as

$$m_{ij}^D = -C_d\overline{\Delta}^2|\sigma(\mathbf{v})|\sigma_{ij}(\mathbf{v}) \;\; ; \;\; M_{ij}^D = -C_d\widehat{\overline{\Delta}}^2|\sigma(\widehat{\mathbf{v}})|\sigma_{ij}(\widehat{\mathbf{v}}) \tag{7.45}$$

The 'optimal' C_d follows from $C_d = \langle R_{ij}\mathcal{B}_{ij}\rangle / \langle \mathcal{B}_{ij}\mathcal{B}_{ij}\rangle$. In order to prevent numerical instability caused by negative values of C_d, the model coefficient C_d is artificially set to zero at locations where the procedure would return negative values. Sometimes, in developing flows, it is beneficial to also introduce a 'ceiling'-value for C_d. This value should be chosen such that once the flow is well-developed in time the actual limitation arising from the ceiling-value is no longer restrictive [35].

The dynamic mixed model employs the sum of Bardina's similarity and Smagorinsky eddy-viscosity model as the base model, i.e.,

$$m_{ij}^{DM} = [L, S](v_i, v_j) - C_d\overline{\Delta}^2|S(\mathbf{v})|S_{ij}(\mathbf{v}) \tag{7.46}$$

Likewise, a mixed nonlinear gradient model can be introduced by

$$m_{ij}^{DG} = \frac{1}{24}\overline{\Delta}^2(\partial_k v_i)(\partial_k v_j) - C_d\overline{\Delta}^2|S(\mathbf{v})|S_{ij}(\mathbf{v}) \tag{7.47}$$

The dynamic procedure has been used in a number of different flows. Compared to predictions using only the constitutive base models, the dynamic procedure generally enhances the accuracy and robustness. Moreover, it responds to the developing flow in such a way that the eddy-viscosity is strongly reduced in laminar regions and near solid walls [56]. This avoids specific modeling of transitional regions and near-wall phenomena, provided the resolution is sufficient. At even coarser resolution one may have to resort to specific models for transition and walls. We will not enter into this problem. Rather, we will focus on the properties of the nonlinear gradient model in the next subsection.

2.4 Analysis of instabilities of the nonlinear gradient model in one dimension

From the discussion of the previous two subsections, it would appear that the nonlinear gradient subgrid model would be very well suited to parameterize the dynamic effects of the small scales in a turbulent flow. This model is part of the full LES-α model and it also emerges as a Taylor expansion of the Bardina similarity model. In this subsection we will analyse this model in the context of the one-dimensional Burgers equation and show that this model gives rise to very strong instabilities. In subsequent sections we will show in what way the explicit filtering and the other terms in the LES-α model, or dynamic eddy-viscosity regularization, alter this peculiar behavior of the nonlinear gradient model.

We will analyse the nature of the instability of the pure gradient model for the one-dimensional Burgers equation [31]. The linear stability of a sinusoidal profile will be investigated. If a flow is linearly unstable then it is nonlinearly unstable to arbitrarily small initial disturbances. The linear analysis thus provides some information on the nonlinear equation.

The Burgers equation with gradient subgrid-model is written as:

$$\partial_t u + \frac{1}{2}\partial_x(u^2) - \nu\partial_x^2 u = -\frac{1}{2}\eta\partial_x(\partial_x u)^2 + f(x) \tag{7.48}$$

The parameter $\eta = \Delta^2/12$. The following analysis shows that smooth solutions of equation (7.48) can be extremely sensitive to small perturbations, leading to severe instabilities. In particular, we consider the linear stability of a 2π-periodic, stationary solution, $U(x,t) = \sin(x)$ on the domain $[0, 2\pi]$ with periodic boundary conditions. The forcing function f is determined by the requirement that U is a solution of equation (7.48). We substitute a superposition of U and a perturbation w,

$$u(x,t) = U(x) + w(x,t) \tag{7.49}$$

into equation (7.48) and linearize around U, omitting higher order terms in w:

$$\partial_t w + (1-\eta)\sin(x)\,\partial_x w + (w + \eta\partial_x^2 w)\cos(x) = \nu\partial_x^2 w \tag{7.50}$$

We use a Fourier expansion for w written as $w = \sum \alpha_k(t)e^{ikx}$. After substitution of this series into equation (7.50) we obtain an infinite system of ordinary differential equations for the Fourier coefficients α_k:

$$\dot{\alpha}_k = \frac{1}{2}k(\eta k - \eta - 1)\alpha_{k-1} - k^2\nu\alpha_k + \frac{1}{2}k(\eta k + \eta + 1)\alpha_{k+1} \tag{7.51}$$

To understand the nature of the nonlinear gradient model for the Burgers equation, we first analyse system (7.51) assuming $\nu = 0$. Instead of the infinite system, we consider a sequence of finite dimensional systems,

$$\dot{\mathbf{z}}_n = M_n \mathbf{z}_n \tag{7.52}$$

where $\mathbf{z}_n$ is a vector containing the $2n+1$ Fourier coefficients $\alpha_{-n}...\alpha_n$ and M_n is a $(2n+1)\times(2n+1)$ tri-diagonal matrix:

$$\mathbf{z}_n = \begin{bmatrix} \alpha_{-n} \\ \cdot \\ \cdot \\ \alpha_{-1} \\ \alpha_0 \\ \alpha_1 \\ \cdot \\ \cdot \\ \alpha_n \end{bmatrix}, \quad M_n = \begin{bmatrix} 0 & l_n & & & & & & & \\ r_n & \cdot & \cdot & & & & & & \\ & \cdot & \cdot & l_2 & & & & & \\ & & r_2 & 0 & l_1 & & & & \\ & & & 0 & 0 & 0 & & & \\ & & & & l_1 & 0 & r_2 & & \\ & & & & & l_2 & \cdot & \cdot & \\ & & & & & & \cdot & \cdot & r_n \\ & & & & & & & l_n & 0 \end{bmatrix} \tag{7.53}$$

with

$$l_k = \frac{1}{2}k(\eta k - \eta - 1) \quad ; \quad r_k = \frac{1}{2}(k-1)(\eta k + 1) \tag{7.54}$$

The eigenvalues of M_n determine the stability of the problem. The system is unstable if the maximum of the real parts of the eigenvalues is positive. We denote the eigenvalues of M_n by λ_j and introduce λ_{max} such that

$$|\lambda_{max}| = \max_j |\lambda_j| \tag{7.55}$$

This eigenvalue problem can be shown to have the following asymptotic properties (for a detailed proof see [31]):

$$\begin{aligned} &1. \quad \text{if } \lambda \text{ is an eigenvalue then } -\lambda \text{ is an eigenvalue} && (7.56)\\ &2. \quad |\lambda_{max}| \sim \eta n^2 && (7.57)\\ &3. \quad |\text{Im}(\lambda_{max})| \leq n-1 && (7.58) \end{aligned}$$

The first point implies that λ_{max} can be chosen such that $\text{Re}(\lambda_{max}) \geq 0$. Hence, the combination of these three properties yields the asymptotic behavior of the maximum of the real parts of the eigenvalues:

$$\text{Re}(\lambda_{max}) \sim \eta n^2 \tag{7.59}$$

This shows that the inviscid system is linearly unstable and that the largest real part of the eigenvalues is asymptotically proportional to n^2, where n is the number of Fourier modes taken into account.

It should be observed that the instability is severe, since the system is not only unstable, but the growth rate of the instability is infinitely large as $n \to \infty$. The instability is fully due to the incorporation of the gradient model, since all eigenvalues of the matrix M_n are purely imaginary in case the inviscid Burgers equation without subgrid-model is considered ($\eta = 0$). In numerical simulations the instability will grow with a finite speed, since then the number of Fourier modes is limited by the finite grid. Moreover, expression (7.59) illustrates that grid-refinement (with η kept constant), which corresponds to a larger n, will not stabilize the system, but rather enhance the instability. The growth rate of the instability of the one-dimensional problem can be expressed in terms of Δ and the grid-spacing h: $\eta n^2 \sim (\Delta/h)^2$. Consequently, the instability is not enhanced if the ratio between Δ and h is kept constant.

Finally, we will consider the more complicated case $\nu \neq 0$. The linear system in equation (7.51) now gives rise to matrices M_n which have a negative principal diagonal. It is known that for every fixed value of n there exists an eigenvalue arbitrarily close to the eigenvalue of the inviscid system (λ_{max}) if ν is sufficiently small [57]. Hence for small values of ν the viscous system for finite n is still linearly unstable. The matrix M_n is strictly diagonally dominant if $\nu > \eta + 1$, while all rows except n and $n+2$ are already diagonally dominant if $\nu > \eta$. If the matrix is diagonally dominant, the real parts of all eigenvalues are negative and, consequently, the system is stable. This indicates that stability can be achieved by a sufficiently large viscosity, which does not depend on n, but only on η. Thus, if the gradient model is supplemented with an adequate eddy-viscosity the instability will be removed as is the case with a dynamic mixed model involving the gradient model.

3. Numerical simulations of a turbulent mixing layer

In this section we first present the numerical methods used to solve the DNS and LES equations (subsection 3.1). We illustrate the accuracy of these methods for turbulent flow in a mixing layer in subsection 3.2.

3.1 Time-integration and spatial discretization

The Navier-Stokes or modeled LES equations are discretized using the so-called method of lines. We consider the compressible formulation and perform simulations at a low convective Mach number which was shown to provide essentially incompressible flow-dynamics. The method of lines allows to treat the spatial and temporal discretization separately and gives rise to a large number of ordinary differential equations for the unknowns on a computational grid.

We write the Navier-Stokes or LES equations concisely as $\partial_t \mathcal{U} = \mathcal{F}(\mathcal{U})$ where $\mathcal{U}$ denotes the state-vector containing, e.g., velocity and pressure, and $\mathcal{F}$ is the total flux, composed of the convective, the viscous, and possibly the subgrid fluxes. The operator $\mathcal{F}$ contains first and second order partial derivatives with respect to the spatial coordinates x_j. The equations are discretized on a uniform rectangular grid and the grid size in the x_j-direction is denoted by h_j. If we adopt a specific spatial discretization around a grid point $\mathbf{x}_{ijk}$, the operator $\mathcal{F}(\mathcal{U})$ is approximated in a consistent manner by an algebraic expression $F_{ijk}(\{U_{\alpha\beta\gamma}\})$ where $\{U_{\alpha\beta\gamma}\}$ denotes the state vectors in all the grid-points, labeled by $\alpha,\ \beta,\ \gamma$. Usually, only neighboring grid points around $(i,\ j,\ k)$ appear explicitly in F_{ijk}, e.g., in case finite difference or finite volume discretizations are considered. After applying the method of lines, the governing equations yield

$$\mathrm{d}_t U_{ijk}(t) = F_{ijk}(\{U_{\alpha\beta\gamma}\}) \quad ; \quad U_{ijk}(0) = U_{ijk}^{(0)} \tag{7.60}$$

where $U_{ijk}^{(0)}$ represents the initial condition. Hence, in order to specify the numerical treatment, apart from the initial and boundary conditions, the spatial discretization which gives rise to F_{ijk} and the temporal integration need to be specified. We next introduce these separately.

The time stepping method which we adopt is an explicit four-stage compact-storage Runge-Kutta method. When we consider the scalar differential equation

$du/dt = f(u)$, this Runge-Kutta method performs within one time step of size δt

$$u^{(j)} = u^{(0)} + \beta_j \delta t f(u^{(j-1)}) \quad (j = 1, 2, 3, 4) \tag{7.61}$$

with $u^{(0)} = u(t)$ and $u(t + \delta t) = u^{(4)}$. With the coefficients $\beta_1 = 1/4$, $\beta_2 = 1/3$, $\beta_3 = 1/2$ and $\beta_4 = 1$ this yields a second-order accurate time integration method [58]. The time step is determined by the stability restriction of the numerical scheme. It depends on the grid-size h and the eigenvalues of the flux Jacobi matrix of the numerical flux f. In a short-hand notation one may write $\delta t = \mathrm{CFL}\ h/|\lambda_{max}|$ where $|\lambda_{max}|$ denotes the eigenvalue of the flux Jacobi matrix with maximal size, and CFL denotes the Courant-Friedrichs-Levy-number which depends on the specific choice of explicit time integration method. For the present four-stage Runge-Kutta method a maximum CFL number of 2.4 can be established using a Von Neumann stability analysis. In the actual simulations we use $\mathrm{CFL} = 1.5$, which is suitable for both DNS and LES, irrespective of the specific subgrid model used.

In order to specify the spatial discretization we distinguish between the treatment of the convective and the viscous fluxes. We will only specify the numerical approximation of the ∂_1-operator; the ∂_2 and ∂_3-operators are treated analogously. Subgrid-terms are discretized with the same method as the viscous terms. Throughout we will use a second order method for the viscous fluxes and both a second order, and a fourth order accurate method for the convective fluxes. All these methods are constructed from (a combination of) first order numerical derivative operators D_j.

The second-order method that we consider is a finite volume method [59]. The discretization of the convective terms is the cell vertex trapezoidal rule, which is a weighted second-order central difference. In vertex (i, j, k) the corresponding operator is denoted by D_1 and for the approximation of $\partial_1 f$ it is defined as

$$\begin{aligned} (D_1 f)_{i,j,k} &= (s_{i+1,j,k} - s_{i-1,j,k})/(2h_1) \\ \text{with} \quad s_{i,j,k} &= (g_{i,j-1,k} + 2g_{i,j,k} + g_{i,j+1,k})/4 \\ \text{and} \quad g_{i,j,k} &= (f_{i,j,k-1} + 2f_{i,j,k} + f_{i,j,k+1})/4 \end{aligned} \tag{7.62}$$

The viscous terms contain second-order derivatives which are treated by a consecutive application of two first order numerical derivatives. This requires for example that the gradient of the velocity is calculated in centers of grid-cells. In center $(i + \frac{1}{2}, j + \frac{1}{2}, k + \frac{1}{2})$ the corresponding discretization $D_2 f$ has the form

$$\begin{aligned} (D_2 f)_{i+\frac{1}{2},j+\frac{1}{2},k+\frac{1}{2}} &= (s_{i+1,j+\frac{1}{2},k+\frac{1}{2}} - s_{i,j+\frac{1}{2},k+\frac{1}{2}})/h_1 \\ \text{with} \quad s_{i,j+\frac{1}{2},k+\frac{1}{2}} &= (f_{i,j,k} + f_{i,j+1,k} + f_{i,j,k+1} + f_{i,j+1,k+1})/4 \end{aligned} \tag{7.63}$$

The second derivative is subsequently calculated with operator D_1; thus we approximate, e.g., $\partial_{11}(f)_{ijk} \approx D_1(D_2(f))_{ijk}$.

The combination of D_1 and D_2 is robust with respect to odd-even decoupling but it is only second order accurate. In a similar manner we may construct a fourth-order accurate method. The corresponding expression for $D_3 f$ has the following form:

$$\begin{aligned} (D_3 f)_{i,j,k} &= (-s_{i+2,j,k} + 8s_{i+1,j,k} - 8s_{i-1,j,k} + s_{i-2,j,k})/(12h_1) \quad (7.64) \\ \text{with} \quad s_{i,j,k} &= (-g_{i,j-2,k} + 4g_{i,j-1,k} + 10g_{i,j,k} + 4g_{i,j+1,k} - g_{i,j+2,k})/16 \\ \text{and} \quad g_{i,j,k} &= (-f_{i,j,k-2} + 4f_{i,j,k-1} + 10f_{i,j,k} + 4f_{i,j,k+1} - f_{i,j,k+2})/16 \end{aligned}$$

This scheme is conservative, since it is a weighted central difference. The coefficients in the definition for $g_{i,j,k}$ are chosen such that $g_{i,j,k}$ is a fourth order accurate approximation to $f_{i,j,k}$ and π-waves in the x_3-direction give no contributions to $g_{i,j,k}$. The definition for $s_{i,j,k}$ has the same properties with respect to the x_2-direction. For convenience, we will refer to a combination of D_3 for the convective, and D_1, D_2 for the viscous fluxes as fourth-order methods, but we remark that the formal spatial accuracy of the scheme is only second-order due to the treatment of the viscous terms.

3.2 The turbulent mixing layer

The flow in a temporally developing turbulent mixing layer is well documented in literature (e.g. [7]), and will be considered here to test the LES-α modeling approach. In this section we review the scenario of the development of the flow that is considered and sketch the type of predictions that can be obtained by traditional LES using the dynamic model. This serves as a point of reference for the next section.

We simulate the compressible three-dimensional temporal mixing layer and use a convective Mach number $M = 0.2$ and a Reynolds number based on upper stream velocity and half the initial vorticity thickness of 50. The governing equations are solved in a cubic geometry of side $l = 59$. Periodic boundary conditions are imposed in the streamwise (x_1) and spanwise (x_3) direction, while in the normal (x_2) direction the boundaries are free-slip walls. The initial condition is formed by mean profiles corresponding to constant pressure $p = 1/(\gamma M^2)$ where $\gamma = 1.4$ is the adiabatic gas constant, $u_1 = \tanh(x_2)$ for the streamwise velocity component, $u_2 = u_3 = 0$ and a temperature profile given by the Busemann-Crocco law. Superimposed on the mean profile are two- and three-dimensional perturbation modes obtained from linear stability theory. Further details may be found in [34].

The DNS is conducted on a uniform grid with 192^3 cells using the fourth order spatial discretization method. Visualization of the DNS data demonstrates the roll-up of the fundamental instability and successive pairings (figure 7.1).

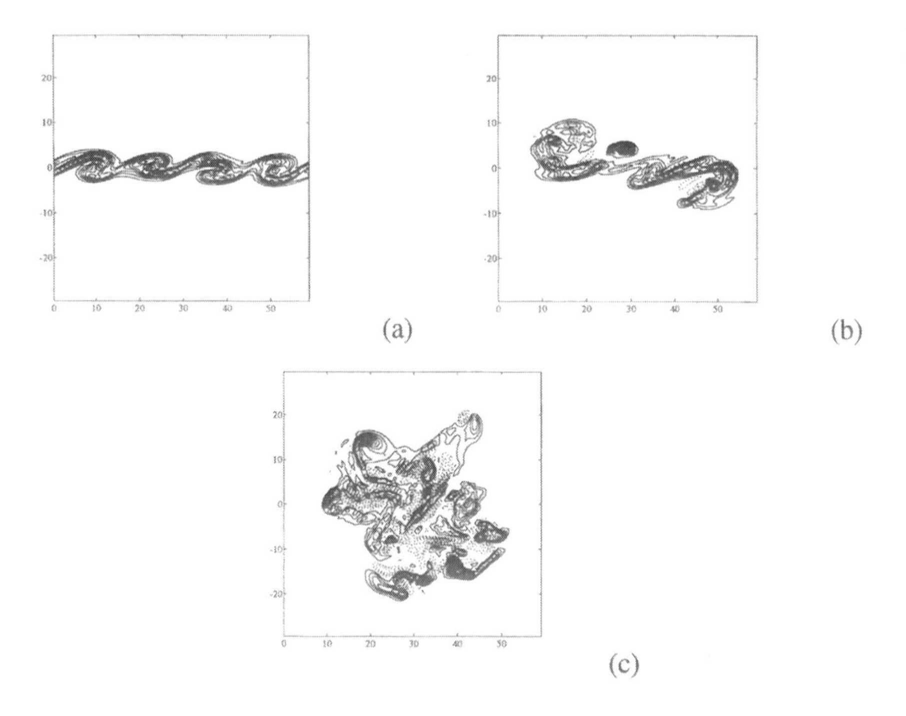

Figure 7.1. Results from a DNS using 192^3 points. Contours of spanwise vorticity for the plane $x_3 = 3L/4$ at (a) $t = 20$, (b) $t = 40$ and (c) $t = 80$. Solid and dotted contours indicate negative and positive vorticity respectively. The contour increment is 0.1.

Four rollers with mainly negative spanwise vorticity are observed at $t = 20$. After the first pairing ($t = 40$) the flow has become highly three-dimensional. Another pairing ($t = 80$), yields a single roller in which the flow exhibits a complex structure.

The accuracy of the simulation with 192^3 cells is satisfactory as is inferred from coarser grid computations on 64^3 and 128^3 cells. The evolution of the momentum thickness

$$\delta(t) = \frac{1}{4}\int_{-L/2}^{L/2} (1 - \langle u_1 \rangle)(\langle u_1 \rangle + 1)dx_2 \tag{7.65}$$

and an instantaneous velocity component at the center of the shear layer are shown in figure 7.2. The 64^3-simulation is inadequate for the prediction of the local instantaneous solution, but the momentum thickness appears quite reasonable.

To illustrate the effect that filtering has on a well-developed DNS solution, vorticity contours for $\Delta = L/16$ are shown in figure 7.3. Comparing this with the corresponding DNS results in figure 7.1 allows one to appreciate the strong smoothing effect that filtering has on the solution. On the right in figure 7.3 we

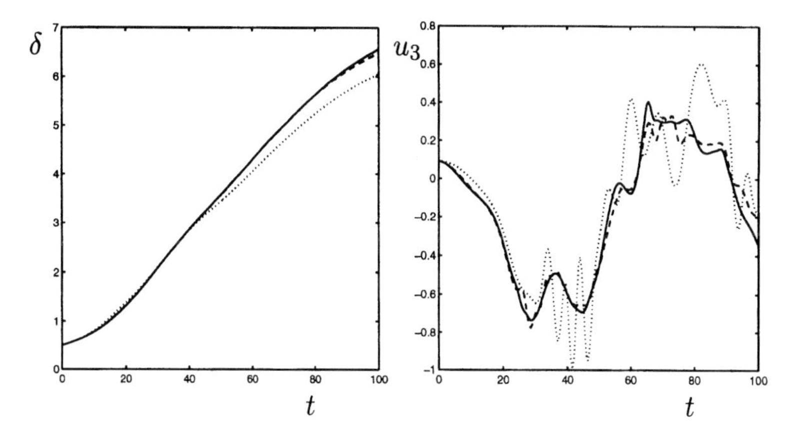

Figure 7.2. Evolution of the momentum thickness (left) and u_3 at $(\frac{1}{4}L, 0, \frac{1}{2}L)$ (right) obtained from simulations which do not involve any subgrid model and employ a sequence of grids: 64^3 (dotted), 128^3 (dashed) and 192^3 (solid).

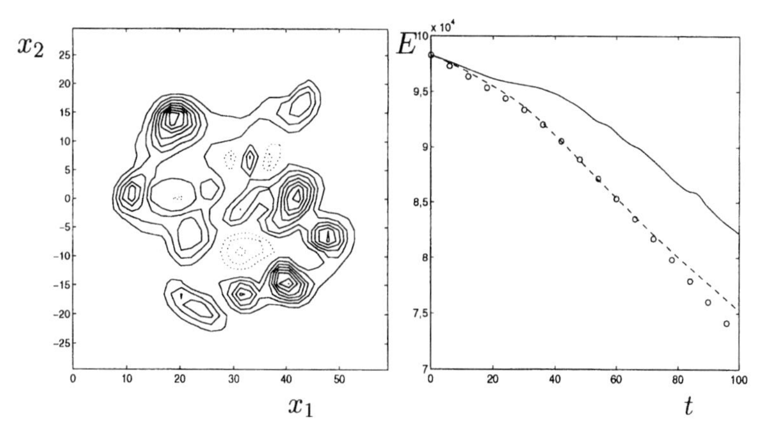

Figure 7.3. Left: Contour-lines of the z-component of the vorticity. The effect of spatially filtering the DNS solution at $t = 80$ in figure 7.1 with a top-hat filter and filter-width $\Delta = L/16$. Right: prediction of the kinetic energy with the dynamic eddy-viscosity model (dashed) compared with the filtered DNS results (markers) and a simulation on the coarse LES grid (32^3) without a model (solid).

included the decay of the resolved turbulent kinetic energy, defined as

$$E = \frac{1}{2}\int_{\Omega}(\overline{u}_1^2 + \overline{u}_2^2 + \overline{u}_3^2)\, d\mathbf{x} \tag{7.66}$$

We observe that the dynamic eddy-viscosity model generates quite a correction of the 'no-model coarse grid simulation'. Other models were also considered in [7], such as Smagorinsky's model, Bardina's scale-similarity model and dynamic mixed models. Roughly speaking, the use of Bardina's model leads to flow predictions which contain somewhat too many small scale features whereas the Smagorinsky model, with eddy-coefficient $C_S = 0.17$ prevents the flow from developing beyond the transitional stage due to excessive dissipation in the early stages of the evolution. Finally, the dynamic mixed models were all shown to perform about equally well and provide accurate predictions.

4. LES-α of a mixing layer

In this section we will consider LES using the LES-α model. Above, in section 2.2, we introduced this model and identified three distinct contributions; in fact, the LES-α model contains the explicitly filtered nonlinear gradient model (m_{ij}^{NG}), the Leray model (m_{ij}^{L}) and the complete LES-α model (m_{ij}^{α}). These are defined as

$$m_{ij}^{NG} = \frac{\Delta^2}{24}\left(\overline{\partial_k v_i \, \partial_k v_j}\right) \equiv \overline{A_{ij}} \tag{7.67}$$

$$m_{ij}^{L} = \frac{\Delta^2}{24}\left(\overline{\partial_k v_i \, \partial_k v_j} + \overline{\partial_k v_i \, \partial_j v_k}\right) \equiv \overline{A_{ij}} + \overline{B_{ij}} \tag{7.68}$$

$$\begin{aligned} m_{ij}^{\alpha} &= \frac{\Delta^2}{24}\left(\overline{\partial_k v_i \, \partial_k v_j} + \overline{\partial_k v_i \, \partial_j v_k} - \overline{\partial_i v_k \, \partial_j v_k}\right) \\ &\equiv \overline{A_{ij}} + \overline{B_{ij}} - \overline{C_{ij}} \end{aligned} \tag{7.69}$$

First we will consider reference LES using these models and compare predictions with those obtained with dynamic subgrid models (subsection 4.1). Then we focus our attention on the resolved kinetic energy dynamics in subsection 4.2. Finally, in subsection 4.3 we consider (nearly) grid-independent LES-α predictions which arise when refining the grid while keeping Δ constant.

4.1 Reference LES of the mixing layer

In order to create a point of reference, we consider LES defined on a resolution of 32^3 grid-points. This choice represents a significant saving compared to the full DNS and places a considerable importance on the subgrid fluxes. This resolution was used previously in a comparative study of subgrid models in [7].

The simulations will be illustrated by considering the evolution of the resolved kinetic energy $E(t)$, defined in (7.66). In addition, we consider the momentum thickness $\delta(t)$, based on filtered variables which quantifies the spreading of the mean velocity profile. We also investigate the Reynolds-stress profiles $\langle q_1 q_2 \rangle$ defined with respect to the fluctuation $q_i = v_i - \langle v_i \rangle$. Finally, we incorporate the streamwise kinetic energy spectrum in the turbulent regime at $t = 80$. In this way a number of essentially different quantities (mean, local, plane averaged) are included in the comparisons in order to assess various aspects of the quality of the models.

For all simulations we will use a LES-filter-width $\Delta = L/16$. On the 32^3 grid this implies that $\Delta/h = 2$, i.e., two grid-intervals cover the filter-width. Moreover, unless explicitly stated otherwise, the explicit filter used in the definition of the LES-α subgrid models will have the same width as the LES-filter, i.e., $\kappa = 1$. The filtering is done using the top-hat filter and we adopt the trapezoidal rule to perform the numerical integrations. The simulations that will be presented in the following subsections correspond to a slightly different initial condition than used in section 3.2. The differences are fairly small, but still prevent a direct comparison with the filtered DNS results presented in section 3.2.

Explicit filtering is essential The proposed subgrid models in the α framework each contain the nonlinear gradient model and also involve an explicit filtering. As analyzed in section 2.4, the nonlinear gradient model, without explicit filtering gives rise to instabilities. These instabilities manifest themselves, e.g., by an increase in the resolved kinetic energy, instead of the monotonic decrease that is characteristic of this relaxing shear layer, cf. figure 7.3.

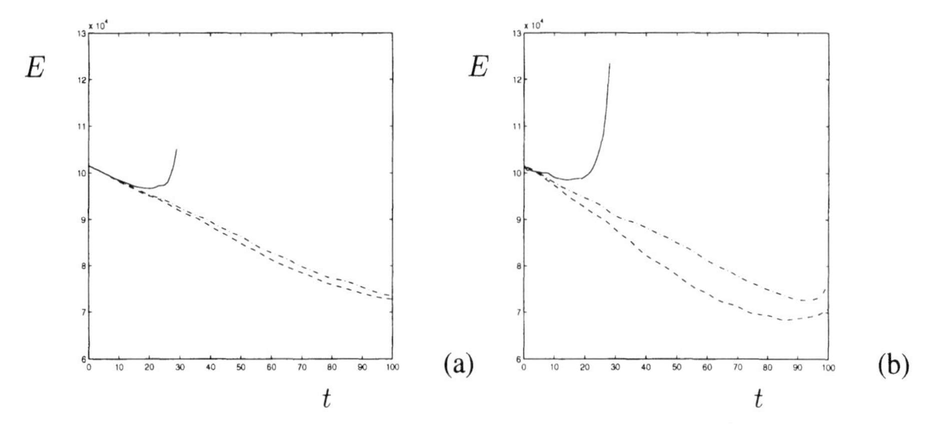

Figure 7.4. Evolution of resolved kinetic energy for the nonlinear gradient model (solid) and the filtered nonlinear gradient model, using $\kappa = 1$ (dashed) and $\kappa = 2$ (dash-dotted) (a). In (b) we show the corresponding results obtained with the unfiltered (solid) and filtered full LES-α model. These instabilities are expected on grounds discussed in subsection 2.4.

The question arises whether the explicit filtering can stabilize the simulations on this reference grid. In figure 7.4 we compiled predictions for the kinetic energy, obtained with the nonlinear gradient and the full LES-α model, both without and with explicit filtering, at different values of the ratio κ. We notice that the explicit filtering is essential in order to maintain stability of the simulation. It appears that the unfiltered LES-α model is even slightly more unstable than the unfiltered nonlinear gradient model. We also considered these models at a higher resolution of 64^3 grid-points. Consistent with the analysis in section 2.4 the instability becomes stronger if the grid is refined while keeping the LES filter-width Δ constant. It is seen that the value of κ, which defines the width of the explicit filter relative to the width of the LES filter, has a comparably small effect on the predictions of the nonlinear gradient model. The instabilities which arise when using the full LES-α model, appear somewhat stronger and, e.g., E even increases in the turbulent regime, despite the explicit filtering. This indicates a marginally unstable simulation, and the situation improves when κ is increased.

Reference LES-α predictions Some basic predictions obtained using the three LES-α models will be presented next. These predictions will contain errors because of shortcomings in the subgrid parameterizations and the numerical treatment. These aspects will be focused upon in the next two subsections respectively; here it is our aim to provide an impression of the predictions under numerical conditions that are fairly common in present-day LES, e.g. $\Delta/h = 2$.

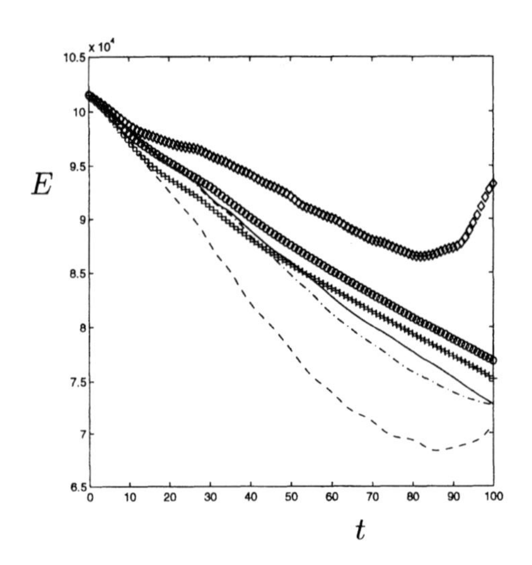

Figure 7.5. Evolution of the resolved kinetic energy E comparing the following models: LES-α (dashed), Leray (solid), filtered nonlinear gradient model (dash-dotted), dynamic mixed (+), dynamic eddy-viscosity (o) and no-model ($\diamond$). We used $\kappa = 1$.

In figure 7.5 we compare the evolution of the resolved turbulent kinetic energy E for a number of subgrid models. We included not only predictions corresponding to the three LES-α models, but also the dynamic mixed model, the dynamic eddy-viscosity model and the simulation without any subgrid model at all. The subgrid models provide a significant improvement compared to the case without a model. From previous simulations we know that a fairly close agreement exists between filtered DNS data and the dynamic models, as shown in figure 7.3 (see [7] for more details). Using the dynamic predictions as point of reference here as well, we notice that the Leray and the filtered nonlinear gradient model provide more accurate predictions than the full LES-α model. We also considered the Bardina model and observed that the predictions are virtually identical to those obtained with the filtered nonlinear gradient model. The Smagorinsky model at $C_S = 0.17$ was used as well and showed too strong dissipation.

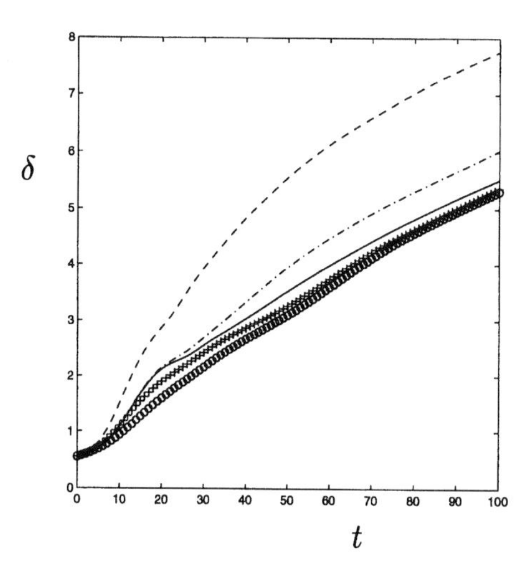

Figure 7.6. Evolution of the resolved momentum thickness δ comparing the following models: LES-α (dashed), Leray (solid), filtered nonlinear gradient model (dash-dotted), dynamic mixed (+), dynamic eddy-viscosity (o). We used $\kappa = 1$.

The momentum thickness δ is shown in figure 7.6. The prediction of δ from the full LES-α model is much higher than those obtained with the other subgrid models and compared to the dynamic model predictions as point of reference, it appears too high. The predictions of the Bardina similarity model again coincide with the filtered nonlinear gradient model, and these predictions are somewhat larger than arise from the Leray model. All the LES-α models predict δ larger than the dynamic models. Since the dynamic predictions slightly underestimate δ according to [7], it appears that the Leray model and the filtered nonlinear gradient model predict δ more accurately, compared to filtered DNS results, than the other models.

In figure 7.7 we collected the Reynolds stress $-\langle q_1 q_2 \rangle$. We observe that all three LES-α models predict a considerably higher level of fluctuations compared to the dynamic models. The full LES-α model predicts levels of fluctuation close to those obtained from the simulation without any subgrid model, suggesting that this model introduces too many small scales into the solution. Likewise, the filtered gradient model generates high levels of fluctuations, while the Leray model is much closer to the levels of fluctuation that are found using the dynamic models.

We consider the streamwise kinetic energy spectrum in the turbulent regime at $t = 80$, in figure 7.8. We observe a clear separation of the predictions in two groups. The two dynamic models show a strong reduction of the smaller

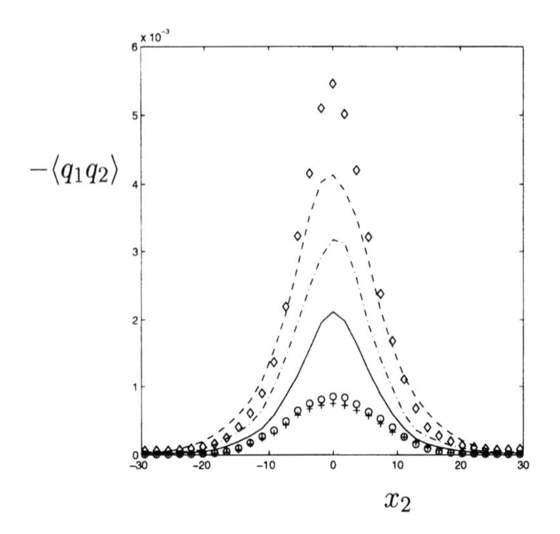

Figure 7.7. Comparison of the Reynolds stress $-\langle q_1 q_2 \rangle$ at $t = 70$: LES-α (dashed), Leray (solid), filtered nonlinear gradient model (dash-dotted), dynamic mixed (+), dynamic eddy-viscosity (o) and no-model ($\diamond$). We used $\kappa = 1$.

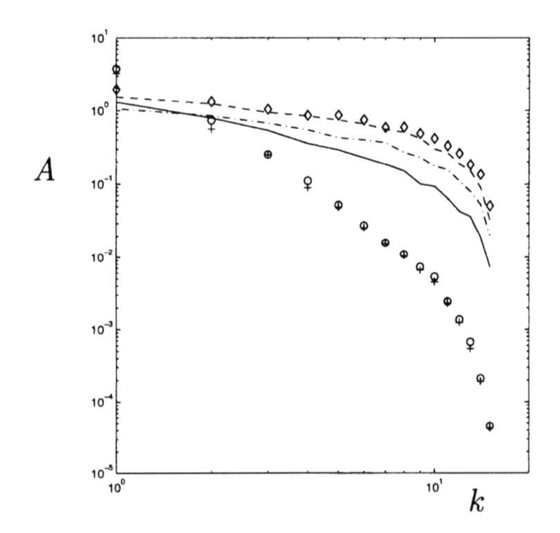

Figure 7.8. Comparison of the streamwise energy spectrum $A(k)$ at $t = 80$: LES-α (dashed), Leray (solid), filtered nonlinear gradient model (dash-dotted), dynamic mixed (+), dynamic eddy-viscosity (o) and no-model ($\diamond$). We used $\kappa = 1$.

scales. In contrast, the full LES-α model displays a spectrum that is quite close to the spectrum of the simulation without any subgrid model. This situation improves significantly for the filtered nonlinear gradient model and finally, the Leray model provides the largest attenuation of the small scales among the three LES-α models.

In summary, the simulations suggest that the full LES-α model does not sufficiently reduce the resolved kinetic energy, leads to too large momentum-

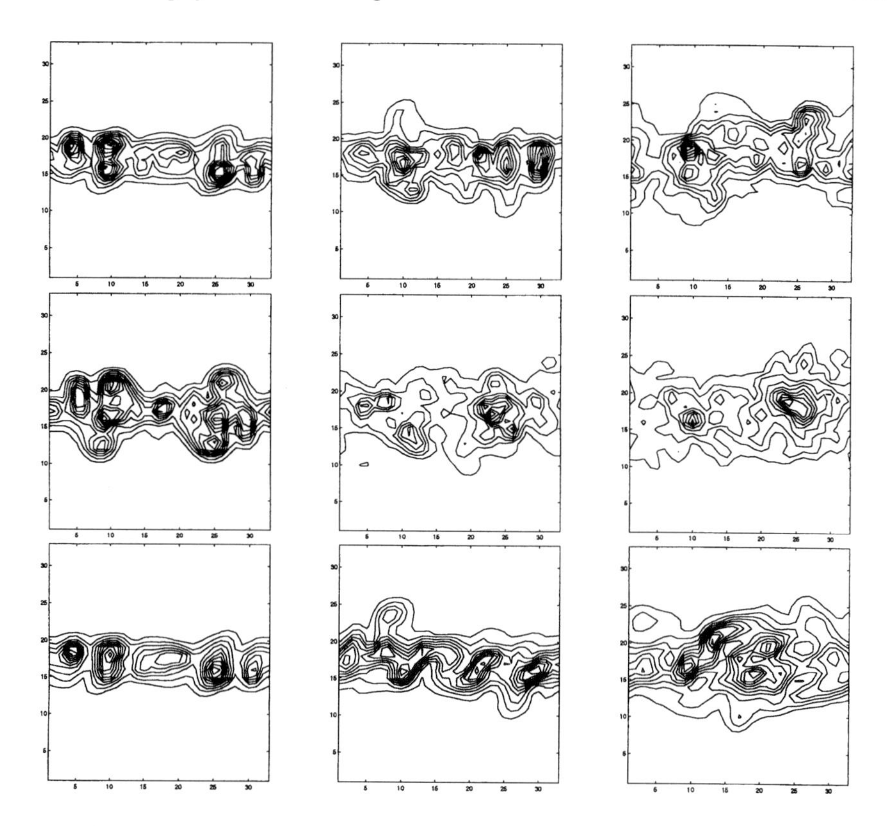

Figure 7.9. Comparison of the evolution of the spanwise vorticity at $x_3 = L/2$ and $t = 20$, $t = 40$ and $t = 80$, from left to right. This is shown for the Leray model (top row), the full LES-α model (middle row) and the dynamic mixed model (bottom row) at $\kappa = 1$.

thickness and too high levels of fluctuation, which is apparent in the spectrum at small scales and snapshots of the solution. The filtered nonlinear gradient model performs better than the full LES-α model but also over-predicts the smaller scales. In contrast to these two models, the Leray model, appears to predict the energy decay properly, shows accurate momentum-thicknesses and apparently reliable levels of turbulence intensities, as shown also in the spectrum and in snapshots of the solution. In order to better understand these predictions we turn to the resolved kinetic energy dynamics in the next subsection and consider the contribution of the individual terms in the models.

4.2 Resolved kinetic energy dynamics

In this section we consider the evolution of the resolved kinetic energy and determine the type and magnitude of the various subgrid contributions. The evolution of E is governed by

$$\partial_t E = \int_\Omega \{ \frac{1}{Re} \overline{u}_i \partial_j \overline{\sigma_{ij}} - \overline{u}_i \partial_j \tau_{ij} \} \, d\mathbf{x}$$

$$= \int_\Omega \{-\frac{1}{2Re}\overline{\sigma_{ij}}\,\overline{\sigma_{ij}} + \tau_{ij}\partial_j\overline{u}_i\}\,d\mathbf{x} \tag{7.70}$$

where use was made of the identity $\overline{\sigma_{ij}}\partial_j\overline{u}_i = \frac{1}{2}\overline{\sigma_{ij}}\,\overline{\sigma_{ij}}$. The predicted kinetic energy evolution, corresponding to a given LES model, emerges by replacing the turbulent stress tensor by its subgrid scale model. We notice that the dynamics of E is governed by a purely dissipative term arising from the molecular dissipation and a term that is associated with the subgrid model. We will consider the resolved energy dynamics both for the coarse reference grid of 32^3 grid points and a much finer simulation in which we use 96^3. The latter simulations use the same filter-width $\Delta = L/16$ but correspond to a much higher subgrid resolution $\Delta = 6h_{LES}$. In this way we can clarify some of the dynamics observed on the coarse grid as well as obtain an impression of the actual dynamical consequences associated with the subgrid model.

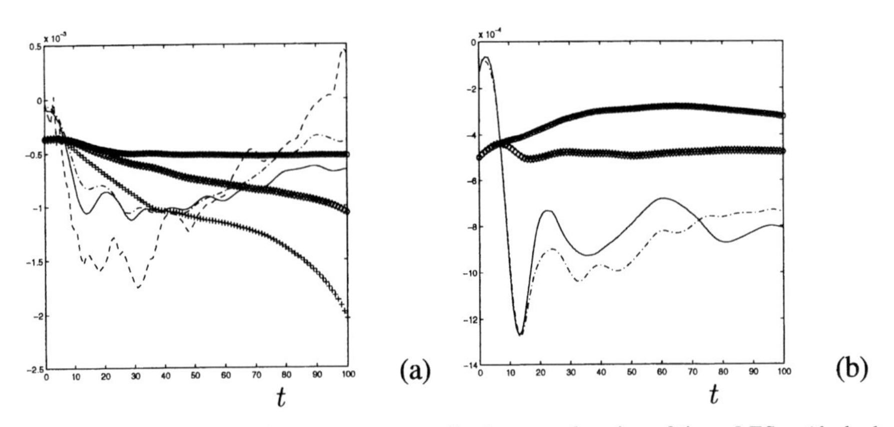

Figure 7.10. Resolved kinetic energy rate contributions as a function of time: LES-α (dashed line $\partial_t E_m$, +: $\partial_t E_v$), Leray (solid line: $\partial_t E_m$, o: $\partial_t E_v$), filtered nonlinear gradient model (dash-dotted line $\partial_t E_m$, $\diamond$: $\partial_t E_v$). We used $\kappa = 1$ and show the results for the 32^3 grid in (a) and for the 96^3 grid in (b).

In figure 7.10 we show the total viscous and subgrid contributions to $\partial_t E$, denoted $\partial_t E_v$ and $\partial_t E_m$, respectively. We notice that on the 32^3 grid the viscous contribution corresponding to the Leray model is quite constant in the turbulent regime and the subgrid contribution gradually becomes of the same order of magnitude. For the filtered nonlinear gradient model we observe a proper dissipation of energy, but slightly less than the Leray model. The corresponding viscous flux contribution increases considerably in the turbulent regime. Finally, for the full LES-α model we observe that the subgrid contribution not only becomes less important in the turbulent regime but even changes sign. This can readily be associated with the overestimated small scale contributions in the solution, as shown in the previous subsection. For the better resolved LES the results of the Leray model and the filtered nonlinear gradient model are quite comparable and appear more predictable. Moreover, all subgrid fluxes are seen to settle and oscillate around some nonzero values, indicating perhaps a more regular self-similar development of the mixing layer in the turbulent regime. The full LES-α model was found to become unstable around $t = 70$, at this high subgrid resolution. Apparently, the explicit filtering, which was found to be essential in the previous section, in order to stabilize the simulation on the coarse grid, is not damping sufficiently well to maintain stability of the LES-α model at increased subgrid resolution.

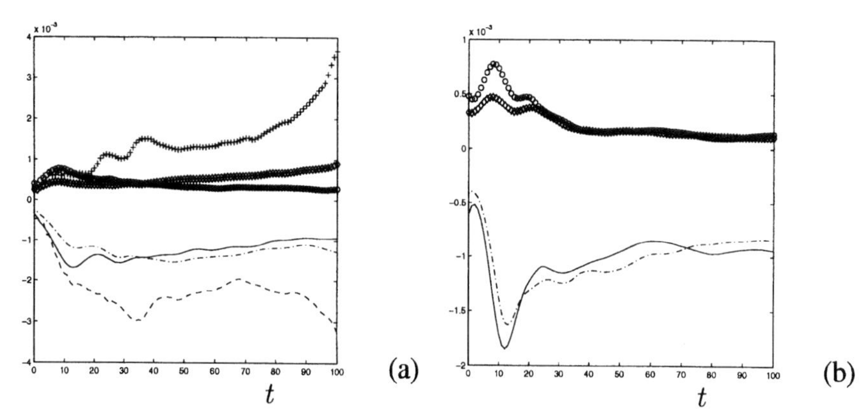

Figure 7.11. Resolved kinetic energy rate contributions: LES-α (dashed line P_f, +: P_b), Leray (solid line: P_f, o: P_b), filtered nonlinear gradient model (dash-dotted line P_f, $\diamond$: P_b). We used $\kappa = 1$ and show the results for the 32^3 grid (a) and the 96^3 grid (b).

To further analyse the dynamical behavior, we can look at splitting the subgrid contribution into a positive, i.e., forward scatter or dissipative, contribution and a negative, i.e., backward scatter or reactive contribution. To formalize this splitting, we introduce

$$P_b(f) = \int_\Omega \frac{1}{2}(f + |f|)\, d\mathbf{x}\ , \quad P_f(f) = \int_\Omega \frac{1}{2}(f - |f|)\, d\mathbf{x} \qquad (7.71)$$

to measure the amount of back-scatter (P_b) and forward scatter of energy (P_f) associated with a term represented by f. In figure 7.11 we collected the forward and backward scatter contributions for the LES-α models. We observe that all these models predict both forward and backward scatter of energy, which sets them apart from simple eddy-viscosity models that only provide forward scatter. On the coarse grid (32^3) the Leray model and the filtered nonlinear gradient model compare fairly well. The full LES-α model, however, shows a large amount of back-scatter in the turbulent regime and a likewise increased importance of forward scatter. On the finer grid the Leray and filtered nonlinear gradient model show a balance between forward and backward scatter in the turbulent regime.

A third decomposition of the total contribution arises in terms of the individual subgrid-terms. If we consider, e.g., the full LES-α model, written as

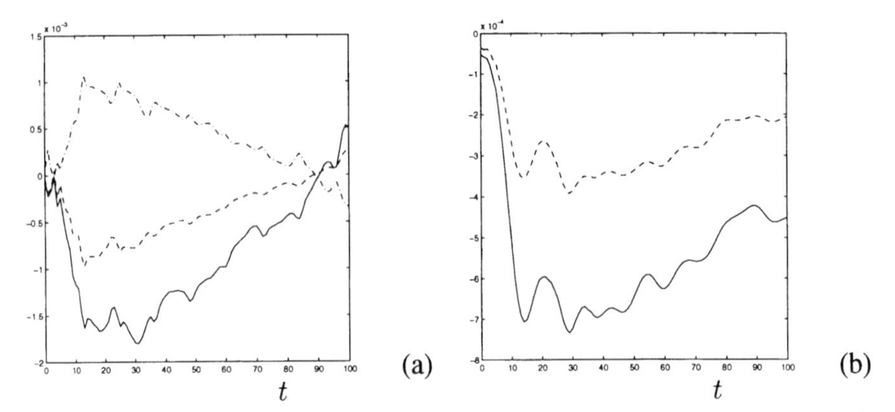

Figure 7.12. Resolved kinetic energy rate contributions: full LES-α shown in (a) (solid $\overline{A}$-term, dashed $\overline{B}$-term, dash-dotted $\overline{C}$-term), Leray model shown in (b):(solid $\overline{A}$-term, dashed $\overline{B}$-term) We used $\kappa = 1$ and show the results for the 32^3 grid.

$m_{ij}^{\alpha} = \overline{A_{ij}} + \overline{B_{ij}} - \overline{C_{ij}}$ we may write

$$\partial_t E = \partial_t E_v + \partial_t E_A + \partial_t E_B - \partial_t E_C \tag{7.72}$$

with individual contributions due to the viscous fluxes, and the $A - B - C$ terms respectively;

$$\partial_t E_v = \int_{\Omega} -\frac{1}{2Re} \overline{\sigma_{ij}}\, \overline{\sigma_{ij}}\, d\mathbf{x}\,, \quad \partial_t E_A = \int_{\Omega} \overline{A_{ij}} \partial_j \overline{u}_i \, d\mathbf{x} \tag{7.73}$$

and similarly for the other terms. In figure 7.12 we collected the detailed energy-dynamics decomposition corresponding to the two terms which make up the Leray model and the three terms that constitute the full LES-α model. Notice that figure 7.10 already contains the single contribution of the filtered nonlinear gradient model. The Leray model is seen to be composed of two terms that both dissipate energy. The full LES-α model behaves less regular and we observe that the dissipative $\overline{B}$ contribution is nearly canceled by the reactive $\overline{C}$ contribution.

From this analysis of the resolved energy dynamics it seems that the Leray model and the filtered nonlinear gradient model provide more accurate results and the internal functioning of these models increases the robustness of the model. We also applied the Leray model to a flow at a ten times higher Reynolds number. Although these latter results are still preliminary, it seems that the Leray model provides reliable results, even in such very turbulent flows. Further analysis of this regime is needed though and this will be published elsewhere [37].

In the next section we use the well resolved LES predictions in combination with the coarse grid simulation results to assess the influence of the spatial discretization scheme on the evolution of the flow.

4.3 Toward grid-independent LES-α

The reference simulations considered above, are executed on a fairly coarse grid which corresponds to a ratio between filter-width Δ and grid-spacing h of $\Delta/h = 2$. In many present-day LES, even a ratio of $\Delta/h = 1$ is frequently used. These choices usually arise from considerations of available computational resources, but at the same time imply that the smallest resolved scales of size on the order of Δ are not accurately represented in the numerical treatment. Hence, there is a strong possibility that the marginal subgrid resolution influences the dynamical properties of the simulated flow.

In order to assess this discretization effect, we will compare the reference LES with simulations at a much higher subgrid resolution. In this way, the effect of subgrid modeling is better represented numerically, while it remains of the same magnitude as in the coarse grid simulation. This allows to isolate the dynamic effects of the spatial discretization in the modeled equations.

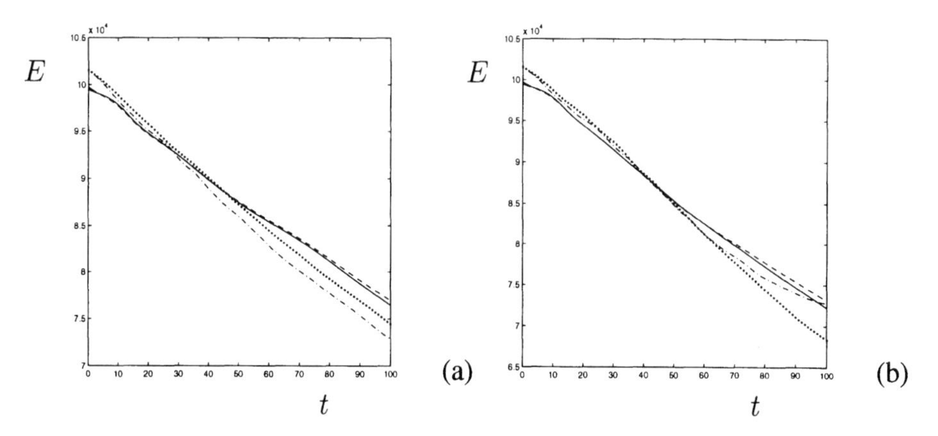

Figure 7.13. Resolved kinetic energy using the Leray model (a) and the filtered nonlinear gradient model (b): solid line (4th order method, resolution 96^3), dashed line (4th order method, resolution 64^3), dash-dotted line (4th order method, resolution 32^3) and dotted-line (2nd order method, resolution 32^3).

We compare simulations on 32^3, 64^3 and 96^3 grid-points and focus our attention on the Leray model and the filtered nonlinear gradient model. In figure 7.13 we compare the predicted resolved kinetic energy obtained with the second and the fourth order accurate spatial discretization method. The subgrid resolution corresponding to these three grids is $\Delta/h = 2$, 4 and 6 respectively. We observe a very close agreement between the predictions using the fourth order accurate method and $\Delta/h = 4$ and 6. This suggests that a mean flow quantity such as the resolved kinetic energy is well represented using $\Delta/h = 4$. Moreover, we notice that on the coarsest grid, the accuracy of the prediction based on the second order method compares closely to that obtained with the fourth order method. Apparently, if the dynamic effects of the spatial discretization errors are quite large, a lower order method can be competitive with a higher order method. For both models the reliability of the predictions on the coarse grid are affected considerably by the coarseness of the subgrid resolution. In both situations, and for both spatial discretizations the effect of the discretization error is seen to enhance the reduction of the resolved kinetic energy.

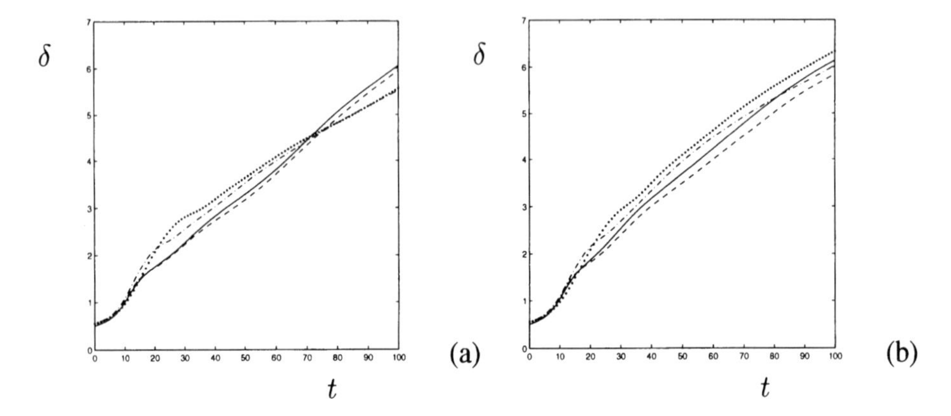

Figure 7.14. Resolved momentum-thickness using the Leray model (a) and the filtered nonlinear gradient model (b): solid line (4th order method, resolution 96^3), dashed line (4th order method, resolution 64^3), dash-dotted line (4th order method, resolution 32^3) and dotted-line (2nd order method, resolution 32^3).

To further establish the convergence, we show the momentum-thickness in figure 7.14. We observe that the convergence is clear for the Leray model and that a value of $\Delta/h = 4$ corresponds to reliable predictions for both subgrid models considered, although the sensitivity of the momentum thickness is larger than that of the resolved kinetic energy. Regarding the results for the best resolved simulations, we observe that the momentum-thickness develops very nearly linearly with time in the Leray model, while a slight reduction of the growth-rate predicted by the nonlinear gradient model is seen in the turbulent regime.

Finally, we show the spectra obtained with the Leray and the filtered nonlinear gradient model on the selected grids in figure 7.15. We notice a general resemblance between the results obtained with both models. As the subgrid resolution is increased, a larger portion of the spectrum is better resolved and correspondingly the spectra obtained on 64^3 and 96^3 grid-points agree for a significant number of modes. Moreover, the differences due to the use of the second order or the fourth order accurate methods are expressed very clearly in the spectra; a strong reduction in the high wavenumbers on the 32^3 grid results when using the second order method. This is consistent with the stronger attenuation of the high wavenumbers in the second order method, compared to the fourth order accurate method.

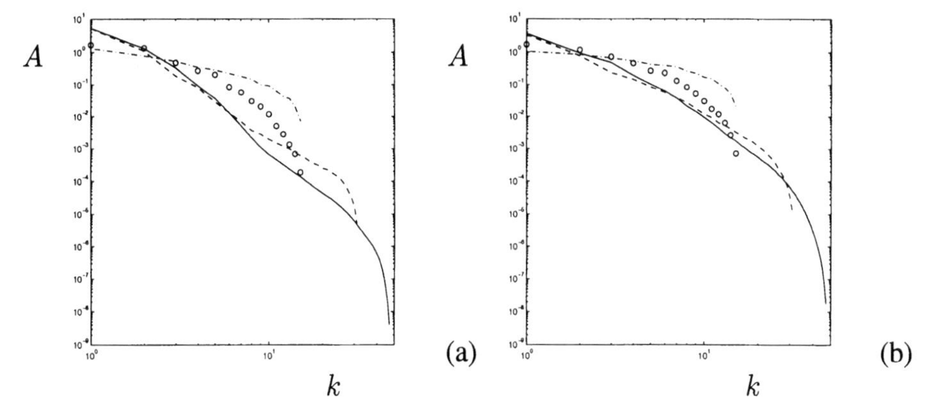

Figure 7.15. Streamwise kinetic energy spectrum $A(k)$ using the Leray model (a) and the filtered nonlinear gradient model (b): solid line (4th order method, resolution 96^3), dashed line (4th order method, resolution 64^3), dash-dotted line (4th order method, resolution 32^3) and marker 'o' (2nd order method, resolution 32^3).

5. Concluding remarks

In the α-framework we derived a new subgrid closure for the turbulent stress tensor in LES, by using the Kelvin theorem applied to a filtered transport velocity. The proposed LES-α subgrid model was shown to contain two other subgrid parameterizations, i.e., the nonlinear gradient model and a model corresponding to the Leray regularization of Navier-Stokes dynamics. Moreover, the LES-α model stress tensor contains an explicit filtering in its definition, which sets it apart from other subgrid models in literature. It was shown that this explicit filtering is essential for the LES-α models; without it, the simulations develop a finite time instability. This instability was also observed in an analysis of the viscous one-dimensional Burgers equation and appears to be associated with the nonlinear gradient term.

The flow in a turbulent mixing layer was considered, in order to test the capabilities of the three 'nested' LES-α models. Through a comparison with dynamic (mixed) models, we inferred that the Leray model provides particularly accurate predictions. The filtered nonlinear gradient model in turn, compares well with the Leray model and corresponds closely to the Bardina similarity model for the flows considered. The full LES-α model was seen to generate too many small scales in the solution and correspondingly poorer predictions, e.g., too large growth-rate, too high levels of turbulence intensities, etc. An analysis of the resolved kinetic energy dynamics showed that the full LES-α model contains two competing contributions which may tend to cancel and, thus, destabilize some simulations involving this model. In particular, this tendency implied that simulations using the full LES-α model were unstable in our simulations of turbulent mixing at high subgrid resolution.

Apart from accuracy and a certain degree of numerical robustness, the computational overhead associated with a subgrid model is an element of importance in evaluating simulation methods. The computational effort associated with the three LES-α models is considerably lower than that of dynamic (mixed) models. This is primarily a result of the reduction in the number of explicit filtering operations required to evaluate the LES-α model. Moreover, the accuracy of the predictions is higher for the Leray model than for *any* of the other subgrid models considered. Consequently, the Leray model is favored in this study and holds promise for applications to even more complex and demanding flow problems. Preliminary results at significantly higher Reynolds number suggest that the Leray model performs well also in this case.

Finally, we also considered the contribution to the dynamics arising from the spatial discretization method at coarse subgrid resolution. In general, the role of numerical methods in relation to LES has not yet been sufficiently clarified to determine unambiguously whether the accuracy of predictions is restricted because of shortcomings in the subgrid model, or whether this inaccuracy is due to spatial discretization effects. Resolving this ambiguity and determining the main sources of error would help in finding the best strategy for employing computational resources in an LES. In one strategy, a grid-independent solution of the modeled LES equations at fixed filter-width is sought and one only assigns computational resources for reducing numerical errors by increasing the subgrid resolution ratio Δ/h. This approach was applied here and used to evaluate the additional dissipation that arises from the spatial finite volume discretizations at either second order, or fourth order accuracy. This additional dissipation is associated with the implicit filtering effect, corresponding to the attenuation of the small flow features represented on the grid.

References

[1] Spalart, P.R.: 1988. Direct simulation of a turbulent boundary layer up to $R_\theta = 1410$. *J. Fluid Mech.* **187**, 61.

[2] Rogallo, R.S., Moin, P.: 1984. Numerical simulation of turbulent flows. *Ann. Rev. Fluid Mech.* **16**, 99.

[3] Lesieur, M.: 1990. Turbulence in fluids. *Kluwer Academic Publishers.*, Dordrecht.

[4] Meneveau, C., Katz, J.: 2000. Scale-invariance and turbulence models for large-eddy simulation. *Ann. Rev. Fluid Mech.* **32**, 1.

[5] Germano, M.: 1992. Turbulence: the filtering approach. *J. Fluid Mech.* **238**, 325.

[6] Geurts B.J.: 1999 Balancing errors in LES. Proceedings *Direct and Large-Eddy simulation III: Cambridge*. Eds: Sandham N.D., Voke P.R., Kleiser

L., Kluwer Academic Publishers, 1.

[7] Vreman A.W., Geurts B.J., Kuerten J.G.M.: 1997. Large-eddy simulation of the turbulent mixing layer, *J. Fluid Mech.* **339**, 357.

[8] Vreman A.W., Geurts B.J., Kuerten J.G.M.: 1994. Realizability conditions for the turbulent stress tensor in large eddy simulation. *J.Fluid Mech.* **278**, 351.

[9] Ghosal, S.: 1999. Mathematical and physical constraints on large-eddy simulation of turbulence. *AIAA J.* **37**, 425.

[10] Holm, D.D., Marsden, J.E., Ratiu, T.S.: 1998. Euler-Poincaré equations and semidirect products with applications to continuum theories. *Adv. in Math* **137**, 1.

[11] Holm, D.D. , Marsden, J.E., Ratiu, T.S.: 1998. Euler-Poincaré models of ideal fluids with nonlinear dispersion. *Phys. Rev. Lett.* **80**, 4173.

[12] Camassa, R., Holm, D.D.: 1993. An integrable shallow water equation with peaked solitons. *Phys. Rev. Lett.* **71**, 1661.

[13] Chen, S.Y., Foias, C., Holm, D.D., Olson, E.J., Titi, E.S., Wynne, S.: 1998. The Camassa-Holm equations as a closure model for turbulent channel flow. *Phys. Rev. Lett.* **81**, 5338.

[14] Chen, S.Y., Foias, C., Holm, D.D., Olson, E.J., Titi, E.S., Wynne, S.: 1999. A connection between Camassa-Holm equations and turbulent flows in channels and pipes. *Phys. Fluids* **11**, 2343.

[15] Chen, S.Y., Foias, C., Holm, D.D., Olson, E.J., Titi, E.S., Wynne, S.: 1999. The Camassa-Holm equations and turbulence. *Physica D* **133**, 49.

[16] Chen, S.Y., Holm, D.D., Margolin, L.G., Zhang, R.: 1999. *Direct numerical simulations of the Navier-Stokes alpha model, Physica D* **133**, 66.

[17] Holm, D.D.: 1999. Fluctuation effects on 3D Lagrangian mean and Eulerian mean fluid motion. *Physica D* **133**, 215.

[18] Marsden, J.E., Shkoller, S.: 2001. The anisotropic Lagrangian averaged Navier-Stokes and Euler equations. *Arch. Ration. Mech. Analysis.* (In the press.)

[19] Shkoller, S.: 1998. Geometry and curvature of diffeomorphism groups with H^1 metric and mean hydrodynamics. *J. Func. Anal.* **160**, 337.

[20] Marsden, J.E., Ratiu, T.S., Shkoller, S.: 2000. The geometry and analysis of the averaged Euler equations and a new diffeomorphism group. *Geom. Funct. Anal.* **10**, 582.

[21] Foias, C., Holm, D.D., Titi, E.S.: 2002. The three dimensional viscous Camassa–Holm equations, and their relation to the Navier–Stokes equations and turbulence theory. *J. Diff. Eqs.* to appear.

[22] Marsden, J.E., Shkoller, S.: 2001. Global well-posedness for the Lagrangian averaged Navier-Stokes (LANS$-\alpha$) equations on bounded domains. *Phil. Trans. R. Soc. Lond. A* **359**, 1449.

[23] Foias, C., Holm, D.D., Titi, E.S.: 2001. The Navier-Stokes-alpha model of fluid turbulence. *Physica D* **152** 505.

[24] Domaradzki, J.A., Holm, D.D.: 2001. Navier-Stokes-alpha model: LES equations with nonlinear dispersion. *Modern simulation strategies for turbulent flow*. Edwards Publishing, Ed. B.J. Geurts. 107.

[25] Mohseni, K., Kosovic, B., Marsden, J.E., Shkoller, S., Carati, D., Wray, A., Rogallo, R.: 2000. Numerical simulations of homogeneous turbulence using the Lagrangian averaged Navier-Stokes equations. *Proc. of the 2000 Summer Program*, 271. Stanford, CA: NASA Ames / Stanford University.

[26] Holm, D.D., Kerr, R.: 2001. Transient vortex events in the initial value problem for turbulence. In preparation.

[27] Holm, D.D.: 1999. Alpha models for 3D Eulerian mean fluid circulation. *Nuovo Cimento C* **22**, 857.

[28] Bardina, J., Ferziger, J.H., Reynolds, W.C.: 1983. Improved turbulence models based on large eddy simulations of homogeneous incompressible turbulence. Stanford University, Report TF-19.

[29] Leonard, A.: 1974. Energy cascade in large-eddy simulations of turbulent fluid flows. *Adv. Geophys.* **18**, 237.

[30] Clark, R.A., Ferziger, J.H., Reynolds, W.C.: 1979. Evaluation of subgrid-scale models using an accurately simulated turbulent flow. *J. Fluid Mech.* **91**, 1.

[31] Vreman A.W., Geurts B.J., Kuerten J.G.M.: 1996. Large eddy simulation of the temporal mixing layer using the Clark model *TCFD***8**, 309.

[32] Winckelmans, G.S., Jeanmart, H., Wray, A.A., Carati, D., Geurts, B.J.: 2001. Tensor-diffusivity mixed model: balancing reconstruction and truncation. *Modern simulation strategies for turbulent flow*. Edwards Publishing, Ed. B.J. Geurts. 85.

[33] Leray, J.: 1934. Sur le mouvement d'un liquide visqueux emplissant l'espace *Acta Math.* **63**, 193. Reviewed, e.g., in P. Constantin, C. Foias, B. Nicolaenko and R. Temam, *Integral manifolds and inertial manifolds for dissipative partial differential equations*. Applied Mathematical Sciences, **70**, (Springer-Verlag, New York-Berlin, 1989).

[34] Vreman A.W., Geurts B.J., Kuerten J.G.M.: 1995. A priori tests of Large Eddy Simulation of the compressible plane mixing layer. *J. Eng. Math.* **29**, 299

[35] de Bruin, I.C.C.: 2001. Direct and large-eddy simulation of the spatial turbulent mixing layer. *Ph.D. Thesis*, Twente University Press.

[36] Germano, M., Piomelli U., Moin P., Cabot W.H.: 1991. A dynamic subgrid-scale eddy viscosity model. *Phys.of Fluids* **3**, 1760

[37] Geurts, B.J., Holm, D.D.: 2001. Leray simulation of turbulent flow. *In preparation.*

[38] Ghosal, S.: 1996. An analysis of numerical errors in large-eddy simulations of turbulence. *J. Comp. Phys.* **125**, 187.

[39] Vreman A.W., Geurts B.J., Kuerten J.G.M.: 1996. Comparison of numerical schemes in Large Eddy Simulation of the temporal mixing layer. *Int.J.Num. Meth. in Fluids* **22**, 297.

[40] Vreman A.W., Geurts B.J., Kuerten J.G.M.: 1994. Discretization error dominance over subgrid-terms in large eddy simulations of compressible shear layers. *Comm.Num.Meth.Eng.Math.* **10**, 785.

[41] Geurts, B.J., Fröhlich, J.: 2001. Numerical effects contaminating LES: a mixed story. *Modern simulation strategies for turbulent flow*. Edwards Publishing, Ed. B.J. Geurts. 309.

[42] Geurts B.J., Vreman A.W., Kuerten J.G.M.: 1994. Comparison of DNS and LES of transitional and turbulent compressible flow: flat plate and mixing layer. Proceedings *74th Fluid Dynamics Panel and Symposium on Application of DNS and LES to transition and turbulence, Crete*, AGARD Conf. Proceedings 551:51

[43] Ghosal, S., Moin, P.: 1995. The basic equations for large-eddy simulation of turbulent flows in complex geometry. *J. Comp. Phys.* **286**, 229.

[44] Du Vachat, R.: 1977. Realizability inequalities in turbulent flows. *Phys. Fluids* **20**, 551.

[45] Schumann, U.: 1977. Realizability of Reynolds-stress turbulence models. *Phys. Fluids* **20**, 721.

[46] Ortega, J.M.: 1987. Matrix Theory. *Plenum Press.* New York

[47] Ghosal, S., Lund, T.S., Moin, P., Akselvoll, K.: 1995. A dynamic localization model for large-eddy simulation of turbulent flows. *J. Fluid Mech.* **286**, 229.

[48] Lilly, D.K.: 1992. A proposed modification of the Germano subgrid-scale closure method. *Phys.of Fluids A* **4**, 633.

[49] Geurts, B.J.: 1997. Inverse modeling for large-eddy simulation. *Phys. of Fluids* **9**, 3585.

[50] Kuerten J.G.M., Geurts B.J., Vreman, A.W., Germano, M.: 1999. Dynamic inverse modeling and its testing in large-eddy simulations of the mixing layer. *Phys. Fluids* **11**, 3778.

[51] Horiuti, K.: Constraints on the subgrid-scale models in a frame of reference undergoing rotation. *J. Fluid Mech.*, submitted.

[52] Germano, M.: 1986. Differential filters for the large eddy numerical simulation of turbulent flows. *Phys Fluids* **29**, 1755.

[53] Smagorinsky, J.: 1963. General circulation experiments with the primitive equations. *Mon. Weather Rev.* **91**, 99.

[54] Stolz, S., Adams, N.A.: 1999. An approximate deconvolution procedure for large-eddy simulation. *Phys.of Fluids*, **11**, 1699.

[55] Domaradzki, J.A., Saiki, E.M.: 1997. A subgrid-scale model based on the estimation of unresolved scales of turbulence. *Phys.of Fluids* **9**, 1.

[56] Wasistho, B., Geurts, B.J., Kuerten, J.G.M.: 1997. Numerical simulation of separated boundary layer flow. *J. Engg. Math* **32**, 179.

[57] Chatelin, F.: 1993. Eigenvalues of matrices. *John Wiley & Sons*. Chichester.

[58] Jameson, A.: 1983. Transonic flow calculations. *MAE-Report 1651*, Princeton

[59] Geurts, B.J., Kuerten, J.G.M.: 1993. Numerical aspects of a block-structured flow solver. *J.Engg.Math.* **27**, 293.

Chapter 8

FORWARD-IN-TIME DIFFERENCING FOR FLUIDS: SIMULATION OF GEOPHYSICAL TURBULENCE

Piotr K. Smolarkiewicz
National Center for Atmospheric Research, P.O. Box 3000, Boulder, Colorado 80307-3000, USA.
smolar@ncar.ucar.edu

Joseph. M. Prusa
Department of Mechanical Engineering, Iowa State University, Ames, Iowa 50010, USA.
prusa@iastate.edu

Abstract The Earth's atmosphere and oceans are essentially incompressible, highly turbulent fluids. Herein, we demonstrate that nonoscillatory forward-in-time (NFT) methods can be efficiently utilized to accurately simulate a broad range of flows in these fluids. NFT methods contrast with the more traditional centered-in-time-and-space approach that underlies the bulk of computational experience in the meteorological community. We challenge the common misconception that NFT schemes are overly diffusive and therefore inadequate for high Reynolds number flow simulations, and document their numerous benefits. In particular, we show that, in the absence of an explicit subgrid-scale turbulence model, NFT methods offer means of implicit subgrid-scale modeling that can be quite effective in assuring a quality large-eddy-simulation of high Reynolds number flows. The latter is especially important where complications such as large span of scales, density stratification, planetary rotation, inhomogeneity of the lower boundary, etc., make explicit modeling of subgrid-scale motions difficult. Theoretical discussions are illustrated with examples of meteorological flows that address the range of applications from micro turbulence to climate.

Keywords: Geophysical Turbulence, Large Eddy Simulation, Finite Difference Methods for Fluids, Subgrid-scale Models

1. Introduction

Prediction of the Earth's climate and weather is difficult in large part due to the ubiquity of turbulence in the atmosphere and oceans. Geophysical flows

D. Drikakis and B.J. Geurts (eds.), Turbulent Flow Computation, 279–312.

evince fluid motions ranging from dissipation scales as small as a fraction of a millimeter to planetary scales of thousands of kilometers — a range of scales spanning ~ 10 decades. The span in time scales (from a fraction of a second to many years) is equally large. Turbulence in the atmosphere and the oceans is generated by heating and by boundary stresses — just as in engineering flows. However geophysical flows are further complicated by planetary rotation and density/temperature stratification, which lead to phenomena not commonly found in engineering applications. In particular, rotating stratified fluids can support a variety of inertio-gravity and planetary waves. When the amplitude of such a wave becomes sufficiently large (i.e., comparable to the wavelength) the wave can break, generating a localized burst of turbulence. If one could see the phenomena that occur internally in geophysical flows at any scale, one would be reminded of familiar pictures of white water in a mountain stream or of breaking surf on a beach.[1]

Because of the enormous range of scales, direct numerical simulation (DNS) of the Earth's weather and climate is far beyond the reach of current computational technology. Consequently, all efforts in numerical simulation must begin with an attempt to truncate the number of degrees of freedom of the atmosphere/ocean system to a magnitude that is tractable on modern computational machines, and to attempt to do so in as physically meaningful a way as possible. One powerful tool in the numerical study of turbulent atmospheric and oceanic flows is large-eddy simulation (LES). Most often, LES is understood as a numerical integration of coarse-grained (filtered) Navier-Stokes' equations, where all scales of motion larger than some multiple of the grid interval ΔX (e.g., $2\Delta X$, [8]) are resolved explicitly, but all finer scales are modeled based on universal properties of fully developed turbulence; cf. [35]. The intent of LES is to account for the effects of subgrid-scale (SGS) motions (i.e., below the available resolution of the numerical model) on those resolved on the grid. The literature on LES is vast and continuously growing — the interested reader is referred to [33, 30, 22, 32, 21] for reviews and discussion.

The idea underlying LES, to represent the effects of the unresolved scales in terms of the resolved scales, is simple and intellectually appealing. Unfortunately, LES is often difficult to realize effectively in practice. For the Navier-Stokes' equations, the formalism of decomposing flow variables into resolved and unresolved scales of motion by spatial filtering leads straightforwardly to altered equations, modified by the appearance of the divergence of the so-called SGS stress tensor. In general, the components of the SGS stress tensor contain products of various components, resolved and unresolved, of the flow variables. Such components are not computable *a priori* for the obvious reason that the unresolved flow variables are not available. To proceed, one must model these components in terms only of the resolved components of the flow variables. That one can do so accurately is the fundamental assumption of the LES approach, an assumption that is only justified by expediency and ultimately by comparison with experiment. In meteorology, the most popular SGS model postulates an SGS stress tensor proportional to the rate of strain (of the resolved flow) via a local eddy-viscosity coefficient. The eddy coefficient itself is assumed to depend on the magnitude of the local strain rate. This model, first proposed by Smagorinsky, is a multidimensional generalization of the artificial viscosity developed by von Neumann to regularize shock waves [44]. The simplicity of this model, mainly the result of assuming isotropy and stationarity of the unresolved scales, has made this a widely used choice for LES. The need to incorporate additional physics has lead to elaborations of the original Smagorinsky model, such as dynamic models that address transient, inhomogeneous behavior — e.g., the *TKE* model that adds an additional equation for turbulent kinetic energy [24], and *double filter* models that allow for backscatter as a negative diffusion [16], [25]. More recently, models have been developed that address anisotropy and backscatter [21] by utilizing a nonlinear relationship between the strain rate and the SGS stress tensor.

As an example of the complexity that can arise in the broad range of geophysical applications, consider the inhomogeneous, anisotropic turbulence that occurs when deep gravity waves break at mesopause altitudes [39, 40, 41]. Figure 8.1 shows the resulting potential temperature, Θ, 155 minutes after the onset of forcing in the lower atmosphere (e.g., below the tropopause located at 15 km altitude). The basic state is one of uniform stability, zonal wind (right to left in the xz field), and density scale height. These results were computed using the nonoscillatory, semi-Lagrangian option of our model on a $544\times80\times291$ grid. The onset of wave breaking occurs approximately 25 minutes earlier, in a very localized region near 100 km altitude and zonal position -90 km. In contrast, the wave breaking and turbulence depicted in Fig. 8.1 cover a very large region with global scales $\sim$ 100 km. The Θ field clearly exhibits inhomogeneity in the zonal and vertical directions, yet is homogeneous in the spanwise (y) direction.

The anisotropy of the turbulence may be demonstrated by examining the velocity derivative skewness, $S_{u^j}(t) \equiv -\langle(\partial u^j/\partial x^j)^3\rangle/\langle(\partial u^j/\partial x^j)^2\rangle^{3/2}$ (no summation is implied by the repeated indices), in each of the coordinate directions ($\langle\ \rangle$ denotes domain averaging). For isotropic turbulence, laboratory experiments with low Reynolds number flows yield values of $S \sim 0.4$, whereas direct numerical simulations at moderate Reynolds numbers give $S \sim 0.5$ [23]. Measurements in the atmospheric boundary layer yield $S \sim 0.8$ [4]. In our breaking gravity wave application, if the averaging domain is taken as the full computational domain, then S_u=1.64, S_v=0.10, and S_w=0.20 result. While clearly documenting anisotropy, a much better appreciation of the complexity of the turbulence and wavefield can be obtained by computing *local* values of S_{u^j} using an averaging domain of smaller size — so as to produce S_{u^j} *fields*. Figure 8.2 depicts the resulting S_u, S_v, and S_w fields using a (7.5 km)^3 (9261

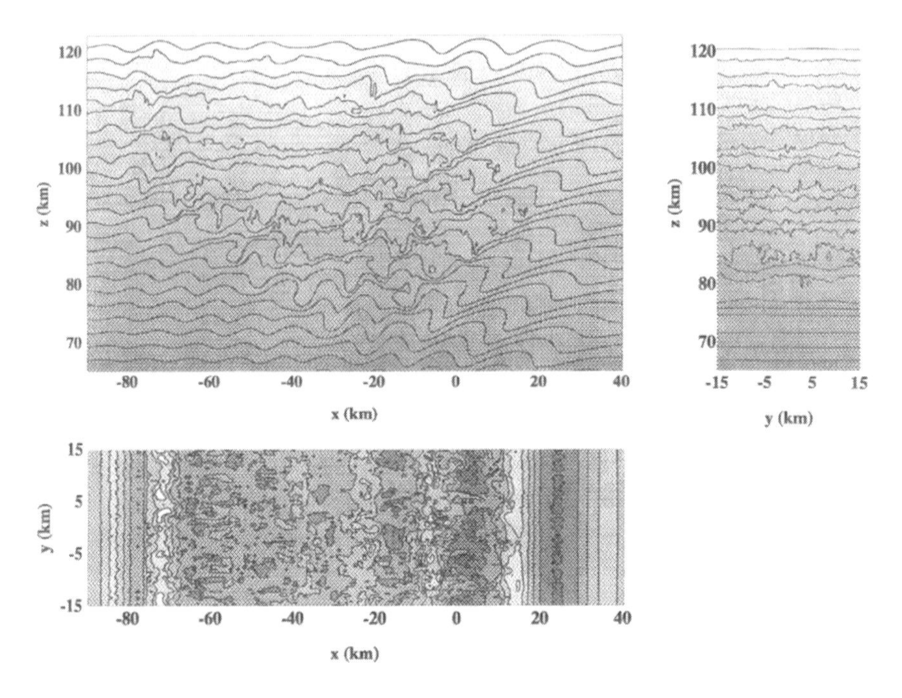

Figure 8.1. Turbulence resulting from breaking of deep atmospheric gravity waves: contour density plots of Θ in vertical xz (zonal) and yz (spanwise) planes at spanwise and zonal locations $y = 0$ km and $x = -35$ km, respectively; and horizontal xy plane at altitude $z = 100$ km. The range in Θ is from 2950 at the bottom to 31,000 K at the top of the xz and yz plots; for the xy plot the average is $\sim$ 12,000 K. Darker grays indicate cooler temperatures. The contour intervals are uniform in $\ln(\Theta)$; for the xz and yz plots they range from $\Delta\Theta$ = 300 to 2900 K at the bottom and top, respectively; for the xy plot $\Delta\Theta \sim$ 450 K. Note that the full computational domain extends zonally from -103 to 103 km and vertically from 15 to 125 km.

grid points) averaging domain. For isotropic turbulence, all three plots shown in the figure should be an identical, uniform shade of gray. Instead, only small, isolated pockets with $S \sim 0.5$ are seen to exist; and these regions do not correlate strongly among S_u, S_v, and S_w.

Another source of difficulty for the practical implementation of formal SGS models occurs when they are combined with nonorthogonal time-dependent geometries, which underlie mesh refinement schemes. Here one lacks the ability to define a global grid-interval-dependent coarsening scale; in addition, the SGS models themselves become overly complicated and cumbersome to program. In operational models for weather or air-quality prediction — which frequently employ the composition of several mappings (e.g., from the Cartesian to terrain-following, to the horizontally stretched coordinates, and to spherical geometry) — it is quite common that formal SGS models are further simplified by writing the *Laplacian* in Cartesian form.

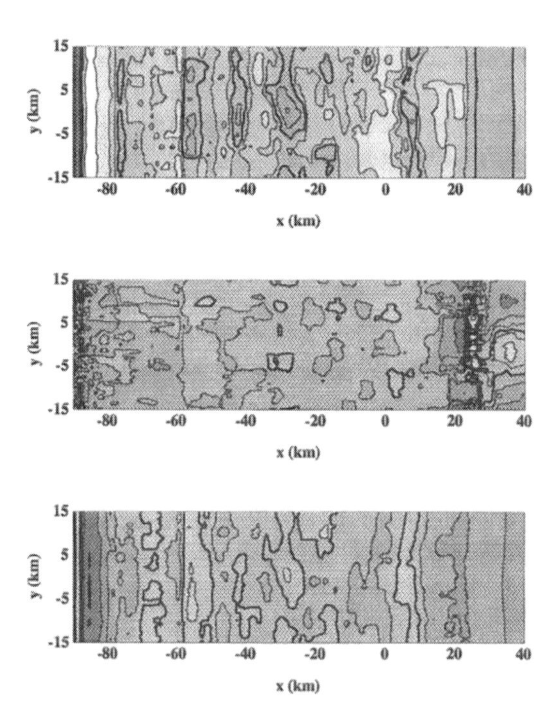

Figure 8.2. Contour density plots of velocity derivative skewnesses, S_u (top), S_v (middle), and S_w (bottom), corresponding to $i = 1, 2, 3$ for coordinate directions x, y, z with velocity components u, v, w. Horizontal xy plane at same altitude ($z = 100$ km) and time (155 minutes) as in bottom plot of Fig. 8.1. Contour lines are at S_{uj} = -2, -1, -.35, 0, 0.35, 0.65, 1, and 2; with bold contour at value 0.35. Darker grays correspond to smaller values of S_{uj}. Note that the lighter gray regions bounded by the bold contour and first regular contour have skewness values in the range 0.35 to 0.65 and thus nominally correspond to isotropic turbulence. The averaging domain for the skewness computations is (7.5 km)^3.

Historically, the development of SGS models for meteorological codes was motivated by the numerical necessity to guarantee the stability of simulations (viz., nonlinear instability problem; [44]). In consequence, SGS models are often appreciated more for their ability to suppress unphysical oscillations rather than for their accurate representation of actual turbulence stresses that might result from the subgrid scales. However, if nonlinear instability is a concern, there are simpler, more effective, and more universal means of prevention (cf. the paragraph concluding the Smagorinsky's [44] historical remarks). In particular, during the past decade a class of finite difference methods — nonoscillatory forward-in-time (NFT)[2] — have been shown to exhibit the remarkable property of representing the effects of unresolved scales of motion without need for any explicit subgrid scale model. Numerical simulations of turbulent flows that abandon the rigorous notion of LES and merely aim at computing explicitly large coherent eddies resolvable on the grid are sometimes referred to as VLES (for very-large-eddy simulations [35, 55]). Hereafter, we shall use the VLES term exclusively to emphasize that no explicit SGS model has been employed.

Successful simulations of turbulent flows, while relying only on the dissipative properties of nonoscillatory advection schemes, have been reported in a variety of regimes and applications. The use of nonoscillatory schemes as an implicit turbulence model *per se* has been discussed in the context of free shear flows [34], strongly compressible turbulence [37], and the development of turbulence at Rayleigh-Taylor unstable interfaces [26]. More recently, Margolin et al. [29] demonstrated that an atmospheric code based on the nonoscillatory advection scheme MPDATA [52] can accurately reproduce (i.e., in close agreement with field and laboratory data and the existing benchmark computations) the structure of the convective planetary boundary layer. When an explicit turbulence model was implemented, MPDATA did not add any unnecessary diffusion. Of greater interest, when no explicit turbulence scheme was employed, MPDATA itself appeared to include an effective SGS model. Finally, using the explicit turbulence model with the eddy viscosity reduced by some factor, MPDATA added just enough dissipation. These experiments document the self-adaptive character of MPDATA and suggest the physically realistic character of its truncation error (i.e., numerical dissipation). Further, the diverse experience of several research groups noted above illustrates the overall effectiveness of the approach, and suggests that it is the general properties of NFT rather than the specific details of any particular scheme that are responsible for its success.

The fact that finite difference advection schemes contain implicit SGS models is by no means a revelation. Consider that any finite-volume (i.e. flux form) advection scheme yields a truncation error in the form of a flux divergence. Thus, when such a scheme is employed to transport momentum, the truncation-error fluxes effectively define a subgrid-scale stress tensor. The remarkable efficacy of the implicit SGS models contained in NFT methods is, nonetheless, surprising. At this point there is no established theory justifying this success. Recently, the similarity between some of the truncation error terms of MPDATA and corrections to the analytic equations that account for finite volume effects has been demonstrated for the one-dimensional Burgers' equation [28]. From this point of view, the effects of turbulence do not arise from the unresolved scales of motion (i.e., from the flow), but rather from the finite scales of length and time that characterize the experiment or the simulation (i.e., from the observer). A further prediction of this development is that it is only those NFT schemes based on flux-form differencing that should represent accurate implicit SGS models.

In this chapter, we will present further evidence in support of the NFT approach for VLES of turbulent geophysical flows. In each of these applications, our VLES approach has worked extremely well. However a comprehensive theory that would, among other things, indicate the limits of applicability of the implicit SGS approach is essential. Brown et al. [3], have performed detailed simulations of convective and shear boundary layers. In general their results are similar to those of [29]. However they note that in very coarsely resolved simulations, the presence of unresolved boundary layers in the flow may require explicit models to account for wall forcings. Moin and Kravchenko [31] are more critical of the approach. However their NFT scheme is based on an advective formulation, and in our understanding should be expected to produce inferior results [28]. In this chapter, we will focus on flows without prescribed boundary layer forcings—a convenient scenario relevant to a broad class of geophysical applications—and will disregard potential difficulties in near-wall regions. The latter problems are the subject of current research.

The chapter is organized as follows. In the following section, we will summarize the NFT formulation that we have found particularly useful in geophysical applications. Although we shall pursue the discussion in abstraction from any specific system of the equations of motion, we will tailor the presentation for the class of anelastic (incompressible type) nonhydrostatic models. In section 3, we shall discuss the accuracy of the NFT schemes in context of turbulent flow simulations. In particular, we shall document the efficacy of NFT methods as effective SGS models, employing results of the decaying turbulence and convective boundary layer simulations. Section 4 culminates our discussion with an example of a relatively complex VLES study — a simulation of an idealized climate of the Earth. At this occasion, we shall also summarize the anelastic model equations used in our research, and outline some details of implementing our NFT approach. The remarks in section 5 conclude the chapter.

2. NFT methods for fluids

In modeling atmospheric/oceanic flows the governing equations can always be viewed in the form of a generalized transport equation

$$\frac{\partial \rho^* \psi}{\partial \bar{t}} + \overline{\nabla} \cdot (\rho^* \overline{\mathbf{v}}^* \psi) = \rho^* R \,, \tag{8.1}$$

where $\overline{\nabla} \cdot \equiv (\partial/\partial\bar{x},\ \partial/\partial\bar{y},\ \partial/\partial\bar{z}) \cdot$ is *defined* as the divergence operator in a time variable, curvilinear coordinate system; and ψ is an intensive dependent fluid variable, e.g., component of specific momentum (i.e. velocity component), potential temperature, water vapor mixing ratio, specific salinity, etc. In (8.1), the coefficients ρ^*, $\overline{\mathbf{v}}^*$, and R are assumed to be known functions of the independent variables $(\overline{\mathbf{x}}, \bar{t})$. In anelastic-flow applications addressed in this chapter, $\rho^* \equiv \rho_b \overline{G}$ plays the role of a reference density multiplied by the Jacobian of the transformation from a stationary, physical coordinate system $\mathbf{x}$ (such as Cartesian) to the curvilinear system $\overline{\mathbf{x}}$ (such as terrain-following coordinates on a rotating sphere);[3] $\overline{\mathbf{v}}^* \equiv (\overline{u^*},\ \overline{v^*},\ \overline{w^*}) \equiv d\overline{\mathbf{x}}/d\overline{t} \equiv \dot{\overline{\mathbf{x}}}$ is the *contravariant* velocity in the curvilinear system; and R combines all forcings and/or sources. In general, both $\overline{\mathbf{v}}^*$ and R are functionals of the dependent variables (see [52] for examples). With these definitions, (8.1) is mathematically equivalent to the Langrangian evolution equation

$$\frac{d\psi}{d\overline{t}} = R\,, \tag{8.2}$$

regardless of the assumed form of the mass continuity equation (i.e., compressible, incompressible, Boussinesq, anelastic, etc.), with $d/d\overline{t} = \partial/\partial\overline{t} + \dot{\overline{\mathbf{x}}}\overline{\nabla}$ for the material derivative.

Our basic NFT approach for approximating either (8.1) or (8.2) on a discrete mesh is second-order-accurate in space and time. The two optional model algorithms, Eulerian [49] and semi-Lagrangian [48], correspond to (8.1) and (8.2). Either algorithm can be written in the compact form

$$\psi_{\mathbf{i}}^{n+1} = LE_{\mathbf{i}}(\widetilde{\psi}) + 0.5\Delta t R_{\mathbf{i}}^{n+1}\,. \tag{8.3}$$

Here and throughout the remainder of this section, we drop the overbar for the transformed time coordinate to simplify notation; all t's are to be understood as $\overline{t}$'s. We denote $\psi_{\mathbf{i}}^{n+1}$ as the solution at the grid point $(\overline{\mathbf{x}}_{\mathbf{i}},\ t^{n+1})$; $\widetilde{\psi} \equiv \psi^n + 0.5\Delta t R^n$; and LE denotes either an advective semi-Lagrangian or a flux-form Eulerian NFT transport operator. In the Eulerian scheme, LE integrates the homogeneous transport equation (8.1), i.e., LE advects $\widetilde{\psi}$ using a fully second-order-accurate multidimensional NFT advection scheme (cf. [46]). In the semi-Lagrangian algorithm, LE remaps transported fields arriving at the grid points $(\overline{\mathbf{x}}_{\mathbf{i}}, t)$ back to the departure points of the flow trajectories $(\overline{\mathbf{x}}_o(\overline{\mathbf{x}}_{\mathbf{i}}, t^{n+1}), t^n)$. Interestingly, the Lagrangian LE is also composed from NFT advection schemes. In contrast to the Eulerian case, however, it exploits directionally split 1D NFT advection schemes. The theory underlying such a design relies on elementary properties of differential forms [48, 47].[4] In technical terms, the semi-Lagrangian LE advects $\widetilde{\psi}$ with a constant advective velocity on a local stencil in the vicinity of the departure point. This advective velocity is really a normalized displacement of the departure point to its nearest grid point, and can be different for each arrival point $(\overline{\mathbf{x}}_{\mathbf{i}},\ t^{n+1})$ in (8.3). Because of the latter, the semi-Lagrangian LE operator is not conservative, but thanks to the constancy of the local advective velocity, it is free of splitting errors. This is particularly convenient, as it allows the construction of fully second-order accurate NFT fluid models based on constant-coefficient 1D advection schemes.

Transporting the auxiliary field $\widetilde{\psi}$ (rather than the fluid variable alone) is important for the accuracy and stability of FT approximations. In the semi-

Lagrangian algorithm, transporting $\widetilde{\psi}$ derives straightforwardly from the trapezoidal-rule approximation for the trajectory integral of (8.2)

$$\psi_{\mathbf{i}}^{n+1} = \psi_o + \int_T R d\tau \approx \psi_o + 0.5\Delta t(R_o + R_{\mathbf{i}}^{n+1})$$
$$\approx (\psi + 0.5\Delta t R)_o + 0.5\Delta t R_{\mathbf{i}}^{n+1} \,, \tag{8.4}$$

where the subscript "o" is a shorthand for the value at the departure point $(\overline{\mathbf{x}}_o(\overline{\mathbf{x}}_{\mathbf{i}}, t^{n+1}),\ t^n)$, and the second approximate equality accounts for eventual nonlinearity of the LE remapping operator.

In the Eulerian algorithm, transporting $\widetilde{\psi}$ is a consequence of correcting for the first-order truncation error proportional to the divergence of the advective flux of R. To show this, we first rewrite (8.1) in a simpler form

$$\frac{\partial \rho^* \psi}{\partial t} + \overline{\nabla} \cdot (\widetilde{\mathbf{u}}\psi) = \rho^* R \,, \tag{8.5}$$

where $\widetilde{\mathbf{u}} \equiv \rho^* \overline{\mathbf{v}}^* = \rho_b\, \overline{G}\, \dot{\overline{\mathbf{x}}}$. We assume a temporal discretization of (8.5) in the form

$$\frac{\rho^{*n+1}\psi^{n+1} - \rho^{*n}\psi^n}{\Delta t} + \overline{\nabla} \cdot (\widetilde{\mathbf{u}}^{n+1/2}\psi^n) = \rho^{*n+1/2} R^{n+1/2} \,, \tag{8.6}$$

where $n + 1/2$ superscript denotes an $\mathcal{O}(\Delta t^2)$ accurate approximation for a field value at $t = t^n + 0.5\Delta t$. The lack of temporal centering of the transported field ψ in the second term on the LHS of (8.6) is the defining property distinguishing FT schemes from CTS methods. With a straightforward (centered) higher-order discretizations of the spatial derivatives, the differencing in (8.6) leads typically to computationally unstable schemes; for example, consider the classical Euler-forward discretization for the constant coefficient, homogeneous case of (8.6). In order to arrive at a stable, fully second-order-accurate, and robust NFT scheme, we derive the modified equation of (8.6)

$$\frac{\partial \rho^* \psi}{\partial t} + \overline{\nabla} \cdot (\widetilde{\mathbf{u}}\psi) = \rho^* R - \overline{\nabla} \cdot \left[\frac{1}{2}\Delta t \frac{1}{\rho^*} \widetilde{\mathbf{u}}(\widetilde{\mathbf{u}} \cdot \overline{\nabla}\psi)\right] + \overline{\nabla} \cdot \left(\frac{1}{2}\Delta t \widetilde{\mathbf{u}} R\right) + \mathcal{O}(\Delta t^2) \,. \tag{8.7}$$

To obtain (8.7), we have evaluated the complete $\mathcal{O}(\Delta t)$ truncation error of (8.6) due to uncentered time differencing (via a second-order Taylor series expansion), represented it in terms of spatial differences, and exploited the general form of the mass continuity equation

$$\frac{\partial \rho^*}{\partial t} + \overline{\nabla} \cdot \widetilde{\mathbf{u}} = 0 \,. \tag{8.8}$$

For details of the derivation procedure the interested reader is referred to [52] and the references therein. Ultimately, to compensate for the error on the RHS of (8.7), we add appropriate terms on the RHS of (8.6).

The $\mathcal{O}(\Delta t)$ truncation error on the RHS of (8.7) has two distinct components. The first is due solely to advection and must be compensated, by design, in any second-order-accurate FT advection scheme. For example, in the special case of 1D flow with constant velocity U, it is compensated by the third term of the classical 1-step Lax-Wendroff scheme

$$\psi_i^{n+1} = \psi_i^n - 0.5\alpha(\psi_{i+1}^n - \psi_{i-1}^n) + 0.5\alpha^2(\psi_{i+1}^n - 2\psi_i^n + \psi_{i-1}^n) \tag{8.9}$$

where $\alpha \equiv U\Delta t/\Delta X$ is the Courant number. The second $\mathcal{O}(\Delta t)$ term on the RHS of (8.7) is related to implementation of the FT schemes in inhomogeneous advection problems. It appears in those approximations to (8.1) that simply combine an FT advection scheme for homogeneous transport with an $\mathcal{O}(\Delta t^2)$ approximation of RHS$^{n+1/2}$. Ignoring this error leads to spurious $\sim \mathcal{O}(\Delta t)$ sinks/sources of "energy" ψ^2 and, eventually, to nonlinear instability (appendix A in [49]). Compensating this error to $\mathcal{O}(\Delta t^2)$ only requires subtracting a first-order-accurate approximation from the RHS of (8.6). This can be further upgraded and/or simplified depending upon a particular approximation adopted for representing RHS$^{n+1/2}$ in (8.6); cf. [49, 52] for discussions. The particular representation that leads to the Eulerian variant of the NFT algorithm in (8.3) assumes trapezoidal-rule approximation

$$\rho^{*n+1/2}R^{n+1/2} = 0.5\left(\rho^{*n}R^n + \rho^{*n+1}R^{n+1}\right), \tag{8.10}$$

The basic MPDATA algorithm, employed in our NFT models for anelastic flows, is based upon a *time-independent* coefficient ρ^*. For time variable ρ^*, it is possible to generalize the algorithm and subtract additional error correction terms from the RHS of (8.6) so as to still maintain an $\mathcal{O}(\Delta t^2)$ approximation [45, 19]. However there is an easier and more elegant alternative that works for any NFT flux-form advection scheme, second-order-accurate for a homogeneous (8.6) with time-independent coefficient. Consider any NFT flux-form advection scheme $\mathcal{A}$ (e.g., MPDATA) such that

$$\psi_{\mathbf{i}}^{n+1} = \mathcal{A}_{\mathbf{i}}(\psi^n,\ \tilde{\mathbf{u}}^{n+1/2},\ \rho^*) \tag{8.11}$$

approximates solutions of a *homogeneous* (8.5) with *time-independent* coefficient ρ^* to the second-order accuracy. Then, the algorithm

$$\psi_{\mathbf{i}}^{n+1} = \frac{\rho^{*n}}{\rho^{*n+1}}\mathcal{A}_{\mathbf{i}}(\psi^n + 0.5\Delta t R^n,\ \tilde{\mathbf{u}}^{n+1/2},\ \rho^{*n}) + 0.5\Delta t R_{\mathbf{i}}^{n+1}, \tag{8.12}$$

approximates solutions of an *inhomogeneous* (8.5) with *time-dependent* ρ^* also to the second-order accuracy. Identifying the first term on the RHS of (8.12) with the LE operator on the RHS of (8.3) closes the derivation of an Eulerian option of the NFT algorithm.

The congruence of the semi-Lagrangian and Eulerian options of the NFT approach in (8.3) is convenient for applications. It allows for a fairly simple yet effective design of fluid models with optional use of either algorithm selectable by the user. Both options have merits, and in fact simulating the same physical problem with two different numerical algorithms helps to assess the significance of the truncation error [54]. Semi-Lagrangian advection schemes are not subject to the Courant-Friedrichs-Lewy (CFL) stability condition, thereby allowing for large-time-step integrations for smooth flows. The smoothness of the flow can be measured with a global maximum of $\mathcal{L} \equiv \| \partial \mathbf{u} \Delta t / \partial \mathbf{x} \|$ — hereafter, the Lipschitz number — that links the size of the time step to the flow topology, or *realizability* [48]. In particular, $\mathcal{L} \leq 1$ suffices to prevent trajectory intersections within Δt, while the necessary condition requires the flow Jacobian $J \equiv \{\partial \mathbf{x} / \partial \mathbf{x}_o\}$ positive and bounded, $0 < J < \infty$ (see chapter 2 in [36], for a discussion). Also, $\Delta X/(1 \pm \mathcal{L})$ provides lower and upper bound estimates for the relative displacements, within Δt, of the trajectories arriving at the two neighboring points on the mesh; see appendix in [48] for details. Because of such topological concerns, the unconditional stability of semi-Lagrangian advection offers little advantage for simulating turbulent flows.[5] Since the flux-form of the Eulerian schemes appears consistent with finite-volume representation of the fluid equations [28], while assuring local conservation and accurate representation of flux boundary conditions [7], it leads, ultimately, to more suitable NFT schemes for simulating high Reynolds number flows; cf. [2]. We shall return to this discussion later in the next section.

3. Accuracy of NFT methods

Many nonoscillatory schemes are available nowadays "on the market", and no one can make claims for them all. In general, however, nonoscillatory advection schemes are dissipative. In practice, this means that a) they contain both even and odd truncation-error terms in their Taylor series expansion (in contrast with, e.g, the CTS leapfrog scheme), and b) they tend to dissipate (rather than conserve) the quadratic integrals ("energy" or "entropy") of the transported variable (in contrast, with, e.g., the CTS Arakawa-type methods [1]). By no means does dissipativity imply low accuracy—although statements in this spirit do occasionally appear in the literature. Implementation issues also exist (some quite subtle) that may shape the opinions of the practitioners of CFD. For example, quite common in the literature are nonoscillatory advection schemes that offer excellent accuracy for uniform 1D flows. However, when applied to variable multi-dimensional flows—by means of directional splitting [56]—such schemes are merely first-order accurate, and may surprise the user with a variety of behavioral errors [42]. These same schemes, however, can be implemented as remapping algorithms in the context of the semi-Lagrangian approach and then they do maintain the accuracy of the 1D constant-coefficient problem [48].

Among the plethora of existing advection schemes, there are nonoscillatory algorithms available that take into account flow variability and multidimensionality. Although such schemes may offer higher-order accuracy for uniform flows, typically they aim at uniformly second-order accurate solutions for arbitrary flows. These are the candidates particularly attractive for VLES modeling of geophysical fluids. The remarkable efficacy (i.e., accuracy versus efficiency) of the nonoscillatory methods has been argued in the literature for the last decade (see Introduction). In order to substantiate such arguments as well as to show the "dissipativity" at work, here we highlight the results of two benchmark simulations of turbulent flows using the NFT incompressible Boussinesq model based on the MPDATA advection scheme [52].

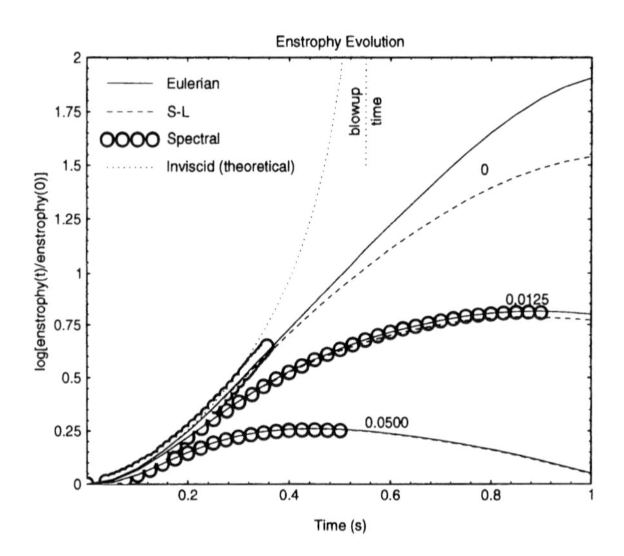

Figure 8.3. DNS and VLES of isotropic, decaying turbulence.

Our first example highlights the results of simulations of the decaying turbulence of a homogeneous incompressible fluid in a triply-periodic cube — a canonical problem in turbulence studies. The assumed homogeneity of the thermodynamics, and the lack of near-wall effects, focus the problem on the nonlinearity of the convective derivatives $\mathbf{u}\nabla\mathbf{u}$ in the momentum equation, i.e., the 'categorical imperative' of the turbulence *per se*. Our NFT simulations with MPDATA (and the nonoscillatory semi-Lagrangian option of our model) follow precisely the 256^3 DNS and inviscid pseudospectral simulations of Herring and Kerr [18].

Figure 8.3 displays the numerical results for the evolution of enstrophy for three values of viscosity, $\nu = 0.0500$, $\nu = 0.0125$, and $\nu = 0\ m^2s^{-1}$ (as indicated). Solid lines are for MPDATA experiments and dashed lines for semi-Lagrangian experiments, whereas Herring-Kerr results are marked with circles. For the sake of a reference, also shown is a theoretical estimate for inviscid flow (dotted lines), based upon the elementary enstrophy relationship (for 3D isotropic turbulence; Chpt. VI.7 in [23])

$$\frac{d\Omega}{dt} = \beta S(t)\Omega^{3/2}\,, \tag{8.13}$$

where $\Omega(t) \equiv \langle\omega^2\rangle$ is the enstrophy; ω is the vorticity; $S(t)$ is the velocity derivative skewness factor (defined earlier in section 1); and β is a constant. Equation (8.13) may be readily integrated, given a phenomenological model for skewness $S(t)$.[6] As there is no unequivocal evidence for physical realizability of the $\sim t^{-2}$ singularity of the Euler equations predicted by (8.13) [20], the theoretical curve shown should be viewed primarily as an upper bound in the ν parameter space.[7]

One striking message from Fig. 8.3 is the remarkable agreement of the NFT and the pseudo-spectral results for DNS ($\nu > 0$). This level of agreement is maintained uniformly throughout all flow characteristics, including spectra (see [18], for discussions and physical insights). Traditionally, pseudo-spectral methods are prized for their accuracy and considered superior tools for studying turbulent flows. Since all convergent methods eventually become accurate when the flow is fully resolved, one may wonder whether the resolution employed is an overkill. This is definitely not the case for $\nu = 0.0125$ where the Kolmogorov scale is about one grid-interval (Kerr, personal communication) and the dissipation of the energy is not well resolved. Apparently, the NFT model employed is at least as accurate as the pseudo-spectral code.

The VLES result ($\nu = 0$) exposes the true power of the NFT approach. Without viscous dissipation, unlimited enstrophy growth is predicted by (8.13), with a finite enstrophy blowup time of t $\sim$ 0.55 s. With a rapid enstrophy growth, the spectral calculations become computationally unstable and must be terminated after $\sim$ 0.35s [18]. Up to this time, the NFT, spectral, and theoretical results agree closely. After the collapse of the spectral model, the NFT computations continue but drop well below the theoretical result.

The quantification of an *effective* viscosity in numerical simulations of turbulent flows has baffled the meteorological community for decades. Typically, the results of numerical simulations are compared with theoretical models of the turbulence; cf. [58]. In this spirit one may evaluate the effective viscosity using

$$-\frac{1}{2}\frac{d}{dt}\langle \mathbf{u}^2 \rangle = 2\nu_{eff}\Omega(t) , \tag{8.14}$$

that relates kinetic energy dissipation rate and the enstrophy for freely-evolving isotropic 3D turbulence; cf. Chpt. VII.7 in [23]. Using (8.14) to evaluate the viscosity for the viscous cases does result in ν_{eff} that *precisely match* the constant explicit viscosity used in the DNS calculations. Equation (8.14) can also be used to evaluate ν_{eff} for the inviscid case but then the interpretation is less clear. The result is that ν_{eff} is a function of time, initially zero and increasing to a maximum value of $\sim 0.0018\ m^2 s^{-1}$ near a time of 1.0 s for the NFT Eulerian scheme. In the semi-Lagrangian simulation, the values of ν_{eff} have similar time dependence, but with approximately twice the amplitude. Continuing either simulation to $t = 2$s (not shown) indicates that ν_{eff} decreases monotonically for $t > 1$. Since physical viscosity is a constant parameter of real fluids (non-Newtonian and temperature dependent complications aside), the ν_{eff} of the inviscid case cannot represent explicit viscosity of a hypothetical, adequately resolved, DNS run. Instead, the ν_{eff} result for the inviscid case should be considered merely a measure of an overall dissipativity of the NFT approximation at hand. That such a measure is quite small may be appreciated from our previous observation that the $\nu = 0.0125 m^2 s^{-1}$ case is barely resolved in the sense of DNS — the ν_{eff} of the inviscid case is an order of magnitude smaller.

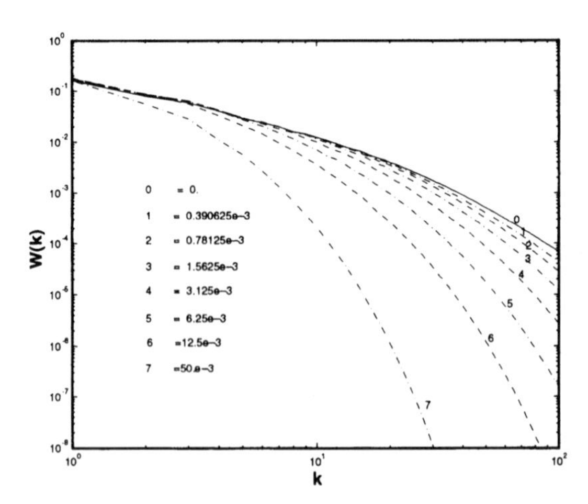

Figure 8.4. Convergence of 256^3 DNS velocity spectra to VLES result.

The results based on (8.14) do not discriminate whether, or in what sense, our VLES results are physically realistic — in principle, they could be dominated by numerical artifacts. The arguments supporting the physicality of VLES come from the simulations and theory of the Burgers' turbulence presented in [28]. According to results there, the transport terms of the fluid equations, properly modified to include finite-scale effects,[8] are renormalizable with respect to the finite-volume averaging; i.e., the finite-volume-averaged fluid equations appear the same regardless of the assumed scales of averaging. Furthermore, the finite-scale terms resulting from the finite-volume averaging closely correspond to the higher-order truncation error of finite-volume NFT schemes. Assuming that these arguments hold precisely for the Navier-Stokes' equations, one would anticipate that our VLES results represent an asymptotic solution in the zero viscosity (i.e., high Reynolds number) limit, projected to a given resolution of the model. Also, one would predict that our $\nu = 0$ simulations at any given resolution will closely match the results generated at other resolutions for the resolved scales. In particular, the spectra of dependent variables should agree at all resolved scales and only diverge at scales below the model resolution. Indeed this is the case.

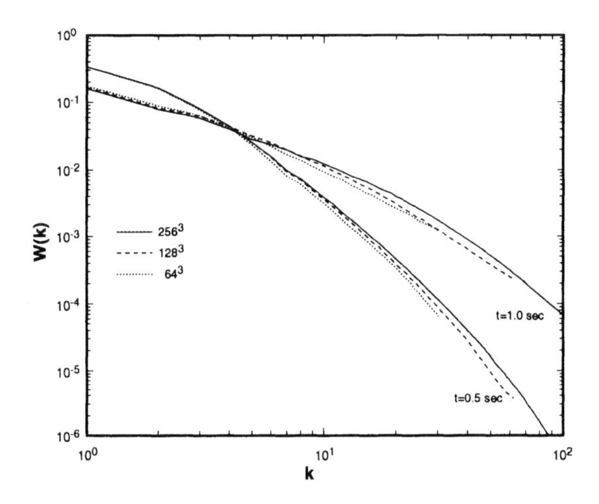

Figure 8.5. VLES velocity spectra for isotropic, decaying turbulence.

Figure 8.4 clearly demonstrates convergence of the 256^3 velocity spectra as $\nu \rightarrow 0$, consistent with our anticipation above. Figure 8.5 shows velocity spectra for the Eulerian VLES at resolutions corresponding to 256^3, 128^3, and 64^3. As predicted by the renormalization argument, the spectra closely match each other for the resolved scales. In other words, the VLES results do appear to approximate the inviscid problem in a physically meaningful way, and the truncation errors of the finite volume NFT schemes provide an accurate as well as effective subgrid-scale model. This becomes particularly clear when applying an NFT approach to LES/VLES simulations of an atmospheric boundary layer, discussed next.

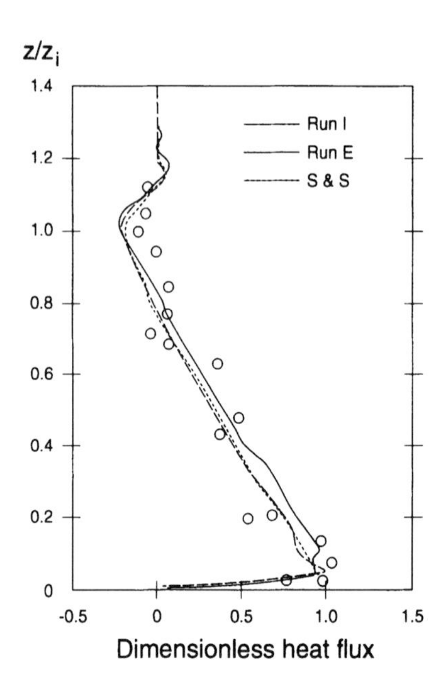

Figure 8.6. LES (run I) and VLES (run E) of convective PBL.

Figure 8.6 highlights the convective planetary boundary layer (PBL) simulations of Margolin et al [29]. The three curves shown in the figure represent mean profiles of the resolved heat flux $\langle \Theta' w' \rangle$ (normalized appropriately) from three different simulations: the short-dashed curve is from LES benchmark simulations of Schmidt and Schumann [43] using a CTS model; the long-dashed curve is from LES simulations with MPDATA, and the solid curve is for MPDATA VLES with no explicit subgrid-scale model. Circles represent field and laboratory data. The comparability of all the results with the data is excellent (for other characteristics of the flow, see [29]). Without the VLES result, one might be tempted to argue that the dissipativity of the employed NFT approach (e.g., LES) is negligibly small. However, the results reveal a more interesting story.

A full appreciation of the results in Fig. 8.6 intertwines with appreciating the mechanics of nonoscillatory schemes. NFT schemes are nonlinear (even for linear problems) as they employ coefficients that depend on the transported variables. In other words, these schemes are *self-adaptive* as they design themselves in the course of the simulation. Thus in contrast to linear CTS methods, different realizations of the same turbulent flow use different numerical approximations to the governing equations of motion. When the explicit SGS model is included (LES) the resolved flow is sufficiently smooth, and the entire machinery assuring nonoscillatory properties of the numerics is effectively turned off (there is no need to limit/adjust linear components of the scheme).

In the absence of an explicit SGS model (VLES) the nonoscillatory machinery adapts the numerics 'smartly' such as to assure solutions that are apparently as smooth as those generated with physically-motivated explicit SGS models. Thus, insofar as the dissipativity *per se* of the NFT methods is concerned, there is no simple scholastic quantification, since the resulting transport scheme can be effectively either non-dissipative or dissipative, depending upon the presence or absence, respectively, of an explicit SGS model. This corroborates the conclusions above, derived from comparing DNS and VLES of the decaying turbulence. Clearly NFT schemes remain accurate and effective tools in DNS, LES, and VLES studies.

Before closing this section, we return to the issue of the relative efficacy of the Eulerian (finite-volume) versus semi-Lagrangian (advective) NFT approximations. In section 2, we indicated that Eulerian schemes appear better suited for simulating high Reynolds number flows, especially in problems dependent on details of flux boundary conditions. In general, however, the overall accuracy of the two options is problem dependent [51], and there is no simple assessment valid throughout the entire range of geophysical applications. Figure 8.3 shows that the VLES simulations of the decaying turbulence appear less dissipative for the Eulerian option of the model, while resolved DNS simulations yield the same results for both options. The semi-Lagrangian velocity spectra (as those in Fig. 8.5; not shown) are in fair agreement with the Eulerian results for the earlier time. But at $t = 1.0$ sec, when smaller scales tend to dominate the flow, semi-Lagrangian spectra for lower resolution runs contain visibly less power in moderate wavelengths than the higher-resolution spectra, while they overshoot higher-resolution spectra at short wavelengths. Apparently, the dissipative properties of the considered Eulerian and semi-Lagrangian NFT schemes are distinct. This is illustrated more clearly in the example below.

Figures 8.7 and 8.8 highlight high-resolution (512×512 grid) simulations of the roll-up of a double-shear layer at $Re = 10^4$ (see [9, 10] and the references therein) using an Eulerian and semi-Lagrangian NFT schemes. The contour plots of the vorticity field in Fig. 8.7 (both panels use the same contour interval) show the two solutions *virtually the same*; however, comparing the centers of the eddies in both panels reveals that the semi-Lagrangian solution rolls up slightly faster. Figure 8.8 displays the dissipation histories, corresponding to the simulations depicted in Fig. 8.7. Solid lines are for the negative of the total kinetic energy decay rate $-\partial\langle \mathbf{u}^2/2\rangle/\partial t$, and dashed lines are for the negative of the viscous dissipation $-\langle Re^{-1}\mathbf{u}\Delta\mathbf{u}\rangle$. For the Eulerian solution, the effective decay rate is uniformly slightly larger than the viscous dissipation, documenting the weak implicit dissipation of the numerical approximation. This is not the

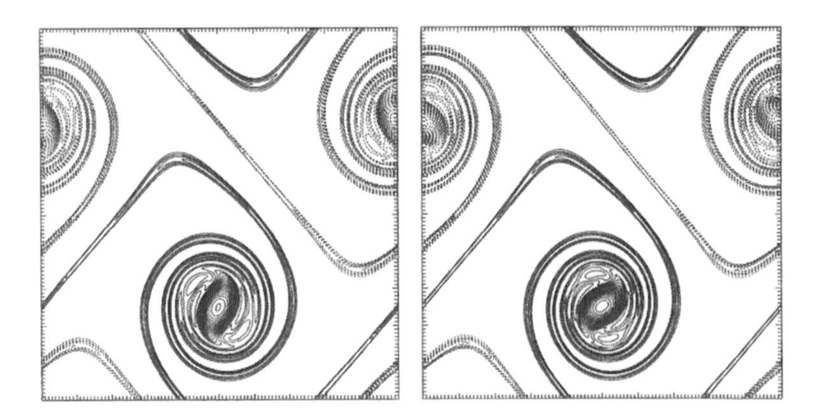

Figure 8.7. Vorticity isolines for Eulerian (left) and semi-Lagrangian (right) simulation of the double shear layer roll-up. Negative values are dashed.

case for the semi-Lagrangian solution, which first generates energy, and then later dissipates it at an increased rate. The unphysical "kink" in the semi-Lagrangian solution appears accentuated in more coarse resolutions, whereas the coarsely resolved Eulerian scheme maintains an energy dissipation path consistent with the more resolved solution.

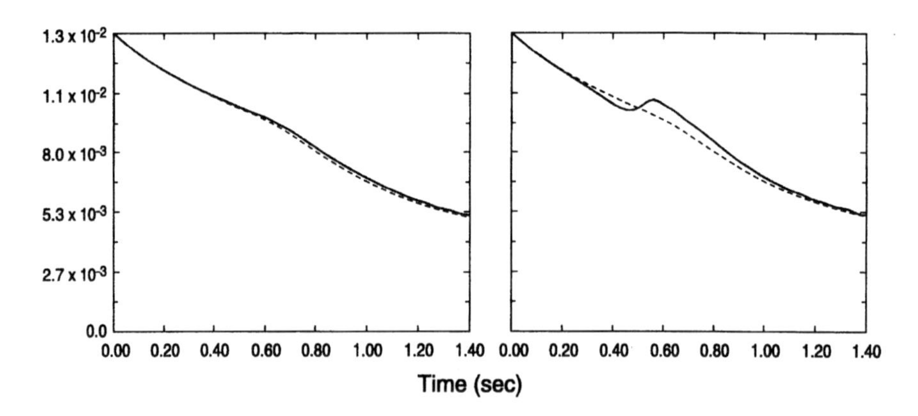

Figure 8.8. Dissipation histories for the solutions in Fig. 8.7. Kinetic energy decay rate (solid lines) versus viscous dissipation (dashed lines).

4. VLES of global geophysical turbulence – an example

4.1 Anelastic Model

Up to now, we have carried out our discussion exploiting rather elementary examples of fluid flows, and without paying much attention to the complexity of theoretical/numerical models underlying geophysical fluids. While the

NFT methods outlined in preceding sections can be directly applied to explicit integrations of a fully compressible Euler system (see [49] for a hydraulic analogy), because of the enormous span of spatial and temporal scales important in geophysical fluids, explicit integrations of generic compressible equations are impractical (viz., prohibitively expensive) for most applications. As a result, meteorological models utilize a variety of analytic/numerical approximations to the fluid equations (hydrostatic, elastic, anelastic, Boussinesq, and so on) and evince many split-explicit or semi-implicit methods for their integrations.

For research studies of all-scale turbulent geophysical fluids, we have found the anelastic nonhydrostatic system optimal. The anelastic equations may be viewed as combining two distinct approximations in the compressible Euler equations: a Boussinesq type linearization of the pressure gradient forces and mass fluxes in momentum and mass continuity equations, respectively; and the anelasticity *per se* equivalent to taking the limit of an infinite speed of sound. Although proven accurate for modeling weakly-stratified deep fluids (from local atmospheric [38] to global solar [13] convection), our recent results [54] document that the anelastic equations can capture adequately a broader range of planetary flows (i.e., including shallow stratified fluids) while requiring relatively minor overhead due to the nonhydrostatic formulation. The adequacy of the anelastic approximation has important practical consequences. The Boussinesq linearization inherent in the anelastic system greatly simplifies the task of designing accurate, flexible, and computationally efficient "all-scale-research" models for meteorological circulations. This is especially important within the class of NFT models, where two-time-level self-adaptive nonlinear numerics leads inevitably to difficult nonlinear elliptic problems for implicit discretizations of fully compressible Euler equations.

Herein, we focus on an inviscid, adiabatic, density-stratified fluid whose undisturbed, geostrophically-balanced "ambient" (or "environmental") state is described by the potential temperature $\Theta_e = \Theta_e(\mathbf{x})$ and the velocity $\mathbf{u}_e = \mathbf{u}_e(\mathbf{x})$. The expanded form of the governing anelastic equations on a mountainous globe, detailed in [53], accounts for the transformation of the anelastic variant of the Navier-Stokes' equations on a rotating sphere to a standard terrain-following coordinates [15]. To facilitate further discussion, here we present only compact, symbolic forms of the governing equations for the anelastic system of Lipps and Hemler [27]:

$$\frac{\mathcal{D}\mathbf{u}}{\mathcal{D}t} = -\mathbf{Grad}\left(\frac{p'}{\rho_b}\right) + \mathbf{g}\frac{\Theta'}{\Theta_b} - \mathbf{f} \times \mathbf{u}' + \mathcal{M}' \,, \tag{8.15}$$

$$\frac{\mathcal{D}\Theta}{\mathcal{D}t} = 0 \,, \tag{8.16}$$

$$\mathbf{Div}(\rho_b \mathbf{u}) = 0 \,. \tag{8.17}$$

Here the operators $\mathcal{D}/\mathcal{D}t$, **Grad**, and **Div** symbolize the material derivative, gradient, and divergence; $\mathbf{u}$ denotes the velocity vector (we employ the standard representation where components of $\mathbf{u}$ are defined in terms of a local tangent Cartesian framework aligned with standard geographical coordinates); $\mathcal{M}$ symbolizes appropriate metric forces (Christoffel terms proportional to products of velocity components, see [53]); Θ, ρ, and p denote potential temperature, density, and pressure; $\mathbf{f}$ and $\mathbf{g}$ symbolize the vectors of the "Coriolis parameter" and gravity. Primes denote deviations from the ambient state, and the subscript $_b$ refers to the basic state, i.e., horizontally homogeneous hydrostatic reference state of the Boussinesq expansion around a constant stability profile (see section 2b in [6], for a discussion).

One useful feature of our anelastic model is that it has the flexibility to solve the model equations in a variety of domains. Consider that all examples discussed throughout this chapter come from different applications of this same numerical model. The capability for such a flexibility comes from formulating our NFT approach (section 2) and the associated model code in terms of generalized time-dependent curvilinear framework [41, 55].[9] Time dependency of the coordinates can be employed for grid adaptation — the computational grid deforms so as to follow features of interest in an evolving solution — and/or to investigate flow responses to time-dependent boundary forcing (e.g., employed in the gravity-wave example discussed in the Introduction).[10]

Equations (8.15) and (8.16) are solved implicitly for the physical velocity, $\mathbf{u}$, and potential temperature perturbation, Θ', respectively; using algorithm (8.3) as given in section 2; cf. [53, 54] for discussions. Acceleration terms arising from rotation, Christoffel symbols, and pressure gradients are all treated implicitly in the $R_{\mathrm{i}}{}^{n+1}$ term of (8.3). SGS terms (if included) are treated explicitly. The implicitness of the potential temperature enhances both stability and accuracy [54]. The implicitness of the pressure gradient forces is an essential feature as it enables the projection of the preliminary values of $LE(\widetilde{\psi})$ onto solutions of the continuity equation (8.17) — see [5]. To make this projection, the system of simultaneous equations resulting from (8.3) are algebraically inverted and the results substituted into a finite-difference form of mass continuity (8.8)[11] to construct an elliptic equation for pressure. This elliptic equation is solved (subject to appropriate boundary conditions) using the generalized conjugate-residual approach — a preconditioned nonsymmetric Krylov solver (see [50, 51]; and the Appendix of [54] for further details).

4.2 Idealized Climate Simulations

An example of simulations of the idealized climates of Held and Suarez [17] culminates our discussion on modeling geophysical turbulence using NFT methods. Although the climate problem *per se* is heavily idealized, it is complex enough to undermine both the practicality and the physicality of explicit SGS modeling. With the physical and numerical complications involved in global modeling, any existing formal SGS model would quickly loose the connection with its roots in universal properties of fully developed turbulence to become merely a numerical filter improving computational stability of the model at hand.

>From a fluid dynamics viewpoint, the Held-Suarez problem represents thermally forced baroclinic instability on the sphere. In a sense, it bears striking resemblance to LES/VLES studies of convective boundary layers (see section 3 of this paper) where small differences in model setups can lead to totally different instantaneous flow realizations, and where different model designs can lead to quite divergent integral flow characteristics. In other words, these simulated flows are both turbulent and stochastic.

Table 8.1. Summary of the four climate simulations. The number of gridpoints refers to zonal $\times$ meridional points; all simulations used 41 vertical nodes with a vertical step size of 800 m, and a uniform time step of 900 s. Symmetry refers to the error in symmetry of the meridional wind field based upon the ratio of maximum to minimum zonally averaged, 2.3-year mean values. $V(\Theta)_{max}$ is the maximum value of the variation of the potential temperature Θ, about its zonal-time average. The CPU time per time step is relative — the "before" and "after" values cited for the TS simulation refer to before and after the time adaptation. During the time adaptation the change was monotone.

simulation	*gridpoints*	*type*	*Symmetry*	$V(\Theta)_{max}$	*CPU time/δt*
U0	64×32	uniform	-8.3%	$42.3K^2$	1
U2	64×64	uniform	1.5%	$49.6K^2$	1.6
SS	130×32	stretched	2.0%	$48.5K^2$	1.6
TS	64×32	adaptive	-1.3%	$46.7K^2$	1 before, 0.8 after

Results from four VLES simulations, that differ only in the design of the horizontal grid, are summarized in Table8.1 and figures 8.9-8.12. U0 and U2 refer to stationary, uniform increment grids in the zonal and meridional directions. These two simulations provide our "coarse" and "fine" resolution control results. Grid SS is stationary with constant zonal increments. In the meridional direction, stretching has been applied using a coordinate transformation such that a broad equatorial region has double the resolution of a uniform grid with the same number of meridional nodes. Grid TS is a time adaptive grid initially with uniform zonal and meridional increments. At 50 days the meridional coordinates begin to adapt to the developing zonal structure, such that at 150 days the region near $\pm 37^o$ latitude has double the resolution of a uniform grid while a narrow equatorial band maintains negligible change in grid resolution. This puts the maximum meridional resolution approximately in the region of the mid-latitude zonal jets. Outside of the time interval 50-150 days, the TS grid is stationary. Following [38], the time dependent coordinate transformation underlying the grid adaptation (as well as the static stretch for case SS) was set up *a priori* in the form of elementary polynomial functions.[12] Each of the four simulations was run for 3 years, beginning with a randomly perturbed no-flow initial state; see [54] for further details of the U0 simulation.

Figure 8.9 illustrates the overall complexity of the flow. It shows instantaneous horizontal cross-sections at 7.2 km altitude of the potential temperature (isentropes) Θ, zonal u, and meridional v wind fields after 3 years of simulated flow from case U2. The results displayed typify the response of an initially stagnant and uniformly stratified fluid to a diabatic forcing that mimics the long-term thermal and frictional forcing in the Earth atmosphere. This diabatic forcing attenuates Θ and $\mathbf{u}$ to the specified equilibrium temperature $\Theta_{EQ}(|y|, r')$ and $\mathbf{u}|_{r'<z_i} = 0$, where z_i represents a height of the boundary layer (see section 2 in [17], for details); y and r' refer to the meridional and the vertical (radial) directions, respectively. The corresponding forcing functions augment the governing equations of motion (8.15, 8.16) with appropriate Rayleigh friction and Newtonian cooling/heating terms.

Figure 8.10 contrasts the complexity of the instantaneous flow in Fig. 8.9 with the display of the resulting "climate", i.e., zonally-averaged 2.3-year means (from 0.7 to 3.0 years) of potential temperature, zonal, meridional and vertical velocities. The top plates of Fig. 8.10 correspond to the results in Figs. 1 and 2 of Held and Suarez [17].[13] The agreement of the two solutions is qualitative, which is not surprising considering the substantial differences between the numerical models employed. For a detailed discussion of the significance of the solutions and their dependence on the model design see [54].

Additional results for the uniform (U0), stretched (SS) and adaptive (TS) grid simulations are presented in Figs. 8.11 and 8.12. The three plates of Fig. 8.11 depict the instantaneous zonal wind field after 3 years of simulation and should be compared to the middle plate of Fig. 8.9 which depicts the corresponding U2 result. The three plates of Fig. 8.12 depict the zonally averaged, 2.3-year means of the zonal wind field and should be compared to the upper right plate of Fig. 8.10 which depicts the corresponding U2 result. While a detailed comparison is outside the scope of this chapter, it is clear that each of the alternative simulations U0, SS, and TS offer some but not all of the features of the U2 simulation. Simulation SS provides the best match to U2 for the instantaneous zonal field at 7.2 km altitude in the equatorial band bounded by $\sim \pm 30^o$ (one may anticipate that it is even better since it has double the zonal resolution of U2), while TS matches better at somewhat higher latitudes but more poorly in the most central $\sim \pm 15^o$ band. At high latitudes the U0 grid does better as the SS and TS grids have lower resolution near the poles. A similar pattern unfolds

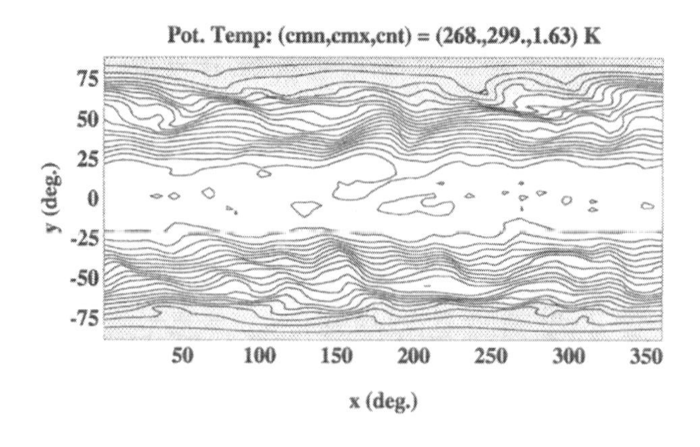

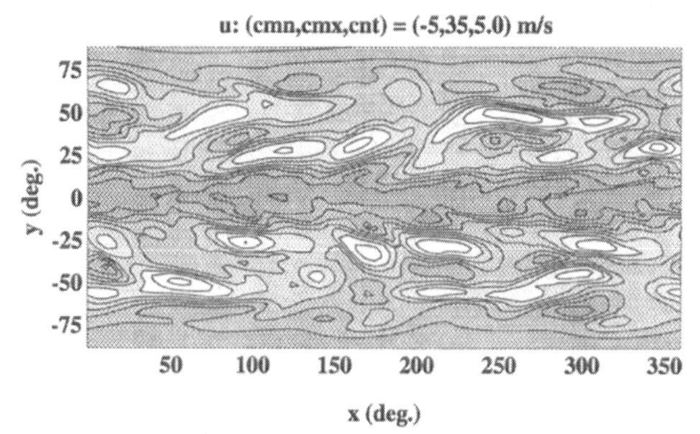

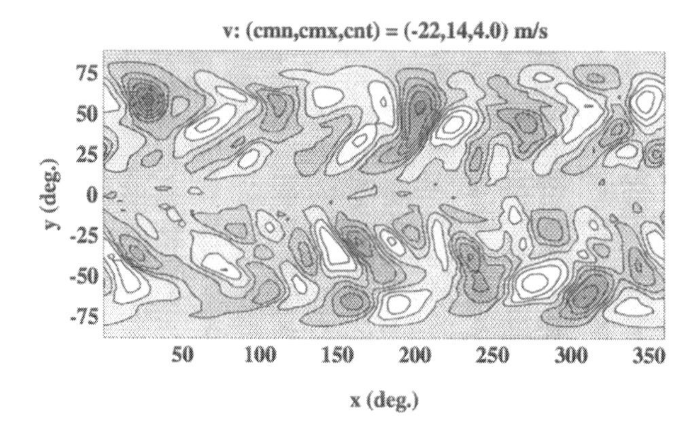

Figure 8.9. Instantaneous solutions of the idealized climate problem after 3 years of simulation. Plates from top to bottom show results based upon U2 simulation for potential temperature, zonal (u), and meridional (v) wind fields, respectively, in the horizontal plane at 7.2 km altitude. Contour extrema (cmx, cmn) and intervals (cnt) are shown above each plate. Darker shades of gray correspond to smaller values of the field variable (e.g, in the potential temperature field, the lightest regions are the warmest and the darkest are the coldest).

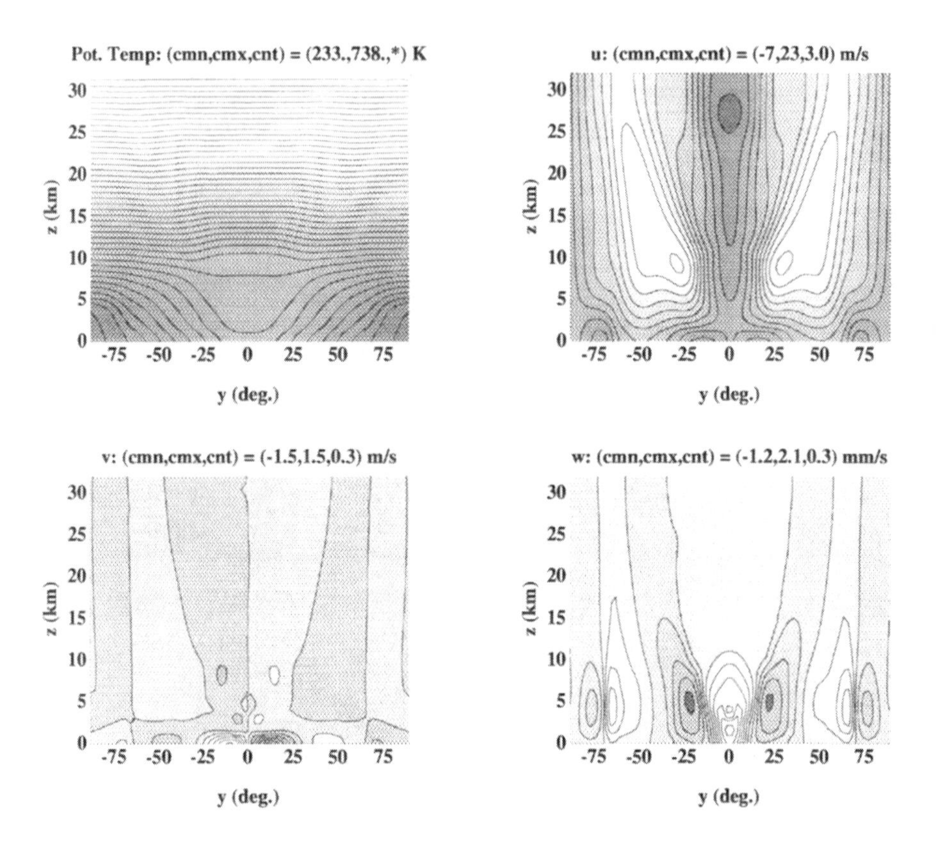

Figure 8.10. The zonally averaged 2.3-year means of potential temperature, zonal (u), meridional (v), and vertical (w) velocities for the simulation U2 highlighted in Fig. 8.9. The contour increments for Θ increase exponentially from 6.6 K at the bottom to 20.8 K at the top of the panel. The vertical velocities have been multiplied by 1000 and thus are in units of mms^{-1}. Contouring convention is similar to that used in Fig. 8.9.

when comparing the zonally averaged 2.3-year means. Only the SS simulation shows the closed, westward jet high over the equator. The lower altitude, mid-latitude easterly jets appear too confined in latitude, however. Simulation ST shows these easterly jets to be broader, though not to the same extent as shown in the U2 result.

Simple estimates of the accuracy of the simulations may be found by considering the symmetries of the flow. Assuming that the climate is stationary, then the zonally averaged 2.3-year means of the meridional wind should be anti-symmetric about the equator. Thus the magnitudes of the maximum and minimum values of meridional wind should be equal. The "symmetry" parameter in column 4 of Table 1 gives the departure of this ratio from unity. Simulations U2, SS, and TS all yield similar values with a departure from symmetry of $\sim 2\%$. The U0 symmetry error is four times larger. Another simple statistic, $V(\Theta)_{max}$ is given in column 5 of Table 1, it is the maximum value of the variation of potential temperature Θ about its zonal-time average. For this statistic, the stretched and time adaptive simulations outperform the coarse, uniform grid simulation. This result generally holds for other global statistics based upon the variation of the flow, whereas statistics based solely on the zonal-time average do not show consistent improvement using SS and TS (with the exception of the vertical wind field).

Comparisons of the four simulations suggest that successful use of stretched and time adaptive grids, to obtain enhanced accuracy over uniform grids for turbulent geophysical flows, is possible but may be highly dependent upon the type of information sought (e.g., local details vs. global statistics, and kind of details and/or statistics) as well as the type of application (e.g., full global climate vs. regional focus such as mid-latitudes). For full global climate simulations, a more sophisticated coordinate transformation than was used in SS and TS is clearly required. In particular, a time adapted stretching could follow the baroclinic eddies. Other applications, such as a global study of equatorially trapped waves would profit immediately from use of a simple stretching such as SS. Consider the computational costs of the simulations as indicated in the last column of Table 1. These data show that by doubling the number of zonal nodes and using stretching to enhance the meridional resolution in the equatorial region by a factor of 2, that the computational time of the SS simulation is increased only by a factor of 1.6 compared to the uniform grid simulation U0. Normally it could be anticipated that such an increase in resolution using a uniform grid (with four times as many nodes as in U0) would cost an order of magnitude more CPU time.

The results illustrated in Figs. 8.9-8.12 exemplify extremely well the value of VLES in modeling geophysical flows. The nonoscillatory machinery of MPDATA proceeds in the stretched and time adapted simulations as easily (if not more so) as in the simulations based upon uniform grids. No compensation

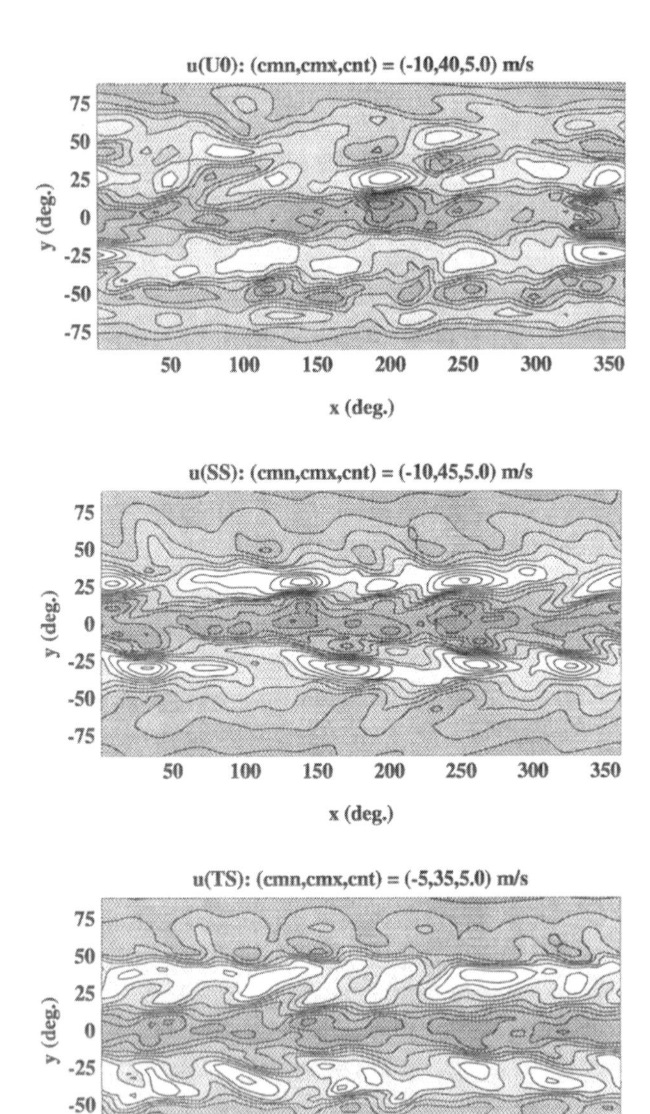

Figure 8.11. Solutions for the instantaneous zonal (u) wind field after 3 years of simulation. Plates from top to bottom show results based upon U0, SS, and TS simulations, respectively, in the horizontal plane at 7.2 km altitude. Contouring convention is the same as that used in Fig. 8.9.

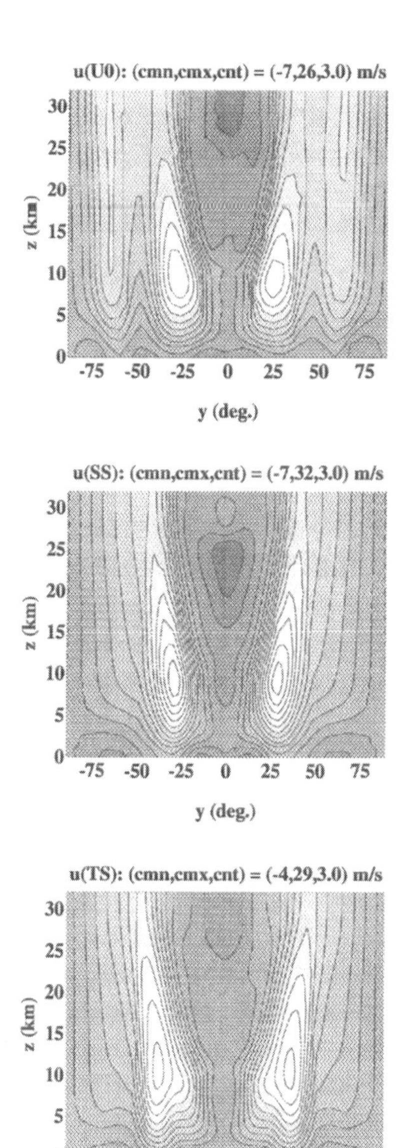

Figure 8.12. The zonally averaged 2.3-year means of zonal winds. Plates from top to bottom show results based upon U0, SS, and TS simulations. Contouring convention is the same as that used in Fig. 8.10.

for grid stretching and motion nor recourse to the actual physical mechanisms at work at sub-grid scales is necessary. Consider that the U0 simulation has model grid boxes $\sim 600 \times 600 km^2$ near the equator — comparable to the size of a small European country. It would be unreasonable to expect that applying any standard turbulence model in the sense of LES could represent an actual turbulence in the Earth's atmosphere at such a low grid resolution. In order to appreciate this issue, note that the processes of convection, terrain forcing, gravity wave breaking, and so on, that are responsible for generating energy at the subgrid scales in the real atmosphere/ocean system, are not accounted for in standard turbulence models.

5. Concluding remarks

Our isotropic, decaying turbulence simulations (section 3) using MPDATA demonstrate that NFT methods are competitive with pseudo-spectral methods for DNS. For high Reynolds number flows, NFT methods are far superior in that they may be used (via VLES) to provide physically meaningful projections of the solution, in the zero-viscosity limit, onto a grid of specified finite resolution. For either viscous or inviscid fluids, the dissipation produced by the nonoscillatory machinery is "sharp", that is, it appears the minimum needed to keep the computations stable.

In geophysical applications, the difficulty of making meaningful LES simulations is exacerbated by the complexity of natural (stratified, rotating, etc.) flow fields. Our gravity wave simulations (section 1) clearly show that correct *a priori* assumptions about the nature of a geophysical turbulence — necessary for LES models — can be difficult to make. In the gravity wave example, the use of standard LES assumptions, based on universal properties of fully-developed turbulence, could be irrelevant as: (i) the flow is strongly non-stationary; (ii) inhomogeneous; and (iii) shows *both* upscale and downscale energy transfer [41]. While regions of isotropic turbulence do exist, they are complicated functions of spatial position and time. By contrast, a VLES implicit SGS model "self adapts" to apply a grid-size filter smartly, only when and where it is needed for the stability and realizability of numerical solutions.

The examples employed throughout the chapter provide evidence that NFT-VLES methods work well in the range of geophysical flows, from decaying isotropic turbulence, through convective planetary boundary layers, free shear layers, gravity-wave-breaking induced stratified turbulence, up to planetary flows. Our last example (section 4.2) details results from an idealized, global climate simulation. It clearly demonstrates that MPDATA, when used in the spirit of VLES, provides a quality solution. This high level of performance is maintained even if *coordinate transformations* are used that stretch and deform the computational grid in time and space.

In each of our VLES simulations, no attempt is made to determine what kind of information may "lie" unresolved at the sub-grid scales (indeed, for the climate simulation example, such an attempt is senseless given the capacity of current machines). This is the defining idea behind VLES. It shifts the goals from developing fine scale turbulence models to effecting stability as "sharply" as possible. Although this may seem a dramatic change of perspective — from a physical viewpoint, for conventional LES models, to a mathematical/computational one, for VLES — our development shows that VLES with NFT methods inevitably appear based in physics (cf. the discussion on flow topology ending section 2, the negative entropy generation observation of [38], and the renormalization argument of [28]).

Acknowledgments

Assistance of Mirek Andrejczuk and Andrzej Wyszogrodzki (both from the Institute of Geophysics, Warsaw University, Warsaw, Poland) with the decaying turbulence simulations discussed in section 3 is gratefully acknowledged. Critical comments from Robert Kerr and Len Margolin of the earlier versions of this chapter helped to improve the presentation. The National Center for Atmospheric Research, NCAR, is operated by the University Corporation for Atmospheric Research under sponsorship of the National Science Foundation. The authors acknowledge partial support from the Department of Energy "Climate Change Prediction Program" (CCPP) while conducting this work.

Notes

1. Complications arising from the thermodynamics and chemistry of the atmosphere and oceans — e.g., due to the ubiquity of water-substance phases and salt — are beyond the scope of this chapter.

2. By "nonoscillatory", we mean all the nonlinear techniques (often referred to as monotonicity or shape preserving, shock capturing, or briefly, monotone schemes; e.g., total variation diminishing, TVD, flux-corrected-transport, FCT, and various flux-limited and sign-preserving schemes) that suppress/reduce/control numerical oscillations characteristic of higher-order linear schemes. After [59], "forward-in-time" (FT) labels a class of generalized one-step Lax-Wendroff type methods. Altogether, "NFT" is meant to distinguish from classical centered-in-time-and-space (CTS) linear methods, notorious for exhibiting spurious oscillations.

3. In elastic flows, it may be more convenient to formulate the archetype problem (8.1) in terms of $\phi \equiv \rho\psi$ instead of ψ; cf. [49] for a discussion.

4. An alternative argument employs the zeroth-order Taylor series expansion with higher-order accurate approximation of the first remainder that takes an integral form of the advection equation [47].

5. In absence of waves, the CFL and $\mathcal{L} \leq 1$ conditions are closely related. For illustration, consider a flow where local velocity U changes sign across the computational cell. Then $\Delta t|U_{i+1/2}-U_{i-1/2}|/\Delta X < 1$ becomes $2\Delta t|U|/\Delta X < 1$, requiring a local Courant number less than 0.5—a familiar condition for the practitioners of CFD.

6. In the present example of decaying, isotropic turbulence, the skewness grows from an initial value of zero (the initial state is Gaussian), to a maximum value of ~ 0.5 ([18, 23]). These characteristics can be modeled using a simple exponential decay function for $S(t)$, with time scale τ. Use of this $S(t)$ model in (8.13) leads to a two parameter enstrophy function $\Omega(t) = \Omega(t : \beta, \tau)$. The parameters were evaluated by matching the model to the numerical results at small time (when viscous effects are negligible).

7. Whether or not the enstrophy truly becomes unbounded in finite time for an inviscid flow is uncertain, as the turbulence may cease to be isotropic at the smallest scales; see chapters 7.8 of [12] and VI.7 in [23] and the references therein.

8. This accounts for the additional terms that appear in the finite-volume-integrated governing fluid equations, solved effectively in finite-volume numerical models.

9. With time variable coordinate mapping, the anelastic model equations (8.15)-(8.17) can be readily written in either of the forms (8.1) or (8.2).

10. The use of coordinate mapping for grid adaptation in the engineering community dates back to a seminal paper by Thompson et. al. [57] and is approaching a status as a mature subfield in computational fluid dynamics. Although the use of coordinate mapping for grid adaptation also dates back to nearly the same time in the meteorological community [15], it has not been as widely embraced; recent works include [11, 14, 19].

11. Actually, the time-derivative of the anelastic reference density, due to the temporal variability of the coordinates, is included under the velocity divergence, by redefining $\widetilde{\mathbf{u}}$ into a "solenoidal" velocity, see [41, 55] for details.

12. Our NFT model is able to handle with ease any time dependent coordinate transformation that preserves the topology of the reference state (spherical coordinates for global applications). Such coordinate transformations are not restricted to specified analytical functions, but in fact may also be determined numerically, say in response to developing flow features.

13. Note that their plots are drawn in the normalized hydrostatic-pressure (viz., mass) coordinate.

References

[1] A. Arakawa, "Computational design for long-term numerical integration of the equations of fluid motions: Two-dimensional incompressible flow", *J. Comput. Phys.*, **1**, 119-143 (1966).

[2] P. Bartello, and S.J. Thomas, "The cost-effectiveness of semi-Lagrangian advection. *Mon. Weather Rev.*, **124**, 2883–2897 (1996).

[3] A.R. Brown, M. K. MacVean, and P. J. Mason, "The effects of numerical dissipation in large eddy simulations", *J. Atmos. Sci.*, **120**, 3337-3348 (2000).

[4] F.H. Champagne,"The fine-scale structure of the turbulent velocity field", *J. Fluid Mech.*,**86**, 67–108 (1978).

[5] A.J. Chorin, "Numerical solution of the Navier-Stokes equations", *Math. Comp.*, **22**, 742-762 (1968).

[6] T.L., Clark, and R. D. Farley, "Severe downslope windstorm calculations in two and three spatial dimensions using anelastic interactive grid nesting: A possible mechanism for gustiness. *J. Atmos. Sci.*, **41**, 329–350 (1984).

[7] M. Cullen, D. Salmond, and P.K. Smolarkiewicz, "Key numerical issues for future development of the ECMWF models", *Proc. ECMWF Workshop on Developments in numerical methods for very high resolution global models*, 5-7 June 2000, Reading, UK, ECMWF, 183–206 (2000).

[8] J.W. Deardorff, "Sub-grid-scale turbulence modelings", *Issues in Atmospheric and Oceanic Modeling, Part B. Weather Dynamics, Advances in Geophysics*. Eds. B. Saltzman and S. Manabe, Academic Press, 337-343 (1985).

[9] D. Drikakis, and P.K. Smolarkiewicz, "On spurious vortical structures", *J. Comput. Phys.*, **172**, 309–325 (2001).

[10] D. Drikakis, L.G. Margolin, and P.K. Smolarkiewicz, "On "spurious" eddies", *Numerical Methods for Fluid Dynamics VII*, Ed. M.J. Baines, Will Print, 289-296 (2001).

[11] G.S. Dietachmayer, and K.K. Droegemeier, "Application of continuous dynamic grid adaptation techniques to meteorological modeling. Part 1: Basic formulation and accuracy", *Mon. Weather Rev.*, **120**, 1675-1706 (1992).

[12] U. Frisch, *Turbulence*, Cambridge Univ. Press, 296 pp. (1995).

[13] J.R. Elliott, and P.K. Smolarkiewicz, "Eddy resolving simulations of turbulent solar convection", *Proc. ECCOMAS computational fluid dynamics conference*, 4-7 September 2001, Swansea, Wales, UK (2001).

[14] M.S. Fox-Rabinovitz, G.L. Stenchikov, M.J. Suarez, L.L. Takacs, and R.C. Govindaraju, "A uniform- and variable-resolution stretched-grid GCM dynamical core with realistic orography. *Mon. Weather Rev.*, **128**, 1883–1898 (2000).

[15] T. Gal-Chen, and C.J. Somerville, "On the use of a coordinate transformation for the solution of the Navier-Stokes equations", *J. Comput. Phys.*, **17**, 209-228 (1975).

[16] M. Germano, U. Piomelli, P. Moin, and W. Cabot, "A Dynamic Subgrid-scale Eddy-Viscosity Model", *Phys. Fluids*, A **3**, 1760-1765 (1991).

[17] I.M. Held, and M.J. Suarez, "A proposal for intercomparison of the dynamical cores of atmospheric general circulation models", *Bull. Amer. Meteor. Soc.*, **75**, 1825-1830 (1994).

[18] J.R. Herring, and R.M. Kerr, "Development of enstrophy and spectra in numerical turbulence", *Phys. Fluids*, A **5**, 2792-2798 (1993).

[19] J.P. Iselin, J.M. Prusa, and W.J. Gutowski, "Dynamic grid adaptation using the MPDATA scheme", *Mon. Weather Rev.*, in press (2001).

[20] R.M. Kerr, "Evidence for a singularity of the three-dimensional, incompressible Euler equations", *Phys. Fluids*, A **5**, 1725-1746 (1993).

[21] B. Kosović, B., "Subgrid-scale modeling for large-eddy simulation of high-Reynolds-number boundary layers", *J. Fluid Mech.*, **336**, 151-182 (1997).

[22] M. Lesieur, and O. Metais, "New trends in LES of turbulence", *Annu. Rev. Fluid Mech.*, **28**, 45-82 (1996).

[23] M. Lesieur, *Turbulence in Fluids*, Kluwer Academic Pub., Dordrecht, 515 pp. (1997).

[24] D.K. Lilly, "The Representation of Small Scale Turbulence in a Numerical Experiment", *Proc. IBM Scientific Computing Symposium on Environmental Sciences*, IBM, White Plains, NY. (1967).

[25] D.K. Lilly, "A Proposed Modification of the Germano Subgrid-scale Closure Method", *Phys. Fluids*, A **4**, 633-635 (1992).

[26] P.F. Linden, J.M. Redondo, and D.L. Youngs, "Molecular mixing in Rayleigh-Taylor instability", *J. Fluid Mech.*, **265**, 97–124 (1994).

[27] F.B. Lipps, and R.S. Hemler, "A scale analysis of deep moist convection and some related numerical calculations", *J. Atmos. Sci.*, **39**, 2192-2210 (1982).

[28] L.G. Margolin, and W.J. Rider, "A rationale for implicit turbulence modeling", *Proc. ECCOMAS computational fluid dynamics conference*, 4-7 September 2001, Swansea, Wales, UK (2001).

[29] L.G. Margolin, P.K. Smolarkiewicz, and Z. Sorbjan, "Large-eddy simulations of convective boundary layers using nonoscillatory differencing", *Physica D*, **133**, 390-397 (1999).

[30] P.J. Mason, "Large-eddy simulation: A critical review of the technique", *Q.J.R. Met. Soc.*, **120**, 1-35 (1994).

[31] P. Moin, and A. G. Kravchenko, "Numerical issues in large eddy simulations of turbulent flows", *Numerical Methods for Fluid Dynamics VI*, Ed. M. J. Baines, Will Print, Oxford, 123-136 (1998).

[32] A. Muschinski, "A similarity theory of locally homogeneous and isotropic turbulence generated by a Smagorinsky-type LES", *J. Fluid Mech.*, **325**, 239-260 (1996).

[33] F.T.M. Nieuwstadt, "Direct and large-eddy simulation of free convection", *Proc. 9th Intl. Heat Transfer Conf.*, Jerusalem, vol.1, Amer. Soc. Mech. Engr., 37-47 (1990).

[34] E.S. Oran, and J.P. Boris, "Computing Turbulent Shear Flows—A Convenient Conspiracy", *Computers in Physics*, **7**, 523-533 (1993)

[35] S.A. Orszag, and I. Staroselsky, "CFD: Progress and problems", *Computer Physics Communications*, **127**, 165-171 (2000).

[36] J.M. Ottino, *The Kinematics Of Mixing: Stretching, Chaos, And Transport*, Cambridge University Press, 364 pp. (1989).

[37] D.H. Porter, A. Pouquet, and P.R. Woodward, "Kolmogorov-like spectra in decaying three-dimensional supersonic flows", *Phys. Fluids*, **6**, 2133-2142 (1994).

[38] J.M. Prusa, P.K. Smolarkiewicz, and R.R. Garcia, "On the propagation and breaking at high altitudes of gravity waves excited by tropospheric forcing", *J. Atmos. Sci.*, **53**, 2186-2216 (1996).

[39] J.M. Prusa, R.R. Garcia, and P.K. Smolarkiewicz., "Three-Dimensional Evolution of Gravity Wave Breaking in the Mesosphere", *Preprints 11th Conf. Atmos. Ocean. Fluid Dynamics*, Tacoma, WA, USA, June 23-27, American Meteorological Society, J3-J4 (1997).

[40] J.M. Prusa, P.K. Smolarkiewicz, and A.A. Wyszogrodzki, "Parallel computations of gravity wave turbulence in the Earth's atmosphere", *SIAM News*, **32**, 1:10-12 (1999).

[41] J.M. Prusa, P.K. Smolarkiewicz, and A.A. Wyszogrodzki, "Simulations of gravity wave induced turbulence using 512 PE CRAY T3E", *Int. J. Applied Math. Comp. Science*, in press (2001).

[42] P.J. Roache, *Computational Fluid Dynamics.*, Hermosa Publishers, Albuquerque, 446 pp. (1972).

[43] H. Schmidt, and U. Schumann, "Coherent structure of the convective boundary layer derived from large-eddy simulation", *J. Fluid Mech.*, **200**, 511-562 (1989).

[44] J. Smagorinsky, J., "Some historical remarks on the use of nonlinear viscosities", *Large Eddy Simulation of Complex Engineering and Geophysical Flows*, Eds. B. Galperin and S. A. Orszag, Cambridge University Press, 3-36 (1993).

[45] P.K. Smolarkiewicz, "A fully multidimensional positive definite advection transport algorithm with small implicit diffusion", *J. Comput. Phys.*, **54**, 325-362 (1984).

[46] P.K. Smolarkiewicz, and T. L. Clark, "The multidimensional positive definite advection transport algorithm: Further development and applications", *J. Comput. Phys.*, **67**, 396-438 (1986).

[47] P.K. Smolarkiewicz, and G.A. Grell, "A class of monotone interpolation schemes", *J. Comput. Phys.*, **101**, 431-440 (1992).

[48] P.K. Smolarkiewicz, and J.A. Pudykiewicz, "A class of semi-Lagrangian approximations for fluids", *J. Atmos. Sci.*, **49**, 2082-2096 (1992).

[49] P.K. Smolarkiewicz, and L. G. Margolin, "On forward-in-time differencing for fluids: Extension to a curvilinear framework", *Mon. Weather Rev.*, **121**, 1847-1859 (1993).

[50] P.K. Smolarkiewicz, and L.G. Margolin, "Variational solver for elliptic problems in atmospheric flows", *Appl. Math. & Comp. Sci.*, **4**, 527-551 (1994).

[51] P.K. Smolarkiewicz, and L.G. Margolin, "On forward-in-time differencing for fluids: An Eulerian/semi-Lagrangian nonhydrostatic model for stratified flows", *Atmos. Ocean Special*, **35**, 127-152 (1997).

[52] P.K. Smolarkiewicz, and L.G. Margolin, "MPDATA: A finite-difference solver for geophysical flows", *J. Comput. Phys.*, **140**, 459-480 (1998).

[53] P.K. Smolarkiewicz, V. Grubišić, L.G. Margolin, and A.A. Wyszogrodzki, "Forward-in-time differencing for fluids: Nonhydrostatic modeling of fluid motions on a sphere", *Proc. 1998 Seminar on Recent Developments in Numerical Methods for Atmospheric Modeling*, Reading, UK, ECMWF, 21-43 (1999).

[54] P.K. Smolarkiewicz, L.G. Margolin, and A.A. Wyszogrodzki, "A class of nonhydrostatic global models", *J. Atmos. Sci.*, **58**, 349–364 (2001).

[55] P.K. Smolarkiewicz, and J.M. Prusa, "VLES modeling of geophysical fluids with nonoscillatory forward-in-time schemes," *Proc. ECCOMAS computational fluid dynamics conference*, 4-7 September 2001, Swansea, Wales, UK (2001).

[56] G. Strang, "On the construction and comparison of difference schemes, *SIAM J. Numer. Anal.*, **5**, 506-517 (1968).

[57] J.F. Thompson, F.C. Thames, and C.W. Mastin, "Automatic Numerical Generation of body-Fitted Curvilinear Coordinate System for Field Containing Any Number of Arbitrary Two-Dimensional Bodies", *J. Comput. Phys.*, **15**, 299-319 (1974).

[58] J. Thuburn, "Dissipation and Cascades to Small Scales in Numerical Models Using a Shape-Preserving Advection Scheme", *Mon. Weather Rev.*, **123**, 1888-1903 (1995).

[59] C.J. Tremback, J. Powell, W.R. Cotton, and R.A. Pielke, "The forward-in-time upstream advection scheme: Extension to higher orders", *Mon. Weather Rev.*, **115**, 540-555 (1987).

Chapter 9

DIRECT NUMERICAL SIMULATIONS OF MULTIPHASE FLOWS

G. Tryggvason, A. Fernández, A. Esmaeeli
Department of Mechanical Engineering, Worcester Polytechnic Institute, 100 Institute Road, Worcester, MA 01609, USA
gretar@wpi.edu, arturo@wpi.edu, aesmae@wpi.edu

B. Bunner
Coventor Inc., 625 Mount Auburn Street, Cambridge, MA 02138, USA
bernard.bunner@coventor.com

Abstract Direct numerical simulations have recently emerged as a viable tool to understand finite Reynolds number multiphase flows. The approach parallels direct numerical simulations of turbulent flows, but the unsteady motion of a deformable phase boundary add considerable complexity. Here, a method based on solving the Navier-Stokes equations by a finite difference/front tracking technique that allows the inclusion of fully deformable interfaces and surface tension, in addition to inertial and viscous effects is described. Studies of homogeneous bubbly flows and the effect of electrostatic fields on the behavior of suspensions are presented in some detail. A parallel version of the method makes it possible to use large grids and resolve flows containing a few hundred bubbles.

Keywords: direct numerical simulations, multiphase flow

1. Introduction

Multiphase and multifluid flows are common in many natural and technologically important processes. Rain, spray combustion, spray painting, and boiling heat transfer are just a few examples. While it is the overall behavior of such flows that is of most interest, this is determined to a large degree by the evolution of the smallest scales in the flow. The combustion of sprays, for example, depends on the size and the number density of the drops. Generally,

D. Drikakis and B.J. Geurts (eds.), Turbulent Flow Computation, 313–338.

these small-scale processes take place on a small spatial and time scale, and in most cases visual access to the interior of the flow is limited. Experimentally, it is therefore very difficult to determine the exact nature of the small-scale processes. Direct numerical simulations, where the governing equations are solved exactly, offer the potential to gain a detailed understanding of the flow. Such direct simulations, where it is necessary to account for inertial, viscous, and surface forces in addition to a deformable interface between the different phases, still remains one of the most difficult problems in computational fluid dynamics.

Here, a numerical method that has been found to be particularly suitable for direct simulations of flows containing moving and deforming phase boundary is briefly described. Various applications of the method are briefly reviewed and studies of bubbly flows and the effect of electrostatic fields of the behavior of suspensions are discussed in some detail.

The method described in this paper is properly described as a hybrid between a front capturing and a front tracking technique. A stationary regular grid is used for the fluid flow, but the interface is tracked by a separate grid of lower dimension. However, unlike front tracking methods where each phase is treated separately, all phases are treated together by solving a single set of governing equations for the whole flow field. Although the idea of using only one set of equations for many co-flowing phases is an old one, the method described here is a direct descendant of a Vortex-In-Cell technique for inviscid multifluid flows described in Tryggvason and Aref (1983) and Tryggvason (1988) and the immersed boundary method of Peskin (1977) developed to put moving boundaries into finite Reynolds number homogeneous fluids. The original version of the method and a few sample computations were presented by Unverdi and Tryggvason (1992) but as our experience in using the method has increased, we have modified and improved the method in several ways.

Before discussing homogeneous bubbly flows and electrostatic effects on suspensions, we will briefly review various applications of the method. We have examine many aspects of bubbly flows in several papers. Unverdi and Tryggvason (1992a,b) computed the interactions of two, two- and three-dimensional bubbles and Jan (1994) examined the motion of two axisymmetric and two-dimensional bubbles in more detail. Ervin and Tryggvason (1997) (see also Ervin, 1993) computed the rise of a bubble in a vertical shear flow and showed that the lift force changes sign when the bubble deforms. The results of Jan and Ervin, which cover a rise Reynolds number range of about a 1 to 100 have yielded considerable insight into the dependency of attractive and repulsive forces between two bubbles on the Reynolds number and bubble deformability. Preliminary studies of the interaction of bubbles with unsteady mixing layers are reported by Taeibi-Rahni, Loth, and Tryggvason (1994) and Loth, Taeibi-Rahni, and Tryggvason (1997). The motion of a few hundred two-dimensional bubbles at O(1) Reynolds number was simulated by Esmaeeli and Tryggvason (1996), who found an inverse energy cascade similar to what is seen in two-dimensional turbulence. Esmaeeli and Tryggvason (1998, 1999) simulated the unsteady motion of several two- and three-dimensional bubbles, examining the dependancy of the rise velocity and the bubble interactions on the Reynolds number. More recently, Bunner and Tryggvason (1999a,b) (see also Bunner, 2000, and Bunner and Tryggvason, 2001abc) used a parallel version of the method to study the dynamics of up to two hundred three-dimensional bubbles. Bunner's results are discussed in more detail later in this chapter. Similar simulations of suspensions of drops have been done by Mortazavi and Tryggvason (2000), who computed the motion of a periodic row of drops in a pressure driven channel flow, and Mortazavi and Tryggvason (2000), who examined the collective behavior of many drops.

We have also examined various aspects of droplet motion and sprays. The head-on collision of two axisymmetric drops was computed by Nobari, Jan, and Tryggvason (1996) and Nobari and Tryggvason (1996) simulated the off-axis collisions of fully three-dimensional drops. Primary focus was on the case where the drops broke up after initial coalescence. The numerical results are in good agreement with available experimental data (see Jiang, Umemura, and Law, 1992, for example) and helped to explain the boundary between the various collision modes. The binary collision of axisymmetric drops was examined again by Qian, Law, and Tryggvason (2000) who focused on the draining of the film between the drops and compared the numerical results with experiments. The capillary breakup of a liquid jet injected into another liquid was examined by Homma *et al.* (1998) and Song and Tryggvason (1999) simulated the formation of a thick rim on the edge of a thin liquid sheet. An extensive study of the secondary breakup of drops has been done by Han and Tryggvason (1999, 2001), and the primary atomization of jets where the drop size is much smaller than the jet diameter, has been examined by Tryggvason and Unverdi (1999), Tauber and Tryggvason (2000), and Tauber, Unverdi, and Tryggvason (2001). Other related problems include simulations of the three-dimensional Rayleigh-Taylor instability by Tryggvason and Unverdi (1990), an examination of the coalescence and mixing of two initially stationary drops by Nobari (1993), and a study of the dissipation of surface waves by Yang and Tryggvason (1998).

In addition to problems where two or more incompressible and immiscible fluids flow together, we have also examined a number of problems where the governing physics is more complex. Simulations of the motion of bubbles and drops with variable surface tension include the examination of contaminated bubbles by Jan and Tryggvason (1991) and the thermocapillary motion of several bubbles by Nas and Tryggvason (1993). See also Jan (1994) and Nas (1995) for details. Other simulations with complex surface forces were presented by Agresar, Linderman, Tryggvason, and Powell (1998) who modeled the response of a biological cell to various flows conditions and Che (1999) who computed the motion and deformation of drops due to electrostatic forces. We will discuss other simulations of the effect of electrostatic forces on the motion of suspensions later. The methodology has been extended in several studies to flows with phase change where the front moves relative to the fluid. Juric and Tryggvason (1996) developed a method for the solidification of pure materials that accounts for the full Gibbs-Thompson conditions at the phase boundary and used it to examine the growth of dendrites. Juric (1996) extended the method to simulate an unstable solidification front in binary alloys. The solidification method has been coupled with a fluid solver to examine the effect of fluid flow on the growth of both two- and three-dimensional dendrites by Al-Rawahi and Tryggvason (2001). A simpler solidification model, freezing the liquid as soon as the temperature drops below the melting temperature, allowed Che (1999) to examine the solidification of multiple hot drops deposited on top of each other. In these computations, both phases had the same density. Generally, however, phase change is accompanied by local expansion at the phase boundary. The collapse of a cavitating bubble in a shear flow was examined by Yu, Ceccio, and Tryggvason (1995), who set the pressure inside the bubble equal to the vapor pressure of the liquid. A simple model of the combustion of a premixed flame, where the local expansion rate and thus the relative expansion velocity is a prescribed constant, was developed by Qian, Tryggvason, and Law (1998), and used to examine the flame generation of vorticity. A more sophisticated method for the complete simulations of boiling was developed by Juric and Tryggvason (1998) who solved the fully coupled fluid and energy equations. The algorithm has been simplified considerably by Esmaeeli and Tryggvason (2001) who used it to examine film boiling.

2. Formulation and Numerical Method

The Navier-Stokes equations govern the fluid motion in all phases and a single vector equation can be written for the whole flow field. For isothermal flow, in conservative form:

$$\frac{\partial \rho \mathbf{u}}{\partial t} + \nabla \cdot \rho \mathbf{u}\mathbf{u} = \\ -\nabla P - \rho \mathbf{g} + \nabla \cdot \mu(\nabla \mathbf{u} + \nabla^T \mathbf{u}) + \gamma \int \kappa_f \mathbf{n}_f \delta(\mathbf{x} - \mathbf{x}_f) dA_f. \quad (9.1)$$

Here, $\mathbf{u}$ is the velocity, P is the pressure, and ρ and μ are the discontinuous density and viscosity fields, respectively. $\mathbf{g}$ is the gravity acceleration. Surface forces are added at the fluid interface. δ is a three-dimensional delta function constructed by repeated multiplication of one-dimensional delta functions. κ is twice the mean curvature and $\mathbf{n}_f$ is a unit vector normal to the interface. γ

is the surface tension. Formally, the integral is over the entire front, thereby adding the delta functions together to create a force that is concentrated at the interface. $\mathbf{x}$ is the point at which the equation is evaluated and $\mathbf{x}_f$ is the position of the front. These equations implicitly enforce the proper mass and momentum conditions at the fluid interface. For fully periodic domains, it is necessary to add a term, $\rho_{av}\mathbf{g}$, where ρ_{av} is the average density, to the equations to prevent uniform downward acceleration of the whole flow field. All phases are taken to be incompressible, so the velocity field is divergence free

$$\nabla \cdot \mathbf{u} = 0. \tag{9.2}$$

which, when combined with the momentum equations leads to a non-separable elliptic equation for the pressure. We also have equations of state for the density and the viscosity:

$$\frac{D\rho}{Dt} = 0 \quad \text{and} \quad \frac{D\mu}{Dt} = 0. \tag{9.3}$$

Here, D/Dt is the material derivative and the last two equations simply state that the density and the viscosity of each fluid remain constant.

This formulation implicitly contains the same conditions at the interface as found in standard references. To demonstrate that, we move to a frame moving with the interface and integrate the equations over a small volume enclosing the interface. As we shrink the volume, most of the terms cancel or go to zero and only gradient terms survive. Integrating the normal component yields

$$\left[\left[-P\mathbf{I} + \mu(\nabla\mathbf{u} + \nabla\mathbf{u}^T)\right]\right] \cdot \mathbf{n} = \sigma\kappa\mathbf{n} \tag{9.4}$$

where the brackets denote the jump across the interface. This is, of course, the usual statement of continuity of stresses at a fluid boundary, showing that the normal stresses are balanced by surface tension. Integrating the tangential component shows that the tangential stresses are continuous and integrating the mass conservation equation across the interface shows that the normal velocities are also continuous.

The formulation described above allows multiphase flows to be treated in a similar way as homogeneous flows. Once the material boundary has been advected and the surface tension found, any standard algorithm based on fixed grids can, in principle, be used to integrate the Navier-Stokes equations in time. Figure 1 summarizes our approach: A fixed grid is used for the conservation equations but a moving grid of lower dimension marks the boundary between the different phases.

To compute the momentum advection, the pressure term, and the viscous forces, several standard discretization schemes can be used. In most of our computations we use a fixed, regular, staggered MAC grid and discretize the momentum equations using a conservative, second-order centered difference

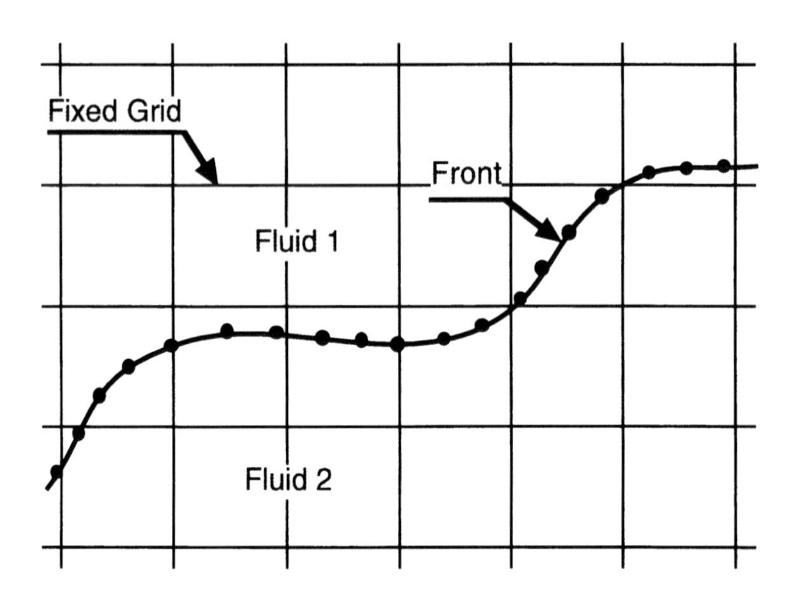

Figure 9.1. Computations of flow containing more than one phase. The governing equations are solved on a fixed grid but the phase boundary is represented by a moving "front," consisting of connected marker points.

scheme for the spatial variables. The time integration is done by an explicit second-order predictor-corrector scheme where the first-order solution at the new time serves as a predictor that is then corrected by the trapezoidal rule.

We generally use a front structure that consists of points connected by elements. Both the points and the elements (the front objects) are stored in linked lists that contains pointers to the previous object and the next object in the list. The order in the list is completely arbitrary and has no connection to the actual order on the interface. The use of a linked list makes the addition and removal of objects particularly simple. For each point, the only information stored are the point coordinates. Each element, on the other hand, knows about the points that it is connected to; the elements that are connected to the same endpoints; the surface tension; the jump in density across the element; and any other quantities that are needed for a particular simulation. The elements are given a direction and for a given front all elements must have the same direction. Three-dimensional fronts are built in the same way, except that three points are now connected by a triangular element. The points, again, only know about their coordinates but the elements know about their corner points and the elements that share their edges.

As the front moves, it deforms and stretches. The resolution of some parts of the front can become inadequate, while other parts become crowded with front elements. To maintain accuracy, additional elements must either be added when the separation of points becomes too large or the points must be redistributed to maintain adequate resolution. It is generally also desirable to remove small elements. In addition to reducing the total number of elements used to represent the front, element removal usually also prevents the formation of "wiggles" much smaller than the grid size.

Since the Navier-Stokes equations are solved on a fixed grid but surface tension is found on the front, it is necessary to convert a quantity that exists at the front to a grid value. Since quantities that exist only at the front are represented by δ-functions, the transfer corresponds to the construction of an approximation to the δ-function on the fixed grid. This "smoothing" can be done in several different ways, but it is always necessary to ensure that the quantity transferred is conserved. The weighting functions are usually written as a product of one-dimensional functions. In three-dimensions, the weight for the grid point (i, j, k) for smoothing from $\mathbf{x}_p = (x_p, y_p, z_p)$ is given by

$$w_{ijk}(\mathbf{x}_p) = d(x_p - ih)\, d(y_p - jh)\, d(z_p - kh) \tag{9.5}$$

where h is the grid spacing. For two-dimensional interpolation, the third term is set to unity. $d(r)$ can be constructed in different ways. The simplest interpolation is the area (volume) weighting:

$$d(r) = \begin{cases} (h - |r|)/h, & |r| < h, \\ 0, & |r| \geq h\,. \end{cases} \tag{9.6}$$

We also often use other smoother distribution function such as those introduced by Peskin (1977) and Peskin and McQueen (1994).

Since the fluid velocities are computed on the fixed grid and the front moves with the fluid velocities, the velocity of the interface points must be found by interpolating from the fixed grid. Although it is not necessary, we generally use the same weighting function to interpolate from the fixed grid as we use to smooth front values onto the grid.

The fluid properties, such as the density and viscosity, are not advected directly. Instead the boundary between the different fluids is moved. It is therefore necessary to reset the properties at every time step. The simplest method is, of course, to loop over the interface elements and set the density on the fixed grid as a function of the shortest normal distance from the interface. Since the interface is usually restricted to move less than the size of one fixed grid mesh, this update can be limited to the grid points in the immediate neighborhood of the interface. This straightforward approach (used, for example by Udaykumar *et al.*, 1997) does have one major drawback: When two interfaces are very close to each other, or when an interface folds back on itself, such that two front segments are between the same two fixed grid points, then the property value on the fixed grid depends on which interface segment is being considered. Since this situation is fairly common, a more general method is necessary. We usually smooth the density jump (or the gradient) onto the grid and then integrate the smoothed jump on the grid to obtain the density function. The integration can be done either line-by-line in the vicinity of the interface or by first taking the numerical divergence of the grid-density gradient field and then solving a Poisson equation. The benefit of working with the density jump, rather than the density itself is that the gradients of the opposite sign simply cancel on the fixed grid.

The accurate computation of the surface tension is perhaps one of the most critical elements of any method designed to follow the motion of the boundary between immiscible fluids for a long time. In our approach the front is explicitly represented by discrete points and elements and while this makes the surface tension computations much more straightforward than reconstructing it from a marker function, there are several alternative ways to proceed. In most of our simulations, we need the force on a front element but not the curvature directly. This simplifies the computations considerably. For a two-dimenisonal system, the force on a short front element is given by:

$$\delta \mathbf{F}_{\gamma} = \gamma \int_{\Delta s} \kappa \mathbf{n} ds. \tag{9.7}$$

Using the definition of the curvature of a two-dimensional curve, $\kappa \mathbf{n} = \partial \mathbf{s} / \partial s$, we can write this as

$$\delta \mathbf{F}_{\gamma} = \gamma \int_{\Delta s} \frac{\partial \mathbf{s}}{\partial s} ds = \gamma (\mathbf{s}_2 - \mathbf{s}_1). \tag{9.8}$$

Therefore, instead of having to find the curvature, we only need to find the tangents of the end points. In addition to simplifying the computation, this ensures that the total force on any closed surface is zero, since the force on the end of one front element is exactly the same as the force on the end of the adjacent element. This conservation property is particularly important for long time computation where even a small error in the surface tension computation can lead to an unphysical net force on a front that can accumulate over time. The strategy for three-dimensional flow is similar. Instead of finding the curvature, we integrate the surface tension around the edges of the elements to find the net force on each element.

The one field approach, where the governing equations are solved simultaneously for both fluids and the interface singularities are approximated on a stationary grid, is also used in the well known volume-of-fluid (VOF) method and in the more recent level set method. See Scardovelli and Zaleski (1999) for a review of the VOF method, and Osher and Fedkiw (2001) and Sethian (2001) for a review of the level set method. However, in those methods, the interface is located by a marker function that is advected on the fixed grid instead of by explicit marker particles as done here. For a detailed description of the method and various validation tests, see Unverdi and Tryggvason (1992), Esmaeeli and Tryggvason (1998), and Tryggvason *et al.* (2001),

Accurate and fast simulations of large, well-resolved, three-dimensional bubble systems can only be obtained with parallel computers. The method was therefore reimplemented for distributed memory parallel computers supporting the Message Passing Interface (MPI) standard. Different strategies are employed for the grid and the front. The Navier-Stokes solver, including the multigrid pressure solver, is parallelized by simple domain decomposition. The flow domain is partitioned into equisized subdomains, each subdomain is supported by a different processor, and boundary data must be exchanged between adjacent subdomains. Our parallel code has mostly been used for simulations of bubbly flows and we use a master-slave technique to parallelize the front. This technique takes advantage of the physical problem and when a bubble is spread over more than one subdomain, one processor is designated as the "master" for that bubble. It handles and centralizes the corresponding data and sends it to the other processors, or "slaves," which then distribute the front data onto the fixed grid. While most of the code parallelizes very efficiently due to the natural load balancing of the physical problem, the parallelization efficiency is somewhat degraded by the multigrid solver. Coarse grain parallelism is therefore employed in most of our computations.

3. Results

Here we will discuss two problems in some detail. We have studied the first one, homogenious bubbly flows in several papers and the simulations have lead to major new insight. The second problem, the effect of electrostatic forces on a suspension of drops is a much more recent investigation and our understanding of the problem is still relatively primitive.

3.1 Homogeneous bubbly flows

Our study of bubbly flows has focused on homogeneous flows where many buoyant bubbles rise together in an initially quiescent fluid. To model such flows, we have done computations of bubbles in fully periodic domains. The simplest case is when there is only one bubble per period so the configuration of the bubbles array does not change. While such regular arrays are unlikely to appear in experiments, they provide a useful reference configuration for freely evolving bubbles. As the number of bubbles in each period is increased, the regular array becomes unstable and the bubbles generally rise unsteadily, repeatedly undergoing close interactions with other bubbles. The behavior is, however, statistically steady and the average motion (averaged over sufficiently

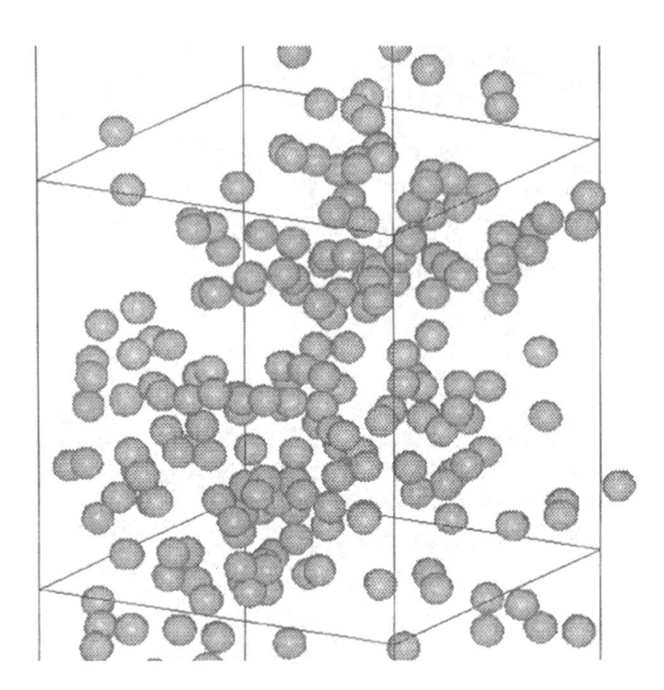

Figure 9.2. One frame from a simulation of the rise of 216 buoyant bubbles in a periodic domain. Here, N= 900, Eo=1, and the void fraction is 6%.

long time) does not change. While the number of bubbles clearly influences the average motion for small enough number of bubbles per period, the hope is that once the size of the system is large enough, information obtained by averaging over each period will be representative of a truly homogeneous bubbly flow.

The goal of computational studies is, first and foremost, to provide insight that is useful to modelers of multiphase flows. In addition to information about the effect of drift Reynolds number, velocity fluctuations, and bubble dispersion on the properties of the system, the computations should allow us to identify how the bubbles interact, whether there is a predominant microstructure and/or interaction mode, and whether flow structures that are much larger than the bubbles are formed. Information about the microstructure is essential for the construction of models of multiphase flows and can also help to identify what approximations can be made. It would, for example, lead to enormous simplification if a dense bubbly flow could be approximated by a regular periodic array of bubbles. Information about the large scale distribution of the bubbles is also critical. Modeling is much easier if the bubbles stay relatively uniformly distributed, than if they form large regions where the bubble density is either very high or very low.

The rise of a single buoyant bubble is governed by four nondimensional numbers. Two are the ratios of the bubble density and viscosity to the ones of

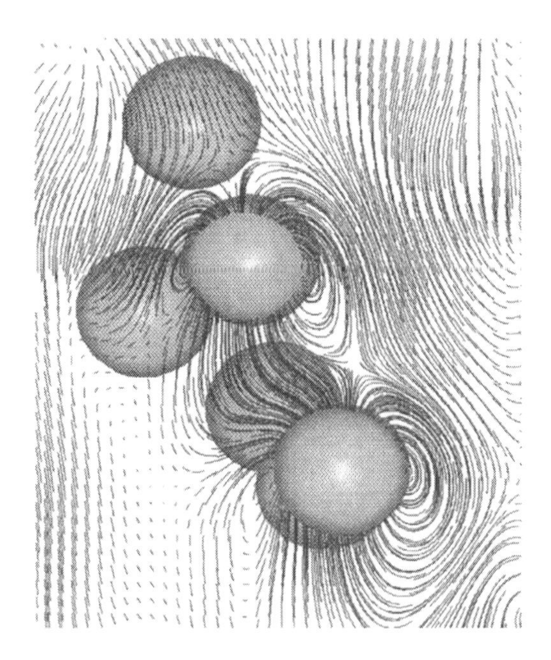

Figure 9.3. A close up of the streamlines around a few bubbles. From a simulation of the buoyant rise of 91 bubbles in a periodic domain. Here, N= 900, Eo=1, and the void fraction is 6%.

the liquid: ρ_b/ρ_o and μ_b/μ_o. Here, the subscript o denotes the ambient fluid and b stands for the fluid inside the bubble. The ratios of the material properties are usually small and have little influence on the motion. The remaining two numbers can be selected in a number of ways. If we pick the density of the outer fluid, ρ_o, the effective diameter of the bubble, d_e, and the gravity acceleration, g, to make the other variables dimensionless, we obtain:

$$N = \frac{\rho_o^2 d_e^3 g}{\mu_o^2}; \qquad \text{and} \qquad Eo = \frac{\rho_o d_e^2 g}{\gamma}. \tag{9.9}$$

The first number is usually called the Gallileo or the Archimedes number (see Clift, Grace and Weber, 1978) and the second one is the Eötvös number. For flow with many bubbles, the void fraction, α, must also be specified.

The motion of nearly spherical bubbles at moderate Reynolds numbers has been examined in a number of papers. Esmaeeli and Tryggvason (1996, 1998) studied a case where the average rise Reynolds number of the bubbles was relatively small, 1-2, and Esmaeeli and Tryggvason (1999) looked at another case where the Reynolds number was 20-30. Most of these simulations were limited to two-dimensional flows, although a few three-dimensional simulations with up to eight bubbles were included. Simulations of freely evolving bubble arrays were compared with regular arrays and it was found that while freely evolving bubbles at low Reynolds numbers rise faster than a regular array (in agreement with analytical predictions for Stokes flow), at higher Reynolds numbers the trend is reversed and the freely moving bubbles rise slower. The time averages of the two-dimensional simulations were generally well converged but exhibited a dependency on the size of the system. This dependency was stronger for the low Reynolds number case than the moderate Reynolds number one. Although many of the qualitative aspects of a few bubble interactions are captured by two-dimensional simulations, the much stronger interactions between two-dimensional bubbles can lead to quantitative differences.

To examine a much larger number of three-dimensional bubbles, Bunner (2000) developed a fully parallel version of the method used by Esmaeeli and Tryggvason. His largest simulation followed the motion of 216 three-dimensional buoyant bubbles per periodic domain for a relatively long time. Figure 9.2 shows the bubble distribution at one time from this simulation. The details of the flow field around a few bubbles from a simulation of 91 bubbles with the same parameters is shown in figure 9.3. The governing parameters are selected such that the average rise Reynolds number is about 20-30, depending on the void fraction. This is comparable to Esmaeeli and Tryggvason, 1999, but not identical, and the deformation of the bubbles are small. Although the motion of the individual bubbles is unsteady, the simulations are carried out for a long enough time so the average behavior of the system is well defined, as in the two-dimensional simulations of Esmaeeli and Tryggvason. Simulations with different number of bubbles have been used to explore the dependency of the various average quantities on the size of the system. The average rise Reynolds number and the Reynolds stresses are essentially fully converged for systems with 27 bubbles, but the average fluctuation of the bubble velocities requires larger systems. Examination of the pair distribution function for the bubbles shows that although the bubbles are uniformly distributed on the average, they tend to line up side-by-side, independent of the size of the system. This trend increases as the rise Reynolds number increases, suggesting a monotonic trend from the nearly no preference found by Ladd (1993) for Stokes flow, toward the strong layer formation seen in the potential flow simulations of Sangani and Didwania (1993) and Smereka (1993). To examine the usefulness of simplified models, the results were compared with analytical expressions for simple cell models in the Stokes flow and the potential flow limits. The results show that the rise velocity at low Reynolds number is reasonably well predicted by Stokes flow based models. The bubble interaction mechanism is, however, quite different. At both Reynolds numbers, two-bubble interactions take place by the "drafting, kissing, and tumbling" mechanism of Joseph and collaborators (Fortes, Lundgren, and Joseph, 1987). This is very different from either a Stokes flow where two bubbles do not change their relative orientation unless acted on by a third bubble, or the predictions of potential flow models

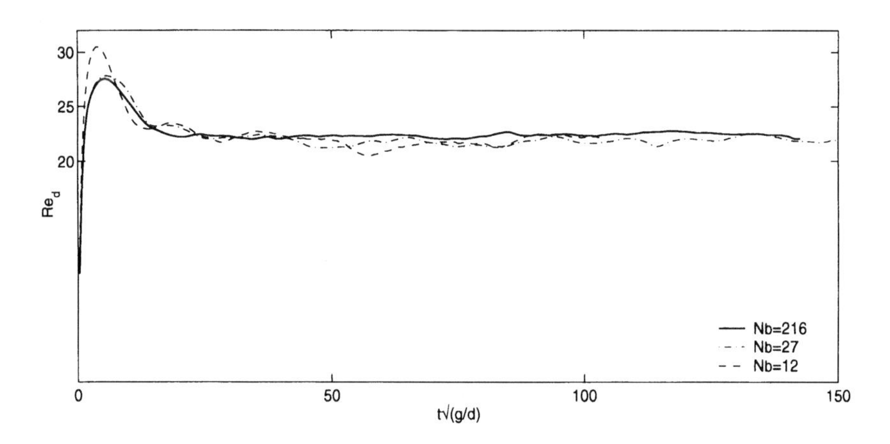

Figure 9.4. The average rise Reynolds number versus time, for simulations of 12, 27, and 216 nearly spherical bubbles.

where a bubble is repelled from the wake of another one, not drawn into it. For moderate Reynolds numbers (about 20), we find that the Reynolds stresses for a freely evolving two-dimensional bubble array are comparable to Stokes flow while in three-dimensional flow the results are comparable to predictions of potential flow cell models. The average rise Reynolds number of bubbles from simulations with 12, 27, and 216 bubbles is plotted versus time in figure 9.4, demonstrating that the rise velocity is insensitive to the number of bubbles simulated.

To examine the effect of deformation of bubbles, Bunner and Tryggvason (2001c) have done two sets of simulations using 27 bubbles per periodic domain. In one set the bubbles are spherical and in the other set the bubbles deform into ellipsoids. The nearly spherical bubbles quickly reach a well defined average rise velocity and remain nearly uniformly distributed across the computational domain. The deformable bubbles initially behave similarly, except that their velocity fluctuations are larger. Figure 9.5 shows the bubble distribution from a simulation of 27 bubbles and a void fraction of 2% at a selected time. The streamlines in a plane through the domain, and the vorticity in the same plane, are also shown. In some cases the nearly uniform distribution seen here changes to a completely different state where the bubbles accumulate in vertical streams, rising much faster than when they are uniformly distributed. This behavior can be explained by the dependency of the lift force that the bubbles experience on the deformation of the bubbles. Although we have not seen streaming in all

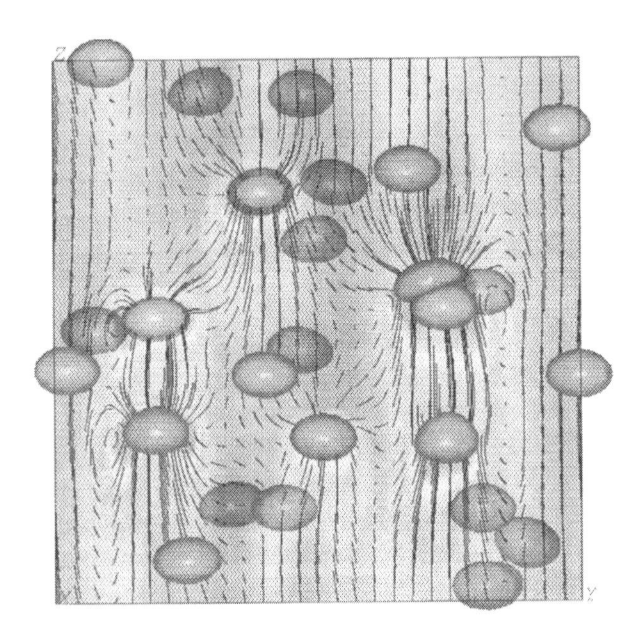

Figure 9.5. One frame from a simulation of 27 ellipsoidal bubbles. The bubbles, along with vorticity contours and streamlines are shown. Here, N= 900, Eo=5, and the void fraction is 2%.

of our simulations with deformable bubbles, we believe that streaming would take place if the computations were carried out for a long enough time or the number of bubbles was larger. Simulations with the bubbles initially confined to a single column show that while the nearly spherical bubbles immediately disperse, the deformable bubbles stay in the column and rise much faster than uniformly distributed bubbles.

The three-dimensional simulations of Bunner & Tryggvason (2000) provided a fairly good picture of the microstructure of a freely moving bubbles at high void fraction. However, the number of bubbles is still relatively small and the question of whether larger scale structures form is still open. We believe, based on the available evidence, that homogeneous systems of nearly spherical three-dimensional bubbles will remain nearly uniformly distributed. This contrasts with the results of Esmaeeli and Tryggvason (1996), who found that a few hundred two-dimensional bubbles at O(1) Reynolds number lead to an inverse energy cascade where the flow structures continuously increase in size. This is similar to the evolution of two-dimensional turbulence, and although the same interaction is not expected in three dimensions, the simulations demonstrated the importance of examining large systems with many bubbles.

Although most of our studies of the motion of many bubbles have been limited to moderate Reynolds numbers, the methodology is, in principle, capable of handling higher Reynolds numbers. The key difficulty is resolution, as in all direct simulations of finite Reynolds number flows. Esmaeeli (1995) conducted several simulations of a few two-dimensional bubbles where the bubbles reached a rise Reynolds number of about 800 which is roughly what one would expect from experimental studies of air bubbles in water at the same nondimensional parameters (and where the bubbles are fully three-dimensional). The high resolution needed and the small time steps required made further studies impractical at that time. A detailed examination of the properties of bubbly flows at these higher Reynolds numbers has therefore not been done, with the exception of a brief examination of two-dimensional regular arrays reported in Esmaeeli, Ervin, and Tryggvason (1993) and Göz, Bunner, Sommerfeld, and Tryggvason (2000). Two-dimensional bubbles start to wobble at much lower rise Reynolds numbers than the three-dimensional counterparts and the results suggest that, in addition to regular periodic wobbly motion, bubbles at very high Reynolds numbers may exhibit chaotic oscillations.

In bubble columns, continuous bubble distributions are often observed It may be expected that small bubbles are drawn into the wake of larger bubbles leading to drag reduction and thus an increase of the average rise velocity of such a group of bubbles. It could also be, however, that this is largely an effect of deformability and occurs for spherical bubbles only if their diameter ratio is sufficiently large. In order to isolate the effect of a bidispersed bubble size distribution, Göz *et al.* (2001) examined systems with an equal number of small and large spherical bubbles. The volume ratio of large to small bubbles was 2. The average rise velocity (of all bubbles) and the average turbulent kinetic energy of the liquid and its dependence on the void fraction turned out to be very close to those of the corresponding monodisperse systems. On the other hand, there are differences in the fluctuation velocities. The average vertical and horizontal fluctuation velocities over each size class are of similar magnitude as the corresponding values in the monodisperse case. Thereby, the averages for the small bubbles tend to be somewhat larger than the averages for the large bubbles, meaning that the small bubbles experience stronger fluctuations about their mean velocity. However, due to the differential rise velocity of the large and small bubbles, the vertical fluctuation velocity averaged over all bubbles is considerably larger in the bidisperse case than in the monodisperse case. Most recently, we have been examining the effect of a larger difference in the bubble sizes, where the large bubbles are deformable and 8 times larger than the small bubbles and the number of small bubbles is 8 times the number of large bubbles. The governing parameters, $N_l = 7200$, Eo_l=4.0, $N_s = 900$, and Eo_s =1.0, allow the large bubbles to deform into ellipsoids with an aspect ratio of about 1.2. Figure 9.6 shows an instantaneous bubble distribution for 40 small and 5 large bubbles at a 6% volume fraction. From a movie produced from this simulation one can clearly see that all bubbles initially move mainly vertically upwards,

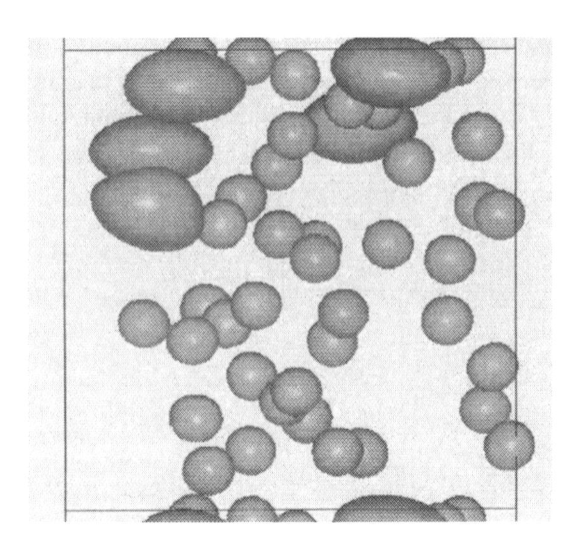

Figure 9.6. One frame from a simulation of a bidispersed bubble system.

without much interaction between the large and the small bubbles except when a faster rising large bubble pushes smaller bubbles out of its way. This is similar to the behavior of the dilute (2%, 6%) bidisperse systems with purely spherical bubbles of volume ratio 2. As the bubbles continue to rise, however, the large bubbles begin to rise unsteadily along a spiral path. The unsteady motion of the large bubbles stirs up the smaller bubbles, leading to large fluctuation velocities for both the small and the large bubbles.

3.2 Electrostatic Effects

Electric fields can have a dramatic effect on multiphase flows, particularly in microgravity. Electric fields have been used to manipulate suspensions to enhance coalescence (Byers and Amarnath,1995), to generate uniform distribution of small drops in sprays (Nawab and Mason, 1958), increase heat and mass transfer from drops (Sadhal, Ayyaswamy, and Chung, 1997), to increase boiling efficiency (Jones, 1987), and to stabilize liquid bridges (Sherwood, 1988). For general reviews see the early article by Melcher and Taylor (1969), and the more recent one by Saville (1997). While some numerical modeling has been done, it is fairly limited in both the physical assumptions used as well as the generality of the numerical method. Here, we are primarily interested in the use of electrostatic forces to modify the behavior of a suspension of bubbles and drops in channel flows.

When the interface between two dielectric fluids is subjected to an external electric field, the dielectric mismatch between the fluids induces a stress at the fluid interface. This stress can be either normal or tangent to the interface, depending on the conduction and dielectric properties of the fluids. In the case of perfect dielectrics and conductors, the electric surface force acts perpendicular to the surface, pushing the interface in the direction of the fluid with the lower dielectric constant. Changes in the shape of the interface and the surface tension balance these induced electric forces at the interface.

For dielectric fluids with small but finite conductivity, Taylor and Melcher (1969) proposed the "leaky dielectric" model. In this approximation, the change in the charge density is assumed to be so small that it can be neglected in the conservation of charge equation. The electric field is therefore obtained from the electric potential by charge conservation equation for the electric potential

$$\nabla \cdot \sigma \mathbf{E} = 0 \tag{9.10}$$

where the electric field is obtained from the electric potential $\mathbf{E} = -\nabla\phi$ and the charge is computed from Gauss' law

$$q_v = \nabla \cdot \varepsilon \nabla \phi. \tag{9.11}$$

Here, σ is the electric conductivity and ε is the permittivity or dielectric constant. Although early experiments showed some disagreement between analytical results from the leaky-dielectric theory and experiments, more careful measurements, particularly accurate determination of the electric properties, appears to have resolved the discrepancy (Saville, 1997). The major approximation in the leaky dielectric fluid is likely to be the neglect of charge advection by the fluid. For drops moving with respect to the continuous fluid, such as a drop falling under gravity, interface particles are continually moved from the front of a drop to the back, possibility leading to a redistribution of the charges.

The force on the fluid is computed directly from the electric field by:

$$\mathbf{f} = q_v \mathbf{E} - \frac{1}{2}(\mathbf{E} \cdot \mathbf{E})\nabla\varepsilon, \tag{9.12}$$

where the charge accumulation is given by equation 9.11. If σ and ε are constant in each fluid the force only acts on the phase boundaries. Once the forces on the interface have been found, they are simply added to the Navier-Stokes equations. For sharp interfaces the force is obviously singular.

The motion of a single drop subjected to an electric field in a stationary fluid is governed by four nondimensional numbers. Those are the ratios of the permittivity of the drop and the ambient fluid, $S^{-1} = \varepsilon_i/\varepsilon_o$, the ratio of the conductivities, $R = \sigma_i/\sigma_o$, the ratio of viscous force to surface tension (the Ohnsorge number $Oh = \varepsilon_o/\sqrt{\rho_o d\gamma}$), and the ratio of surface tension to electric force (the electric capillary number $Ca = \varepsilon E_\infty^2 a/\gamma$). When a fluid flow is added we must also specify the channel Reynolds number, the void fraction, and the nondimensional size of the drops. We note, for reference, that for castor

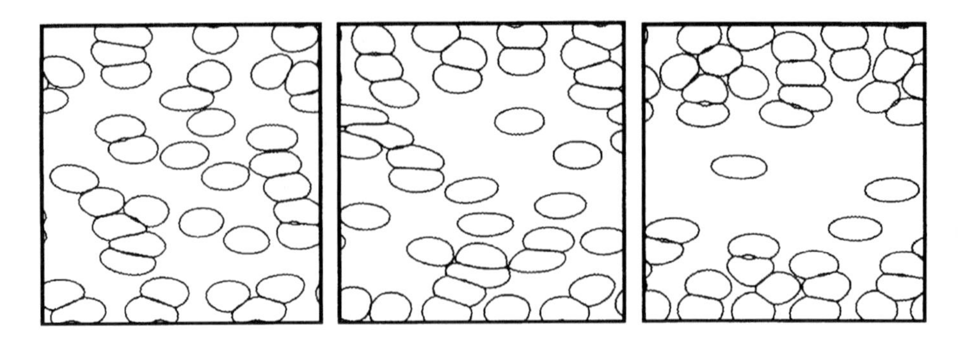

Figure 9.7. The drops distribution at three times for a simulation of two-dimensional drops.

oil drops in water, $S^{-1} = 0.057$ and $R = O(10^4 - 10^6)$. A wide range of values for S and R can be obtained by using different fluid combinations and oil pairs, in particular, can yield values close to unity.

The Navier-Stokes equations are solved in the same way as discussed previously. The new aspect are the solution of a Poisson equation for the electric field, equation 9.10, and the addition of new forces to the momentum equations. The Poisson equation is solved by the same multigrid method that is used for the pressure equation directly on a fixed grid. Alternatively, the electric field could be computed by a boundary integral equation over the fluid interface and the external boundary. There are two main reasons for using the fixed grid: First, this allows a very straight forward extension to fluids with variable dielectric permeability and conductivity. The second is that for complex fluid interfaces it is much faster to solve a Poisson equation on a fixed grid than to compute all interactions directly as in boundary integral calculations. This approach confirms also better to the rest of the numerical method and results in a simpler code. Since the conductivity is generally different in the different fluids, an elliptic solver suitable for a domain with nearly discontinuous properties must be used.

If σ and ε are constant in each fluid, they are simply set by an indicator function constructed from the location of the interface. If they vary, an advection equation for these properties must be solved.

To compute the electric forces and add them to the Navier-Stokes equations, it is possible to proceed in two ways by using either equation 9.12 or by finding normal and tangent component on the front and spreading them onto the fixed grid. The former approach is particularly simple, since it only involves differentiation on the fixed grid. This is used in the calculations presented here.

The method outlined above has been implemented in axisymmetric geometry and used to examine the deformation of single drops and the interactions of two drops (Che, 1999). The deformation of a single drop was computed for various permittivity and conductivity ratios, resulting in both oblate and prolate drops.

The results were in good agreement with the finite element computations of Feng and Scott (1996) for finite Reynolds number flows. The two-drop problem has been examined in the Stokes flow limit by Baygents *et al.* (1998) using a boundary integral technique and Che's results showed that moderate Reynolds numbers had only a small effect. The interaction of two drops is controlled by two effects. The drops are driven together due to the charge distribution on the surface. Since the net charge of the drops is zero, the drops see each other as dipoles. This dielectrophoretic motion always leads to drops attraction. The second effect is fluid motion driven by tangential stresses at the fluid interface. The fluid motion depends on the relative magnitude of the permittivity and conductivity ratios. When the permittivity ratio is higher than the conductivity ratio, the tangential forces induce flow from the poles of the drops to the equator. If the center of two such drops lies on a line parallel to the electronic field, the flow drains from the region between the drops and they attract each other. When the ratios are equal, no tangential motion is induced and the drops attract each other by dielectrophoretic motion. For drops with the permittivity ratio lower than the conductivity ratio, the tangential forces induce flow in the opposite direction (from the equator to the poles of the drops) and if the induced fluid motion is sufficiently strong, the drops repell each other.

When an electric field is applied to many drops suspended in a channel flow the distribution of the drops across the channel cross section can be changed greatly. This problem has been examined by Che (1999), and Fernández *et al.* (2001), using both two- and three-dimensional simulations. An extensive set of two-dimensional simulations has allowed us to explore the effect of the conductivity and permittivity ratios in some detail. For drops that attract each other the initial evolution is governed by the pairwise interactions of drops. For drops with a permittivity ratio that is lower than the conductivity ratio, two drops whose centers lie on an axis parallel to the electric field attract each other and drops are attracted to the walls of the channel by the same mechanism. When the electric field is turned on drops first attract each other pairwise and some drops move to the wall. If the forces are strong (compared to the fluid shear) the drops can form columns, spanning the channel and blocking the two-dimensional flow. If the attractive forces are weaker the columns are immediately broken up.

Figure 9.7 shows a few frames from a two-dimensional simulation of the motion of 36 drops in a periodic channel. The electric field is generated by applying a potential difference between the top and the bottom of the channel and the flow is subject to a constant pressure gradient. Here, S=0.125, R=2.0, Re=20, We=0.0625, the density and viscosity ratios are 1.0, and the void fraction is 0.44. These parameteres result in oblate drops and two drops whose centers are on a line parallel to the electric field attract each other. After the electric field is turned on, the pairwise interactions of drops initially lead to the formation of drop pairs and columns of drops parallel to the field. The drops are also attracted to the walls by the same mechanism that drops are attracted to each other and eventually all the drops migrate to the walls. In this case the drops are not allowed to coalesce, but in reality we would expect the drops to form fluid layers next to the walls.

If the permittivity and the conductivity ratios are comparable, no tangential fluid motion is generated but the drops still attract each other pairwise and columns can form if the pressure drop is low and the drops are not allowed to coalesce. For permittivity ratios lower than the conductivity ratios the electric field induces tangential flow from the equator to the poles. In two-dimensions, this results in interactions between the drops that are similar to the previous case, except that the attractions take place perpendicular to the electric field. The drops therefore tend to form rows aligned with the flow, or "slugs" where many drops clump together. When the conductivity ratio is much higher than the permittivity ratio, the drops become prolate, expell each other and are spread more uniformly across the channel.

While the two-dimensional simulations have allowed us to conduct a large number of simulations relatively inexpensively, and explore a large range of conductivity and permittivity ratios, it is clear that we need to simulate fully three-dimensional systems for quantitative predictions. Figure 9.8 show one frame from a simulation of a three-dimensional system. The parameters are the same as in figure 9.7, but the void fraction is lower. The drops and the electric field is shown after several drops have moved to the walls, but before any significant pairwise interaction has taken place.

Our examination of the effect of electric fields on suspensions is still in its early stage and several aspects of the flow remains poorly understood. We have only done a limited number of simulations to examine the effect of the flow, for example. For flows where the drops tend to form columns across the channel, strong flow (or weaker electric field) breaks up the columns. In some cases this promotes drop accumulation at the walls, as seen in figure 9.7, but in other case the result is a statistically steady state where drop pairs and short drop chains continuously form and break up. When drops are driven together to form columns or pairs, the film between them would eventually drain completely and the drops coalesce. While detailed simulations of the draining of the film and its rupture require both exceedingly fine resolution and the includion of short range attractive forces, it is likely that the overall effect of coalescence can be assessed by a much simpler model when the interfaces are simply reconnected when the film thickness is of the order of the grid-spacing. This is easily acomplished, particularly for two-dimensional flows when the front is simply a chain of connected marker points and is essentially identical to the "automatic" coalescence of interfaces in methods where the interface is followed by advecting a marker function (as in the Volume of Fluid and Level Set methods). We have not yet done such simulations.

Figure 9.8. The drop distribution for a fully three-dimensional simulation.

4. Conclusions

Attempts to simulate multiphase flows go back to the early days of computational fluid dynamics at Los Alamos. While a few successful simulations can be found in the early literature, major progress has been made in the last few years. The "one field" formulation is the key to much of this progress. The method described here is one of the most successful implementations of the "one field" formulation but impressive results have also been obtained by improved VOF methods, level set methods, phase field methods, and the CIP method. The key difference between these methods and the technique described here is our use of a separate "front" to mark the phase boundary, instead of a marker function. While explicit front tracking is generally more complex than the advection of a marker function, we believe that the increased accuracy and robustness is well worth the effort. The explicit tracking of the interface not only reduces errors associated with the advection of a marker function and surface tension computations but the flexibility inherent in the explicit tracking approach should also be important for application to problems where complex interface physics must be accounted for.

The results presented here have been selected to show both a problem where direct numerical simulations havc already had some impact as well as one where the investigation is just starting. Although improvement in accuracy, efficiency, and robustness are always possible, it is probably fair to say that simulations of problems where two incompressible fluids, separated by a constant surface tension interface, flow together are relatively well under control. Future work for such fluids is likely to focus mainly on obtaining an understanding of diverse physical problems. With the continuing development of advanced computers, the next few years are likely to see a large number of such simulations. It is now feasible to conduct simulations with up to about billion grid points (1000^3). If we assume that each bubble is resolved by 20 grid points per diameters and that the bubbles are separated by about a bubble diameter, then we should be able to do simulations of systems with over 15,000 bubbles. While the presence of walls, the inclusion of a distribution of bubble sizes, lower void fraction, and higher Reynolds numbers may lower this estimate it is clear that we are in the position of simulating systems where we should be able to obtain very reliable statistical data. Simulations of that magnitude should also allow us to expand our studies beyond dispersed systems and to examine flows that undergo rapid change in topology, such as in atomization and churn flow. Perhaps the biggest challenge is to understand how to incorporate the knowledge obtained by large-scale direct simulations into models useful for routine engineering predictions.

From a numerical point of view, the main action is likely to be in the development of robust methods for systems with complex physics. Simulations of problems where the fluid motion is coupled with electric, magnetic, or thermal effects are still in their infancy. Similarly, simulations of problems with phase changes, such as solidification, boiling, and evaporation, are just emerging. The possibility to conduct detailed simulations of such systems will transform how research on multiphase flow is conducted and revolutionize the ability to predict the behavior of such systems.

Acknowledgments

This work was supported by the National Science Foundation and the National Aeronautics and Space Administration. The computations were done on the IBM SP2 parallel computers at the Maui High Performance Computing Center and at the Centers for Parallel Computing at the University of Michigan and at WPI.

References

G. Agresar, J. J. Linderman, G. Tryggvason, and K.G. Powell, "An Adaptive, Cartesian, Front-Tracking Method for the Motion, Deformation and Adhesion of Circulating Cells," *J. Comput. Phys.* **43** (1998), 346-380.

N. Al-Rawahi and G. Tryggvason, "Numerical Simulations of Dendritic Solidification with Convection—Two Dimensional Geometry." Submitted for publication (2001).

J.C. Baygents, N.J. Rivette, and H.A. Stone, "Electrohydrodynamic Deformation and Interaction of Drop Pairs," *J. Fluid Mech.* 368 (1998), 359-375.

B. Bunner, "Large Scale Simulations of Bubbly Flow," *Ph.D. Dissertation*, The University of Michigan, 2000.

B. Bunner and G. Tryggvason, "Direct Numerical Simulations of Three-Dimensional Bubbly Flows," *Phys. Fluids*, **11** (1999), 1967-1969.

B. Bunner and G. Tryggvason, "An Examination of the Flow Induced by Buoyant Bubbles," *Journal of Visualization*, 2 (1999), 153-158.

B. Bunner and G. Tryggvason, "Dynamics of Homogeneous Bubbly Flows: Part 1. Rise Velocity and Microstructure of the Bubbles." Submitted for publication (2001).

B. Bunner and G. Tryggvason, "Dynamics of Homogeneous Bubbly Flows. Part 2, Fluctuations of the Bubbles and the Liquid." Submitted for publication (2001).

B. Bunner and G. Tryggvason, "Effect of Bubble Deformation on the Stability and Properties of Bubbly Flows." Submitted for publication (2001).

C.H. Byers and A. Amarnath, "Understand the potential of electro-separations," *Chem. Engng. Prog.* **91** (1995), 63-69.

J. Che, "Numerical Simulations of Complex Multiphase Flows: Electrohydrodynamics and solidification of droplets." *Ph.D. Dissertation*, University of Michigan, 1999.

R. Clift, J.R. Grace, and M.E. Weber, *Bubbles, Drops, and Particles*. Academic Press, 1978.

E. A. Ervin, "Full Numerical Simulations of Bubbles and Drops in Shear Flow." *Ph.D. Dissertation*, The University of Michigan, (1993).

E.A. Ervin and G. Tryggvason, "The Rise of Bubbles in a Vertical Shear Flow," *ASME J. Fluid Engineering 119* (1997), 443-449.

A. Esmaeeli, "Numerical Simulations of Bubbly Flows." *Ph.D. Dissertation*, The University of Michigan, (1995).

A. Esmaeeli, E.A. Ervin, and G. Tryggvason, "Numerical Simulations of Rising Bubbles." In *Proceedings of the IUTAM Conference on Bubble Dynamics and Interfacial Phenomens*. Birmingham, U.K., 6-9 Sept. 1993. Ed.: J.R. Blake, J.M. Boulton-Stone and N.H. Thomas. 247-255.

A. Esmaeeli and G. Tryggvason, "An Inverse Energy Cascade in Two-Dimensional, Low Reynolds Number Bubbly Flows," *J. Fluid Mech.* **314** (1996), 315-330.

A. Esmaeeli and G. Tryggvason, "Direct Numerical Simulations of Bubbly Flows. Part I—Low Reynolds Number Arrays," *J. Fluid Mech.* **377** (1998), 313-345.

A. Esmaeeli and G. Tryggvason, "Direct Numerical Simulations of Bubbly Flows. Part II—Moderate Reynolds Number Arrays," *J. Fluid Mech.* **385** (1999), 325-358.

A. Esmaeeli and G. Tryggvason, "Direct Numerical Simulations of Boiling Flows," *Proceedings of the Fourth International Conference on Multiphase Flows* (2001).

J.Q. Feng and T.C. Scott, "A Computational Analysis of Electrohydrodynamics of a Leaky Dielectric Drop in an Electric Field," *J. Fluid Mech.* 311 (1996), 289-326.

A. Fernández and G. Tryggvason. "Effect of Electrostatic Forces on the Phase Distribution in Droplet Suspension." ASME Fluids Engineering Division Summer Meeting, New Orleance, LA, May 29-June 1, 2001

A. Fortes, D.D. Joseph, and T. Lundgren, "Nonlinear mechanics of fluidization of beds of spherical particles," *J. Fluid Mech.* **177** (1987), 467-483.

M.F. Göz, B. Bunner, M. Sommerfeld, and G. Tryggvason, "The Unsteady Dynamics of Two-Dimensional Bubbles in a Regular Array." Proceedings of the ASME FEDSM'00 Fluids Engineering Division Summer Meeting June 11-15, 2000, Boston, Massachusetts

M.F. Göz, B. Bunner, M. Sommerfeld, and G. Tryggvason, "Simulation of bubbly gas-liquid flows by a parallel finite-difference/front-tracking method." In: Lecture Notes in Comput. Science and Engineering, Springer, in print.

J. Han and G. Tryggvason, "Secondary Breakup of Liquid Drops in Axisymmetric Geometry—Part I, Constant Acceleration," *Phys. Fluids*, **11** (1999), 3650-3667.

J. Han and G. Tryggvason, "Secondary Breakup of Liquid Drops in Axisymmetric Geometry—Part II. Impulsive Acceleration," Submitted to *Phys. Fluids*, **13** (2001), 1554-1565.

S. Homma, G. Tryggvason, J. Koga, and S. Matsumoto, "Formation of a Jet in Liquid-Liquid System and Its Breakup into Drops, " FEDSM98-5216. Proceedings of the 1998 ASME Fluids Engineering Division Summer Meeting, Washington, D.C., June 21-25, 1998.

Y. J. Jan, "Computational Studies of Bubble Dynamics," *Ph.D. Dissertation*, The University of Michigan, (1994).

Y.-J. Jan and G. Tryggvason, "Computational Studies of Contaminated Bubbles," Symp on Dynamics of Bubbles and Vortices Near a Free Surface," AMD Vol. 119 (Ed. Sahin and Tryggvason), ASME (1991), 46-59.

Y.J. Jiang, A. Umemura, and C.K. Law, "An experimental investigation on the collision behavior of hydrocarbon droplets." *J. Fluid Mech.* **234** (1992), 171-190.

D. Juric, "Computations of Phase Change," *Ph.D. Dissertation*, The University of Michigan, (1996).

D. Juric and G. Tryggvason, "A Front Tracking Method for Dentritic Solidification," *J. Comput. Phys.* **123** (1996), 127-148.

D. Juric and G. Tryggvason, "Computations of Boiling Flows," *Int'l. J. Multiphase Flow*, **24** (1998), 387-410.

A.J.C. Ladd, "Dynamical simulations of sedimenting spheres," *Phys. Fluids* A, **5** (1993), 299-310.

E. Loth, M. Taeibi-Rahni, and G. Tryggvason, "Deformable Bubbles in a Free Shear," *Int'l. J. Multiphase Flow*, **23** (1997), 977-1001.

J.R. Melcher and G.I. Taylor, "Electrohydrodynamics: A Review of the Role of Interfacial Shear Stresses," *Ann. Rev. Fluid Mech.* 1 (1969), 111-147.

S. Mortazavi and G. Tryggvason, "A numerical study of the motion of drops in Poiseuille flow. Part 1. Lateral migration of one drop," *J. Fluid Mech.* **411** (2000), 325-350.

S. Mortazavi and G. Tryggvason, "A Numerical Study of Drops Suspended in a Poiseuille Flow. Part II. Many drops," In preparation.

M.A. Nawab and S.G. Mason, "The preparation of uniform emulsions by electric dispersion," *J. Colloid. Sci.* 13 (1958), 179-187.

S. Nas, "Computational Investigation of Thermocapillary Migration of Bubbles and Drops in Zero Gravity," *Ph.D. Dissertation,* The University of Michigan, (1995).

S. Nas and G. Tryggvason, "Computational Investigation of the Thermal Migration of Bubbles and Drops." In AMD 174/FED 175 Fluid Mechanics Phenomena in Micrograviry, Ed. Siginer, Thompson and Trefethen. ASME (1993), 71-83.

M. R. H. Nobari, "Numerical Simulations of Drop Collisions and Coalescence," *Ph.D. Dissertation*, The University of Michigan, (1993).

M. R. Nobari, and G. Tryggvason, "Numerical Simulations of Three-Dimensional Drop Collisions," *AIAA Journal* **34** (1996), 750-755.

M. R. Nobari, Y.-J. Jan and G. Tryggvason, "Head-on Collision of Drops–A Numerical Investigation," *Phys. Fluids* **8**, (1996), 29-42.

S. Osher and R.P. Fedkiw, "Level Set Methods: An Overview and Some Recent Results," *J. Comput. Phys.* **169** (2001), 463-502.

C. S. Peskin, "Numerical Analysis of Blood Flow in the Heart," *J. Comput. Phys.* **25** (1977), 220-.

C. S. Peskin and D.M. McQueen, "A General Method for the Computer Simulation of Biological Systems Interacting with Fluids." In: *SEB Symposium on Biological Fluid Dynamics,* Leeds, England, July 5-8.

J. Qian, G. Tryggvason and C.K. Law, "A Front Tracking Method for the Motion of Premixed Flames," *J. Comput. Phys.* **144** (1998), 52-69.

J. Qian, G. Tryggvason and C.K. Law, "An Experimental and Computational Study of Bounching and Deforming Droplet Collision." Submitted for publication.

S.S. Sadhal, P.S. Ayyaswamy, and J.N. Chung. *Transport Phenomena with Drops and Bubbles*, Springer 1997.

A.S. Sangani and A.K. Didwania, "Dynamic simulations of flows of bubbly liquids at large Reynolds numbers," *J. Fluid Mech.* **250** (1993), 307-337.

D.A. Saville, "Electrohydrodynamics: The Taylor-Melcher Leaky Dielectric Model," *Ann. Rev. Fluid Mech.* 29 (1997), 27-64.

R. Scardovelli and S. Zaleski, "Direct numerical simulation of free-surface and interfacial flow," *Ann. Rev. Fluid Mech.* **31** (1999), 567-603

J.A. Sethian, "Evolution, Implementation, and Application of Level Set and Fast Marching Methods for Advancing Fronts," *J. Comput. Phys.* **169** (2001), 503-555.

J.D. Sherwood, "Breakup of Fluid Droplets in Electric and Magnetic Fields," *J. Fluid Mech.* 188 (1988), 133-146.

P. Smereka, "On the motion of bubbles in a periodic box." *J. Fluid Mech.* **254** (1993), 79-112.

M. Song and G. Tryggvason, "The Formation of a Thick Border on an Initially Stationary Fluid Sheet," *Phys. Fluids*, **11** (1999), 2487-2493.

M. Taeibi-Rahni, E. Loth and G. Tryggvason, "DNS Simulations of Large Bubbles in Mixing Layer Flow," *Int'l. J. Multiphase Flow*, **20** (1994), 1109-1128.

W. Tauber, S.O. Unverdi, and G. Tryggvason, "The nonlinear behavior of a sheared immiscible fluid interface," Submitted for publication, (2000).

W. Tauber and G. Tryggvason, "Direct Numerical Simulations of Primary Breakup," *Computational Fluid Dynamics Journal* **9** (2000).

G. Tryggvason and H. Aref, "Numerical Experiments on Hele Shaw Flow with a Sharp Interface," *J. Fluid Mech.* **136** (1983), 1-30.

G. Tryggvason, "Numerical Simulation of the Rayleigh-Taylor Instability," *J. Comput Phys.* **75** (1988), 253-282.

G. Tryggvason and S. O. Unverdi, "Computations of Three-Dimensional Rayleigh-Taylor Instability," *Phys. Fluids A* **2** (1990), 656-659.

G. Tryggvason and S. O. Unverdi, "The Shear Breakup of an Immiscible Fluid Interface." In *Fluid Dynamics at Interfaces*, W. Shyy and R. Narayanan, editors. Cambridge University Press, 1999, 142-155.

G. Tryggvason, B. Bunner, A. Esmaeeli, D. Juric, N. Al-Rawahi, W. Tauber, J. Han, S. Nas, and Y.-J. Jan, "A front tracking method for the computations of multiphase flow," *J. Comput. Physics*, 169 (2001), 708 759.

H. S. Udaykumar, H. C. Kan, W. Shyy and R. Tran-Son-Tay, "Multiphase Dynamics in Arbitrary Geometries on Fixed Cartesian Grids," *J. Comput. Phys.* **137** (1997), 366-405.

S. O. Unverdi and G. Tryggvason, "A Front-Tracking Method for Viscous, Incompressible, Multi-Fluid Flows," *J. Comput Phys.* **100** (1992), 25-37.

S. O. Unverdi and G. Tryggvason, "Computations of Multi-Fluid Flows," *Physica D* **60** (1992), 70-83.

Y. Yang and G. Tryggvason, "Dissipation of Energy by Finite Amplitude Surface Waves," *Computers and Fluids,* **27** (1998), 829-845.

P.-W. Yu, S.L. Ceccio and G. Tryggvason, "The Collapse of a Cavitation Bubble in Shear Flows-A Numerical Study," *Phys. Fluids* **7** (1995), 2608-2616.

Chapter 10

EXAMPLES OF CONTEMPORARY CFD SIMULATIONS

S. R. Chakravarthy, U. C. Goldberg
src@metacomptech.com, ucg@metacomptech.com

P. Batten
batten@metacomptech.com
Metacomp Technologies, Inc., Westlake Village, California

Abstract This chapter is intended to convince the reader, through a number of relevant and diverse examples, that modern CFD with turbulence modelling is a practical tool for fast and accurate prediction of flow problems of engineering interest. A general introduction to modern CFD and a preface to the general-purpose CFD solver, CFD++, are followed by a substantial number of flow examples.

Keywords: CFD, turbulence closure, accuracy, engineering applications

1. Introduction

Computational Fluid Dynamics (CFD) has evolved much over the last three decades. In years past, aerospace applications were the primary driver for, and client of, CFD technology. Aerospace continues to be a significant focus, but the application areas have now expanded into automotive, marine, environmental, biomedical and industrial processing. This places a strenuous demand on contemporary CFD tools. Various physical phenomena must be accurately and efficiently predicted, including boundary and interior layers, other high gradient regions, discontinuities (e.g. shock waves), chemical reactions, presence of multiple species and phases and their interactions, very high speed and very low speed flows and everything in between. Very complex geometries must be treated, including movement of objects and meshes around multiple objects. The computational power that was only available to a few and which required

D. Drikakis and B.J. Geurts (eds.), Turbulent Flow Computation, 339–369.

specially constructed rooms and buildings is now available at an individual engineer's desk and is at the disposal of groups, large and small. Multi-CPU machines and networked computers are the norm, not the exception. Contemporary CFD tools must fit this computational environment. In the early decades of CFD development, the developer of the algorithm and code was also the user. Today, there is an expectation that a myriad of practical fluid dynamics simulations be achievable by engineers who are not CFD experts. Often CFD is expected to be part of early design and not just a tool to help trouble-shoot undesirable fluid dynamic effects observed on an existing design. Practical CFD must deliver value in this tough environment.

1.1 Algorithmic Elements of Practical CFD

In the early nineteen-eighties, the evolution of the upwind scheme resulted in a step forward in being able to compute more problems more routinely. In today's environment, we find that the following algorithmic and discretization elements are conducive to high fidelity, robust and efficient simulations in a variety of situations:

1 Unstructured grid treatment for full flexibility in dealing with complex topologies and physics, including various types of adaptation.

2 Ability to deal with multi-block meshes with various types of inter-block connectivities.

3 Multidimensional interpolation that more accurately represents local behavior of flow-dependent variables. While formal order of accuracy need not be any higher, this approach leads to practically higher accuracy on relatively coarse meshes. The multidimensional interpolation framework helps deal easily with inter-block connectivities also.

4 The preconditioning approach to dealing with low speed and incompressible fluid flows. We have extended the formulations to include the treatment of both single species and multi-species flows. In the case of single species flows, we have also extended the formulation to successfully treat multi-speed flows (where low and high speed flows may coexist within the same problem).

5 The use of implicit relaxation approaches to avoid factorization errors and time step restrictions often seen in approximate factorization methods.

6 The use of the algebraic multigrid approach for achieving good convergence acceleration for implicit schemes even on dense meshes. There is some disadvantage with this approach being relatively memory intensive for a large coupled system of equations.

7 The use of dual-time stepping to achieve accuracy in simulating transient (not asymptotically steady state) flows. The dual time stepping is particularly helpful in mitigating one side-effect of using preconditioners which is to alter the characteristic speeds of solution evolution. Dual time stepping is similarly effective with other convergence acceleration techniques, such as spatially varying time step. In conjunction with the multigrid approach and relaxation, dual time stepping provides the ability to compute transient periodic behavior with very few physical time steps representing each period of oscillation (on the order of one or two hundred time steps).

8 The use of a pointwise implicit relaxation methodology helps deal with implementations that must exploit parallel computing.

9 The use of an implicit boundary condition treatment whenever possible. This is fundamental to improving robustness as well as convergence (especially for Neumann boundary conditions).

10 The use of upwind formulations including a "Riemann solver" that is not yet very popular. While necessitating a larger effort when developing a computer code, it leads to less uncertainty about the level of numerical dissipation that is practical for robustness and accuracy. The use of the upwind formulation also makes the development of codes based on preconditioners more straightforward. More thoughts on Riemann solvers are presented in a later section.

11 Good turbulence models and modelling approaches. More on this is discussed later.

While none of these elements is, by any means, revolutionary, a survey of the present state of the art will find that these elements are not often present, in combination, in any single piece of software. We have built these features into the CFD++ software package. As a result of struggling hard with many varieties of fluid dynamic problems over a long period of time, and learning from each experience, we believe there is not one single element that represents progress in practical CFD, but rather a combination of those items presented above that leads to satisfactory results (this is not, however, to deny the validity of other approaches or algorithmic choices). Early discussions of our ideas and implementations can be found in[1, 2]. Some of the turbulence models we have developed and used are discussed in[3, 4, 5, 6, 7].

1.2 Progress In Turbulence Modelling

We now turn our attention to more specific information about turbulence closure, which is still considered a pacing item in CFD. We have achieved progress in two sub-fields:

1 Improvements to classical models

2 Hybrid RANS/LES methods

1.2.1 Improvements to classical models. From the outset, our approach to turbulence closure has been based on topography-parameter-free formulations. These models are ideally suited to unstructured book-keeping and massively parallel processing thanks to their independence from constraints related to the placement of boundaries and/or zonal interfaces. Recent contributions to these models include:

- improved behavior of the dissipation-rate transport equation by explicit sensitization to non-equilibrium flow regions.
- enhanced near-wall characteristics and elimination of ad-hoc formulations through introduction of time-scale realizability.
- improved heat-transfer prediction capability in high-speed flow by adopting a dual-dissipation approach[5].

1.2.2 Hybrid RANS/LES methods. While LES is an increasingly powerful tool for unsteady turbulent flow prediction, it is still prohibitively expensive for most problems. To bring LES closer to becoming a design tool, a hybrid RANS/LES approach has been recently developed at Metacomp[8], called LNS (for Limited Numerical Scales). With this method a regular RANS-type grid is used everywhere, except in isolated flow regions where denser, LES-type mesh is used to resolve critical unsteady flow features. The hybrid model transitions smoothly between an LES calculation and a cubic k-e model, depending on grid fineness. It reverts to RANS in the vicinity of elongated near-wall cells (those typically used in RANS calculations), thus avoiding the near-wall modelling problems associated with LES. Fig. 10.1 shows a 2D cut of a square cylinder in crossflow, where a fine grid was placed around the body and in the near-wake; the rest of the flowfield being occupied by a coarser, RANS-like mesh.

1.3 A Brief History of the Riemann Solver

We felt it relevant to present a perspective on the "Riemann solver" which has become a mainstay of CFD. Much of our modern upwind technology arose from

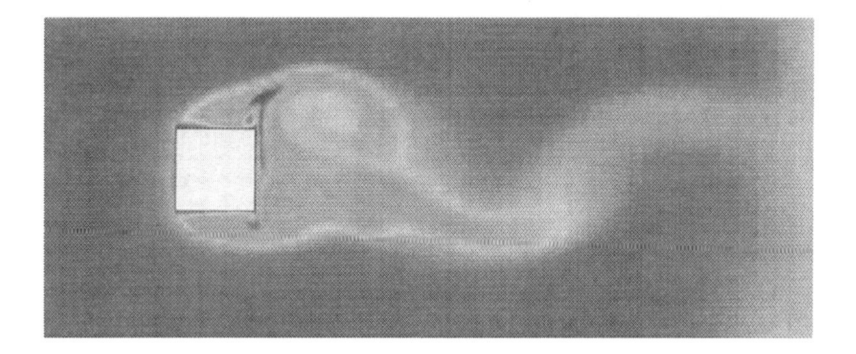

Figure 10.1. LNS: turbulence intensity around a square cylinder in crossflow

developments in the early 1980's. With the advent of these new, parameter-free, scalar transport schemes based on hybridized first- and higher-order upwind schemes (for example, TVD[9], UNO/ENO[10]), the significance of the underlying first-order method became apparent. This generated renewed interest in the work of Godunov[11] who, decades earlier, had devised a first-order approach based on the exact solution to a two-state wave interaction, or Riemann, problem. Godunov's pioneering work enabled these new upwind schemes to be applied directly to hyperbolic systems such as the Euler equations. A flurry of activity led to the development of approximate upwind fluxes, for example, the well-known flux-vector split methods of Steger-Warming[12] and van Leer [13] and the well-known flux-difference-splitting methods of Osher-Solomon [14] and Roe[15]. The original justification for these approximations was the premise that only limited information from the exact solution was needed by most numerical schemes and the practical observation that approximate solutions appeared to give similar predictions, at least in simple test cases.

Ironically, much of the work in the late 1980's and early 1990's (for example, Einfeldt et al.[16], Toro et al.[17], Batten et al.[18], Moschetta and Pullin [19], Liou[20]) was then spent attempting to remedy the host of problems that were subsequently discovered with these early approximate upwind-flux methods. Common problems encountered were a loss of positivity, large glitches near sonic points in expansion waves, kinked Mach stems, excessive growth of disturbances at grid-oblique shear layers or the introduction of strong artificial smoothing in the direction normal to shear or boundary layers. Many of these problems were hard to detect and were often incorrectly attributed to coding errors. Even now, the various known failings of Roe's solver[15] have done little to diminish its popularity.

One could legitimately ask what was gained by abandoning the exact Riemann solver. Its' introduction in a large CFD-code calculation today would likely have only a minor impact on CPU time, given the other complexities that are now handled on a routine basis. The most practical use of approximate Riemann solvers now appears to lie in the implementation of implicit schemes, where Jacobians need to be obtained rapidly, or in the modelling of more complex physics, where exact solutions are currently unknown. In these situations, there appear to be considerable advantages in the (less well-known) approximate Riemann solver framework introduced by Harten, Lax and van Leer[21]. In their inspirational 1983 paper, Harten et al. put forward an approach based on the generation of integral-average states. These integral averages would be exact if the signal velocities in the Riemann problem were known in advance. Based on the work of Roe[15], Einfeldt[16] and Toro[17], Batten et al.[18] devised a set of wave-speeds that ensured a positivity-preserving version of the Toro et al.[17], 'HLLC' solver. This appears to have been the first and only approximate Riemann solver to simultaneously guarantee positivity and entropy conditions, whilst recognizing all isolated discontinuities. The latter property is now widely regarded as significant for viscous flows (see Allmaras[22], Batten [26] and McNeil[23]).

The strategy of ensuring realizability in the numerics complements recent efforts to achieve realizability in the mathematics, for example in the modelling of other physical phenomena, such as turbulence. Modern turbulence closures which ensure non-negative normal stresses and Schwarz inequalities on shear stresses tend to improve both robustness and the quality of predictions. Mathematically, an interesting aspect of these improvements in realizability is the prospect of an absolute guarantee of robustness, at least within some local time-step limit. For example, Perthame and Shu[24] recently demonstrated how positivity-preserving schemes, such as HLLC, can be extended to arbitrary orders of spatial accuracy using an additional gradient constraint. Other questions inevitably remain unanswered and continue to spark new ideas. One curious phenomenon is an instability sometimes seen polluting the bow shock upstream of blunt bodies in hypersonic flows. This has a myriad of other manifestations, but is most commonly known as the 'carbuncle phenomenon'. Depending on grid resolution and orientation, the carbuncle may or may not appear, but one important point which is often ignored, is that the exact Riemann solver also suffers from this 'deficiency'. What is truly at fault is the way in which these exact solutions are defined and used. There have been claims in the literature that certain schemes do not give this effect (HLLC gives carbuncles identical to those of an exact Riemann solver). Some of these claims subsequently proved false, with the carbuncle reappearing using a different grid or flow speed and there remains the curious question of how one can improve upon an exact solution. Nonetheless, new schemes tend to provide new insight and often this is a valuable contribution in itself.

1.4 Preface to the Current CFD Solver Methodology

The combination of modern numerical methods and modern turbulence closures enables increasingly reliable prediction of both internal and external, steady and unsteady aerodynamic flow problems across the Mach number range. By prediction we mean not only the velocity and pressure fields, but also surface phenomena such as skin friction and heat transfer. Contemporary CFD codes can predict such flows on either structured or unstructured meshes. The solver used here, CFD++, is an example of one such code, in which structured, unstructured, hybrid and complex overset grids can be handled within a single unified framework. Further flexibility and computational power are realised through the use of massively parallel computers. Both these attributes are crucial for successful prediction of real-life flow problems of engineering interest.

In order to take full advantage of modern concepts such as unstructured bookkeeping and massively parallel computer architectures, turbulence models must adapt and comply with the requirements dictated by these contemporary approaches. Topography-parameter-free turbulence closures are ideally suited for the task thanks to their local nature and the absence of explicit wall distance requirements. A variety of recently-developed single-, two- and three-equation closures have been incorporated into the code in order to successfully tackle aerodynamic flow problems of engineering importance. These model equations can be integrated directly to solid surfaces or in conjunction with sophisticated wall functions which take into account compressibility, heat transfer and pressure gradient effects.

Contemporary approaches have extended the realm of turbulent flow prediction in two important ways: on the one hand, methods are now available to predict turbulent flows on coarse grids; on the other hand, hybrid RANS/LES approaches can effectively predict unsteady flows on "engineering" meshes with only local refinements. Both these extensions to the classical RANS framework have an appreciable impact on practical CFD.

Flow prediction examples will be given for automotive cases, environmental simulation, aircraft aerodynamics, hypersonic-flow heat transfer, free-shear mixing, reacting flow, acoustics, and the emerging field of active flow control.

1.4.1 Highlights of the Numerical Approach. The CFD solver used here is a very versatile and powerful, modular Computational Fluid Dynamics (CFD) software suite which implements a unified-grid, unified-physics and unified-computing framework. Some of its characteristics include:

1 Unsteady compressible and incompressible fluid flow using the Favre-averaged Navier-Stokes equations with turbulence modelling. The code can be used to investigate flow regimes from incompressible and very low subsonic to hypersonic (Unified Physics).

2 Unification of Cartesian, structured curvilinear, and unstructured grids, including hybrids (Unified Grid).

3 Unification of treatment of various cell shapes including hexahedral, pyramid, tetrahedral and triangular prism cells (3-d), quadrilateral and triangular cells (2-d) and linear elements (1-d). Other special cells for self-similar flows and surface manipulations are also available (Unified-Grid).

4 Treatment of multiblock patched-aligned (nodally connected), patched-nonaligned and overset grids. Interblock connectivity is automatically determined (Unified-Grid).

5 Total Variation Diminishing discretization based on a new multi-dimensional interpolation framework. This results in an extremely versatile discretization formulation that can handle above-mentioned cell and grid topologies.

6 Riemann solvers to provide proper signal propagation physics. These Riemann solvers have been adapted to the various preconditioned forms of the governing equations.

7 Consistent and accurate discretization of viscous terms using the same multi-dimensional polynomial framework and memory-saving implementation. Non-TVD derivatives computed for inviscid terms are reused to compute viscous terms and turbulence model discretization. This provides for additional computational efficiency.

8 Topography-parameter-free turbulence models that do not require knowledge of distance to walls. This formulation is particularly well suited for general book-keeping and multi-CPU implementations.

9 Versatile boundary condition implementation includes a rich variety of integrated boundary condition types to cover most problems encountered in practice.

10 Modern implementation encompasses all computing platforms including all UNIX and Linux platforms, Windows 2000/NT, single-CPU and multi-CPU platforms from PCs to mainframes (Unified-Computing).

11 The implementation on MPP computers is based on the distributed-memory message-passing model. It can use native message-passing libraries or MPI, PVM, etc. The code has been implemented and tested on a variety of parallel computer platforms including products from SGI, Compaq, HP, SUN, IBM as well as Linux clusters and Windows NT/2000 machines (Unified-Computing).

Further details regarding the numerical methodology can be found in[1, 2, 25, 26].

1.4.2 Numerical Accuracy. In general, large complex 3D flow computations, using wall functions, cannot be shown to yield strict grid independence. Furthermore, a one million size grid, for example, may require an increase to ten million cells before grid independence can be achieved. This would render CFD a useless tool from a practical viewpoint. Instead, the approach commonly adopted is to observe the effect of refining the grid in selected subdomains on the most desired predictions, such as drag. Experience plays an important role in determining the flow regions where increased mesh is beneficial. Often just a few local refinements suffice to converge the essential predictions to within a few percent, thus providing a reasonable level of confidence in the overall results.

2. Flow Examples

2.1 DLR-F4 Wing/Body Configuration

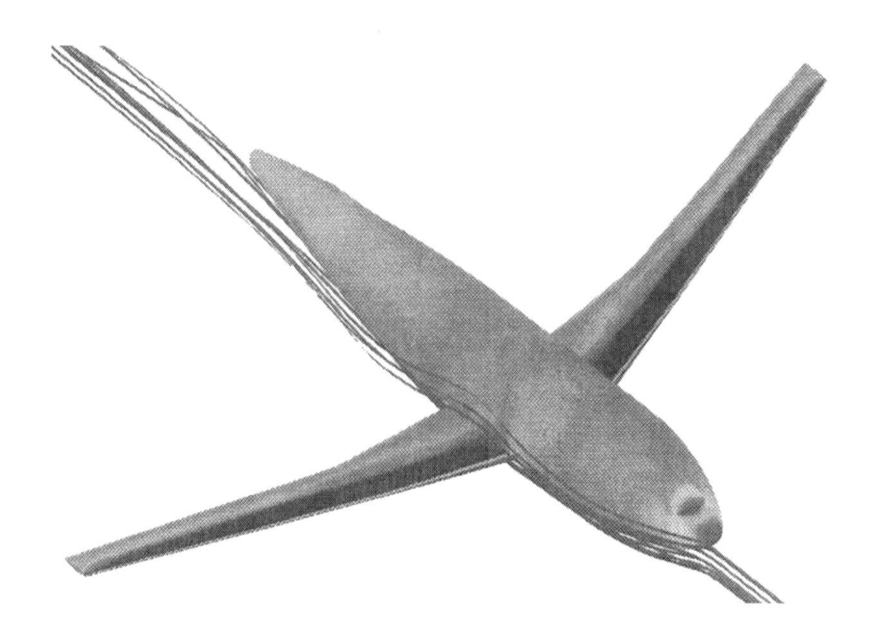

Figure 10.2. DLR-F4: surface pressure contours

In an effort to evaluate CFD tools in predicting aerodynamic drag, experimental data were collected in three European wind tunnels[27] for transonic flow over the DLR-F4 wing/body configuration at several angles-of-attack. The

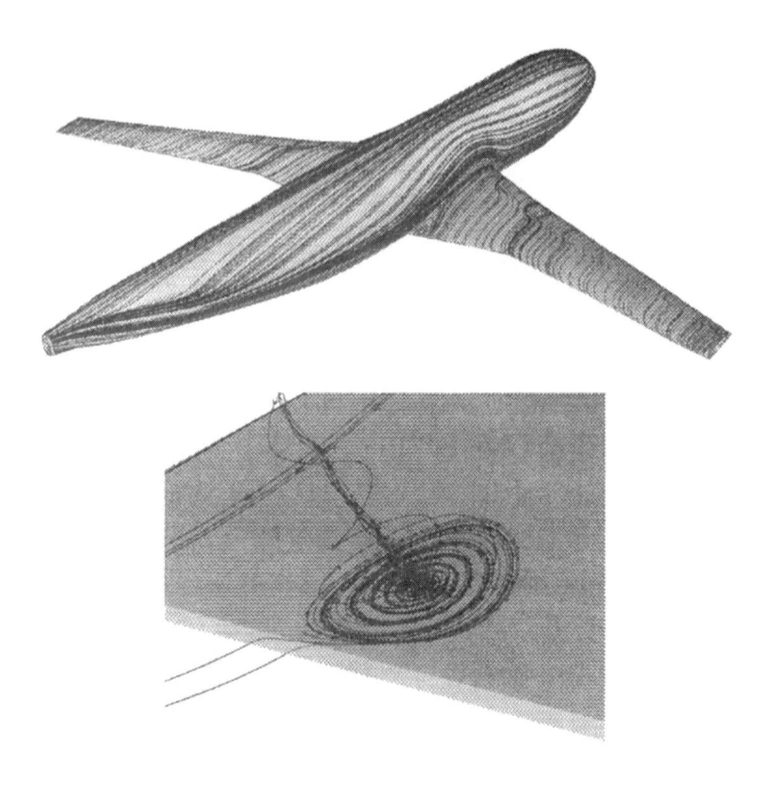

Figure 10.3. DLR-F4: Oil flow pattern (upper), junction vortex (lower)

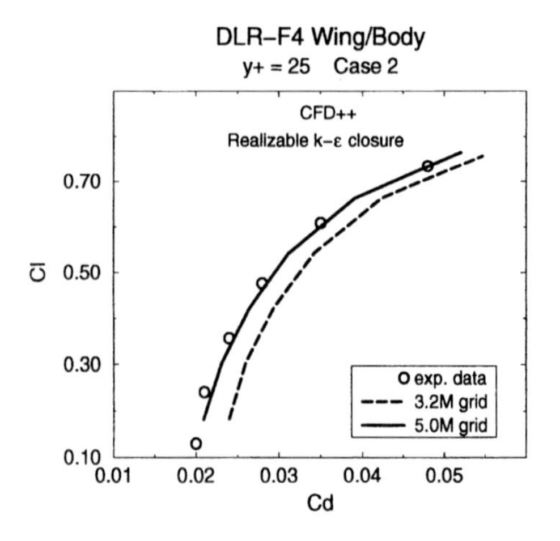

Figure 10.4. DLR-F4: drag polar

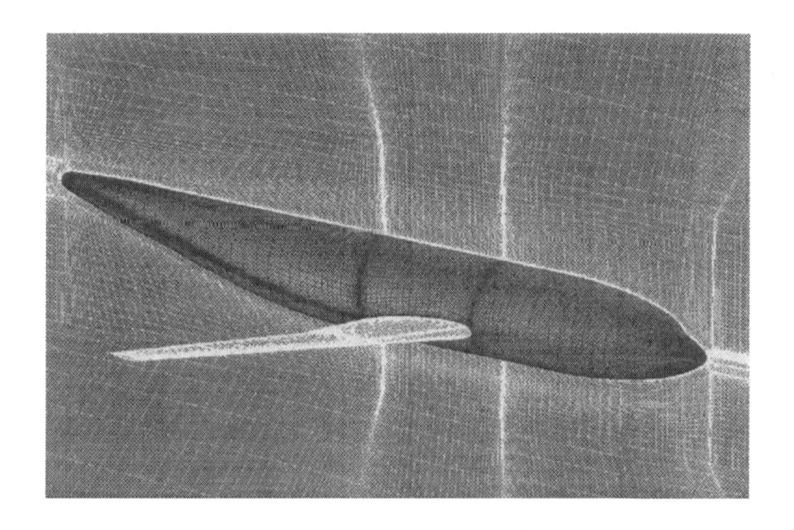

Figure 10.5. DLR-F4: partial grid view

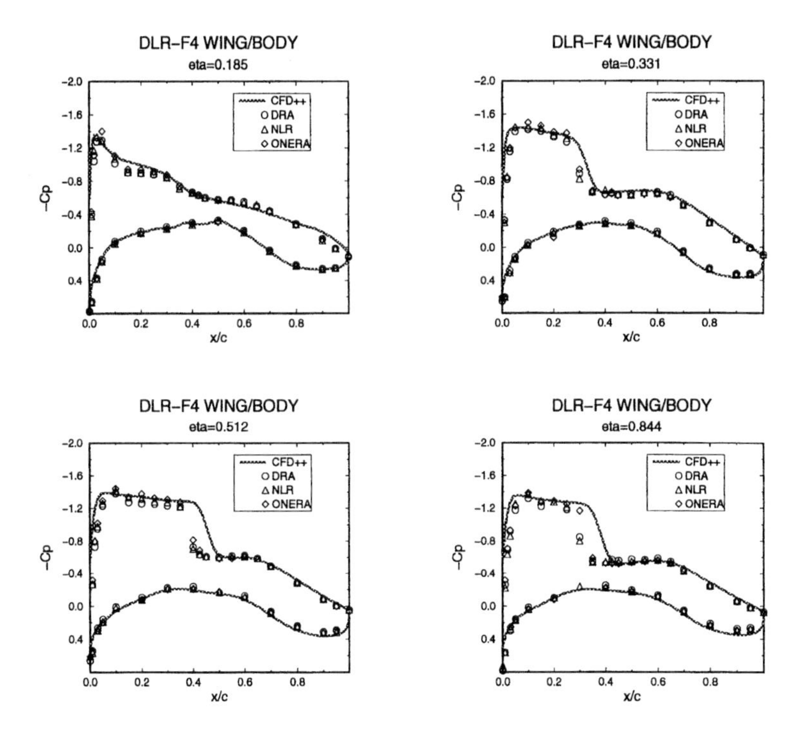

Figure 10.6. DLR-F4: pressure profiles at various wing spanwise stations

current CFD solver was used to predict the flow at M=0.75, employing the realizable wall-distance-free k-e model[3] with a wall function. The level of freestream turbulence was measured at 0.2%. The calculations were performed initially on a 3.2 million hexahedral element mesh and later on a refined 5.0 million grid, both with y^+=25 at the first centroids away from walls. Figs. 10.2-10.4 show surface pressure, surface oil flow pattern, indicating flow separation over the wing at the shock foot, a detail of the vortex generated near the junction of the fuselage and the rear upper wing surface (Figs. 10.3 courtesy of CEI) and the predicted drag polar compared to measurements. The sensitivity of predictive quality to the mesh is clearly seen in the drag polar plot and it points out typical uncertainties and challenges in computing complex 3D turbulent flows. Specifically, the CFD user must be knowledgeable and experienced enough to avoid too many iterations before an adequate grid is produced. Recent advances in CFD methodology enable automatic grid adaptation but, in practice, the resulting mesh is often too large for repetitive calculations of complex 3D viscous flows, necessary within the design cycle. Thus a combination of a few adaptation cycles and engineering judgement may often be necessary to arrive at a grid which is fine enough to obtain correct flow trends yet coarse enough to retain CFD within the design cycle.

In the present flow case, however, even the refined mesh (partial view seen in Fig. 10.5) was insufficient to capture the precise location of the shock foot along the wing's suction side, as seen in Figs. 10.6 which show pressure profiles at several spanwise stations. Except for a shift in the shock location toward the wing tip, very good agreement with the experimental data is observed.

2.2 Two Hole Transverse Injection into a Supersonic Flow

Scramjet propulsion relies on injection and burning of fuel in a supersonic flowfield. The residence time of the fuel inside the combustion chamber is typically a few milliseconds, therefore rapid mixing of the fuel with the freestream air is critical to ensure a working design. One promising configuration being used is transverse injection from an array of holes located at the combustor's lower wall.

The design of high speed propulsion systems is increasingly relying on CFD simulations to predict the complex flowfields present in these devices. The main challenge here is to avoid creating an excessive eddy viscosity field which would force premature diffusion of the mixing vortex core. This leads to the conclusion that the turbulence model is a key ingredient in rendering CFD a reliable tool to predict scramjet flows: a good model must not introduce excessive levels of turbulence in the core flow. This suggests a closure which is based primarily on predicting the Reynolds stresses while using eddy viscosity only in a secondary level of importance. In addition, the model should be able to account for the high levels of turbulence anisotropy and for the strong streamline curvature present in the two-hole injection flowfield. One of the best choices is the cubic k-ϵ closure[4] which is economical yet accounts for Reynolds stress anisotropy, streamline curvature and swirl.

McDaniel et al[28] conducted a spatially complete set of measurements for a complex 3-D unit combustor flowfield, consisting of transverse air injection from two holes into a Mach 2 flow downstream of a backward-facing step. Two non-intrusive optical techniques were employed: (a) laser-induced-iodine fluorescence (LIIF) and (b) laser-doppler anemometry (LDA). The figures below show experimental data using both techniques. The cubic k-ϵ model was employed here together with a wall function. The mesh consisted of half a million cells, using a symmetry plane. The calculation converged in two hours on an SGI parallel workstation, using eight R10000 CPUs.

Fig. 10.7 shows details of the flow topology in the vicinity of the injectors. The horseshoe vortices wrapping around the normal jets are clearly seen, including the lift-off of the downstream vortex due to the impingement of the first jet on the second.

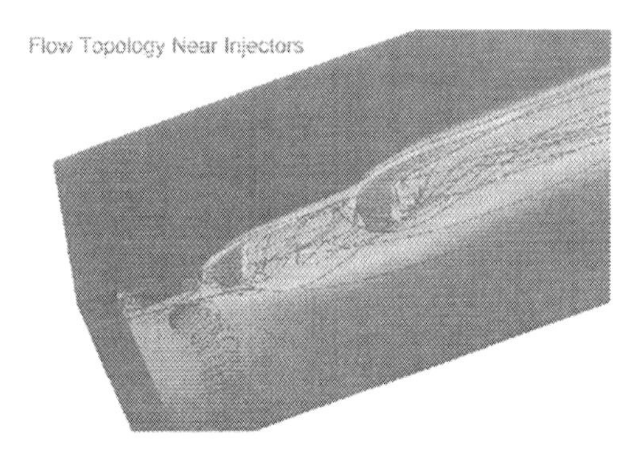

Figure 10.7. Two-hole injection: flow topology features in vicinity of jets

Figs. 10.8-10.9 show detailed profile comparisons on the symmetry plane at the injector locations. A high level of agreement between predictions and data is observed, indicating that the CFD approach, using the cubic closure, was able to avoid premature diffusion of the mixing vortex core.

Finally, Fig. 10.10 is the convergence history plot. A remarkable level of solution convergence was achieved in just 300-400 time steps, demonstrating the computational tool's readiness for the design cycle.

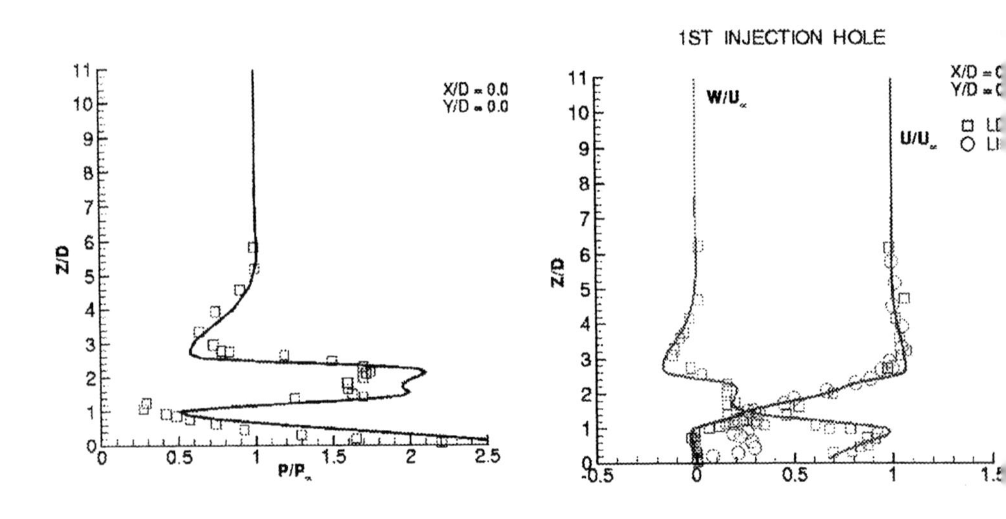

Figure 10.8. profiles at 1st injector hole

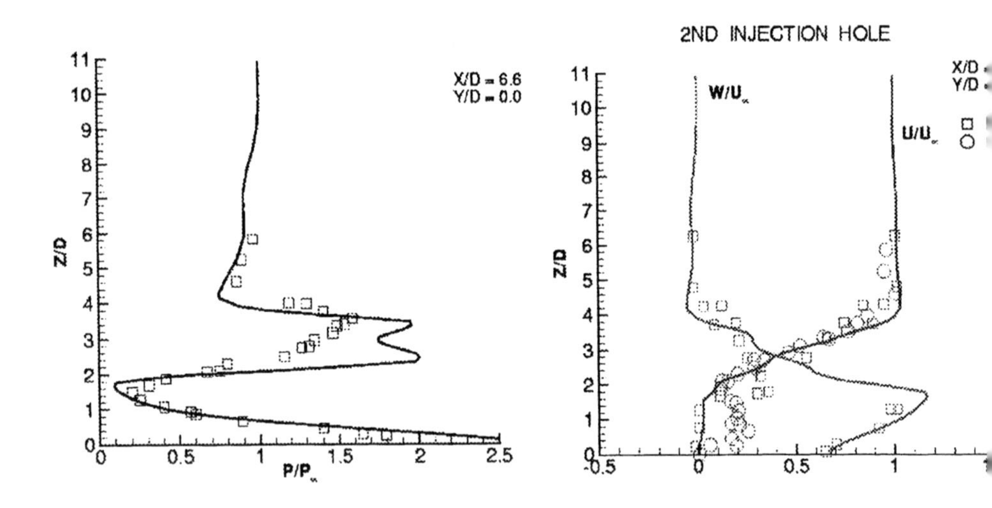

Figure 10.9. profiles at 2nd injector hole

2.3 Hypersonic Flow in a Double-Wedge Inlet

In recent years there has been renewed interest in hypersonic flight vehicles. The engine inlets of these vehicles typically involve compression ramps which, through a series of shocks, reduce the engine inflow Mach number to supersonic levels to enable supersonic combustion. Such a shock system imposes, however,

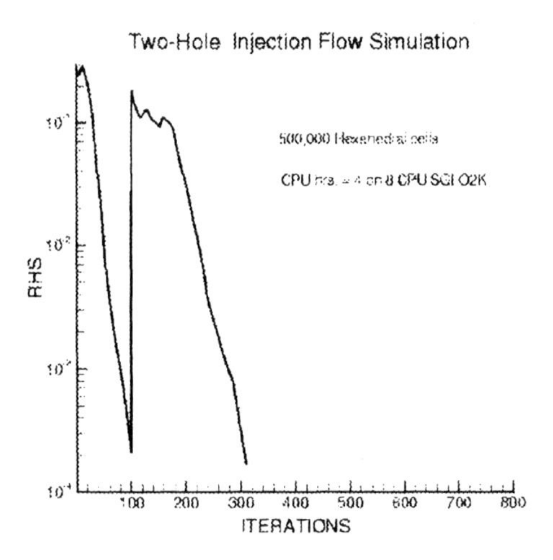

Figure 10.10. convergence history

a severe penalty in terms of surface heating, requiring careful attention to the choice of materials and/or cooling devices to avoid the possibility of local melting of the vehicle's skin. It is important, therefore, to be able to predict hypersonic flow over ramp and wedge configurations, including surface heating characteristics, with the aim of using this capability for analysis and design purposes of vehicle components such as engine inlets. The ability to predict turbulent hypersonic flows with high level of confidence carries much broader benefits, namely entire vehicle external/internal flow prediction capability for preliminary design and, later, for various analysis purposes.

Hypersonic wall-bounded flows pose a difficult challenge to CFD due to severe velocity and temperature gradients adjacent to solid surfaces, the presence of laminar to turbulent flow transition and strong shock/boundary-layer interactions with attendant massive flow separation.

Kussoy et al.[29] performed extensive experimental measurements on a Mach 8.3 flow in a wedge inlet configuration, with $T_{\rm w}/T_0$=0.27. To predict this complex 3D flowfield, involving crossing shock/boundary-layer interactions, the CFD solver was used on a structured mesh consisting of approximately 250,000 cells. First centroidal locations away from walls were at $y^+ \approx 60$ to avoid a much larger grid size. A wall function which accounts for compressibility, heat transfer and pressure gradient effects was employed. This wall function uses the van Driest transformed velocity (see[30]) in conjunction with a version of the Launder-Spalding Law-of-the-Wall[31] which is based on $\sqrt{k}$ rather than on the friction velocity, u_τ, to avoid problems in separation and reattachment zones where the latter vanishes. The wall-law was used to determine momentum and energy fluxes at walls.

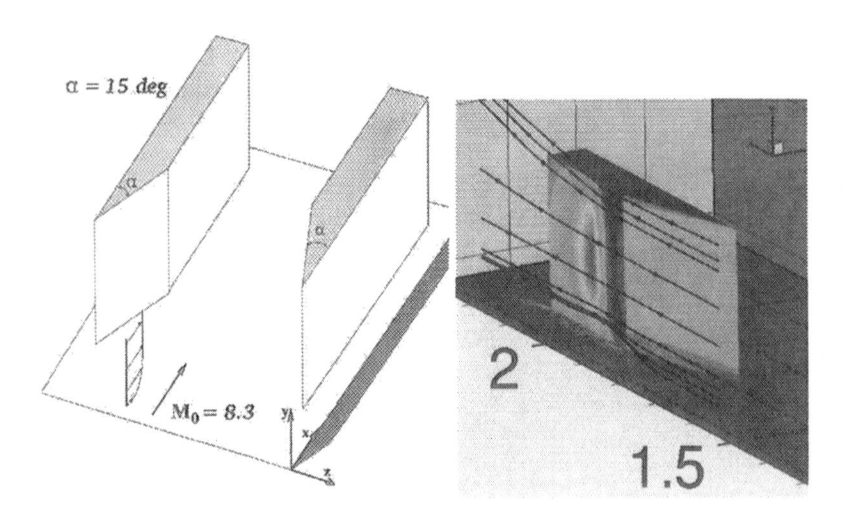

Figure 10.11. Mach 8 inlet: geometry (left), main flow features (right).

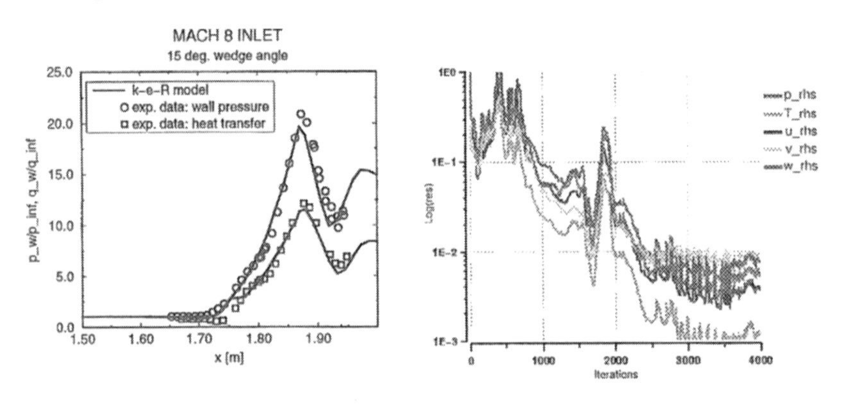

Figure 10.12. Mach 8 inlet: wall pressure and heat transfer comparison (left), convergence history (right).

Fig. 10.11 is a sketch of the topography and an overview of the flow in the region of the wedges, showing streamlines and pressure contours on one wedge surface. It is observed that the flow in the mid-region of the wedge maintains an approximately two-dimensional flavour. The high-pressure region downstream of the shoulder is due to shock impingement from the other wedge. The streamlines at the wall/wedge juncture clearly show streamwise separation due to the adverse pressure gradient downstream of the wedge shoulder. Flow spillage at the upper end of the wedge, due to cross-stream pressure gradient, is also observed. Fig. 10.12(L) compares predicted wall pressure and heat transfer, along the symmetry line, with corresponding measurements. The 3-equation model

[5], used in the present example, yields very good agreement with both pressure and heat transfer data. Finally, Fig. 10.12(R) shows the convergence history plot. These results demonstrate the capability of modern CFD in predicting complex 3D hypersonic turbulent flows of engineering interest.

2.4 Room Ventilation by Forced Convection

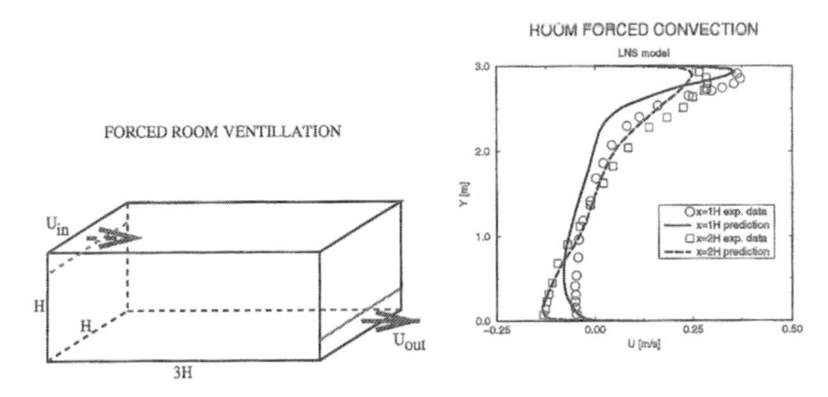

Figure 10.13. Room ventilation: geometry (left), midplane velocity profiles (right).

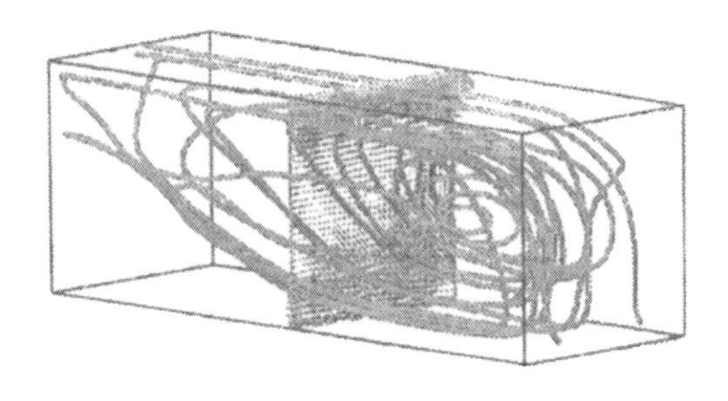

Figure 10.14. Room ventilation: instantaneous streamlines

Healthy and comfortable design of indoor environments necessitates knowledge about the distributions of air velocity and temperature fields, humidity, and contaminant concentrations. Indoor flows are typically unsteady, turbulent, and driven by pressure gradients and thermal effects such as buoyancy. These flows involve natural convection (baseboard heating), forced convection (free cooling) and mixed convection (air conditioning). Predicting these complex flows is beyond the realm of RANS solvers because of the low frequency unsteadiness involved. On the other hand, experimental approaches require expensive, full scale test chambers due to the scaling problems of non-isothermal flows. In addition, taking such measuremets is a lengthy process, typically stretching over a period of several months. This renders CFD, using LES or hybrid RANS/LES, as the only practical alternative currently available to the environmental engineer. 3D time-dependent predictions are, however, a major challenge to CFD even when using massively parallel computers. To minimise uncertainties in predictions, hundreds of time steps must be computed for each period representative of the flow, and several sub-iterations are necessary for each time step to reduce numerical errors to acceptable levels. On top of that, often many periods must be computed before meaningful time averages can be obtained. Because of the extensive computations involved, unsteady CFD is, generally speaking, not yet ready for the design cycle even though the computational capability exists.

The example below shows the potential of hybrid RANS/LES in predicting a typical room ventilation case. Nielsen et al.[32] measured forced convection in a room (see Fig. 10.13(L)) with U_{in}=0.455 m/s. Calculations were performed in unsteady mode, using the current solver's hybrid RANS/LES (LNS) model [8]. Time-averaged predictions are compared with data at the midplane, seen in Fig. 10.13(R). The level of agreement is very encouraging. Fig. 10.14 is an instantaneous streamline plot, showing the complex, vortical, highly three-dimensional flow inside the room.

2.5 Car Aerodynamics

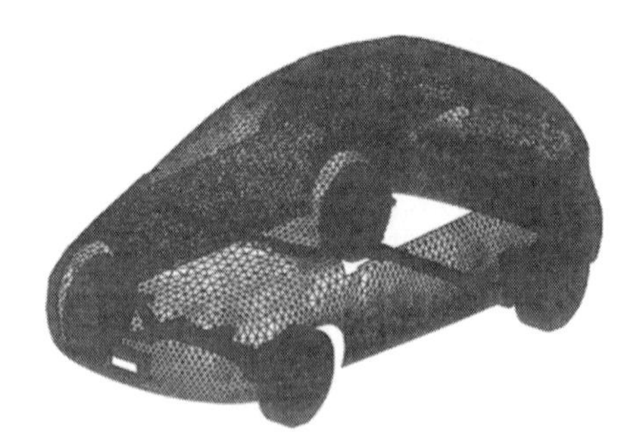

Figure 10.15. Fiat Punto topology: partial view of mesh

Due to a steady rise in oil prices, car manufacturers have been putting much emphasis on enhancing fuel efficiency through improvements in both external and internal aerodynamics as well as enhanced engine performance. While windtunnel testing is commonplace, the automotive industry is increasingly relying on CFD tools to analyse a variety of automobile-related flows. With rising cost of windtunnel operation, car manufacturers are inclined to use CFD also for

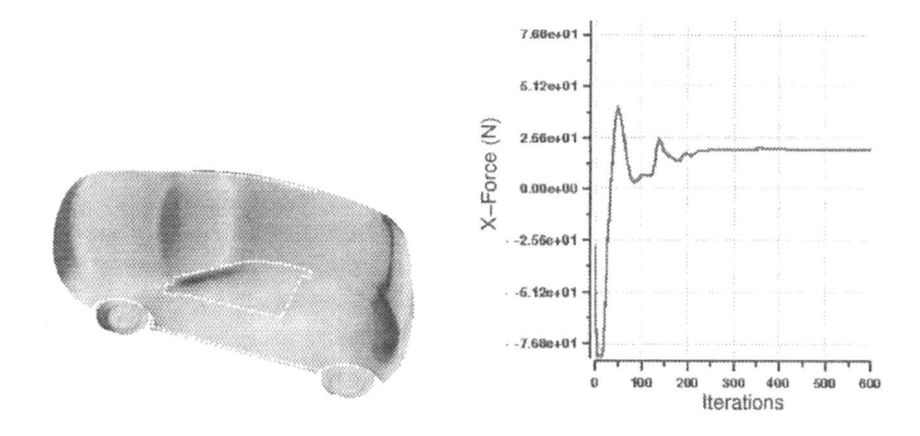

Figure 10.16. Fiat Punto predictions: surface pressure (left), model drag (right).

the design cycle. The ability of modern CFD to predict complex 3D turbulent flows in a timely manner convinces an increasing number of auto manufacturers to make the switch from windtunnel measurements to CFD predictions.

A computation of the external flow around a wind-tunnel model of the Fiat Punto (courtesy of FIAT Auto and FIAT CRF) was carried out using the current CFD code on 587,000 tetrahedral elements, with $y_1^+ \approx 75$. The k-ϵ model was invoked together with a wall function treatment. Fig. 10.15 shows an overview of the topology and mesh, including fine details of the undercarriage. Fig. 10.16 shows surface pressure contours and convergence behaviour in terms of the drag force. Thanks to the rapid turnaround time of these calculations, automotive engineers are able to use this tool in the design cycle using realistic, detailed car geometries.

2.6 Active Flow Control

Dynamic flow control is emerging as an important aerodynamic tool for both external and internal flows. Proper application of control devices can, for example, laminarize turbulent flows; delay or even eliminate flow separation; weaken shocks; serve as an economical vehicle steering device for rockets; and reduce combustion instabilities. These and other efficiency-enhancing benefits render flow control devices increasingly desirable in a wide range of fluid flow applications.

Micro-scale actuators have been used successfully to establish flow control over large extents of flow domains, such as the suction side of wings and the wake downstream of a cylinder. Here we present results from a numerical simulation of separation control of the flow over a blunt aerofoil at a large angle-

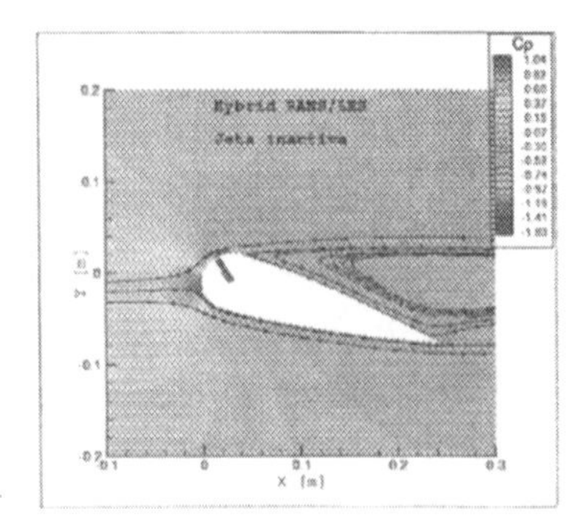

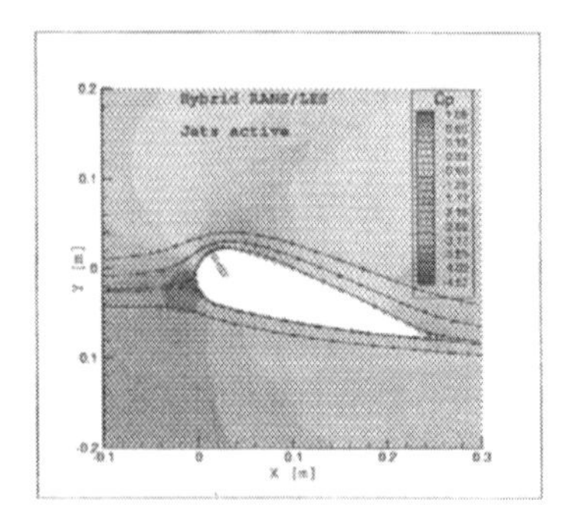

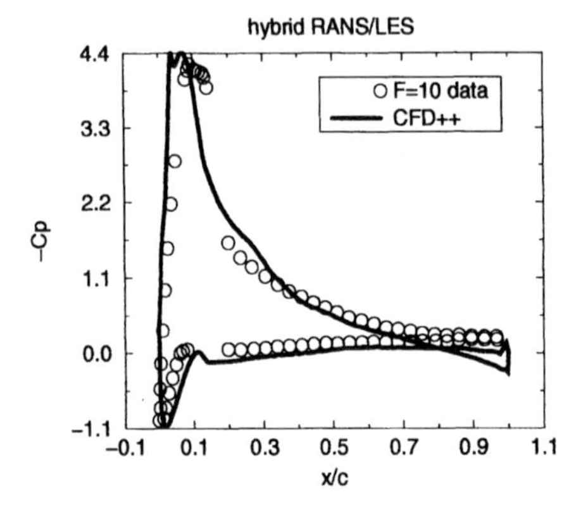

Figure 10.17. Synthetic jet: jets inactive (upper-left), jets active (upper-right), Cp-profile (lower).

of-attack, using micro-actuators. Experimental data were taken at Georgia Institute of Technology[33]. A pair of synthetic jet actuators was mounted on the suction side of the aerofoil near the cylindrical leading edge. The wind-tunnel conditions were $\alpha = 17.5°$ and $Re = 310,000$ based on freestream velocity and chord length. At this angle-of-attack the flow is massively separated on the suction side (see Fig. 10.17(UL)). When the actuators were turned on at 720 Hz the flow completely reattached (Fig. 10.17(UR)). The computational results reproduce the experimentally observed behaviour, including the pressure distribution around the aerofoil with active control (Fig. 10.17(B)). When the actuators are turned on, small-scale, high-frequency turbulent perturbations are created in the flow. These perturbations break up the separating shear layer and enable the flow to reattach. These small-scale fluctuations become excessively diffused using conventional RANS (which is designed to resolve the mean motion only). Since DNS remains too time-consuming to be a practical tool and LES faces unresolved issues relating to near-wall modelling, the current solver's LNS[8] model, a hybrid RANS/LES approach, was the natural closure for this unsteady flow. In LNS, small-scale flow features are resolved wherever the computational mesh is fine enough, with the RANS solution automatically recovered on coarser regions of the grid. Thus, the mesh can be tailored to capture the fine-scale motion where this is of interest, while coarser grid density is employed elsewhere in order to save computational time. The results clearly show the feasibility of the LNS model in predicting flowfields subject to active flow control. As mentioned before, 3D unsteady flows remain a severe challenge for practical CFD due to the large amount of real time necessary to obtain reliable predictions.

2.7 Acoustics

Traditional CFD methods face a number of interesting challenges in the developing field of aeroacoustics, since the prediction of flow noise requires a faithful mechanism for simulating both the source of acoustic disturbances and the long-range propagation of these disturbances. Neither of these phenomena are handled adequately by the classical Reynolds-Averaged Navier-Stokes (RANS) approach, where fine-scale unsteady sources are represented only in a statistical sense and where long-range wave propagation tends to be heavily diffused as a result of numerical and effective (eddy) viscosities. Acoustics methods being developed at Metacomp Technologies include the use of hybrid LES/RANS methods and low diffusion, non-linear disturbance equation solvers, which make a separation of scales in order to isolate acoustic-wave propagation from the effect of turbulent mixing on the mean flow. In both areas, the concept of synthetic, or artificially-reconstituted, turbulence is expected to play a significant future role in describing the fine-scale noise sources and providing a conversion between statistically-represented and directly-resolved kinetic energy in acoustics simulations.

2.7.1 Cavity Flow. The CFD solver was used to investigate a number of cavity flows with free-stream Mach numbers ranging from low subsonic to supersonic. The examples shown here are for a constant length-to-depth ratio of 3.0. A fully-developed turbulent boundary layer profile was imposed upstream of the cavity leading edge, such that the boundary layer thickness was approximately $\delta_s = D/3$ at separation. Fig. 10.18(U) shows an instantaneous snapshot of the resonant flow within the cavity, computed using the hybrid LNS model. Fig. 10.18(B) compares the predicted dominant modes with the theoretical predictions of Heller[34] and experimental data of Zhang[35].

2.7.2 Flow around a Car Wing Mirror. The CFD solver was used to simulate the flow around a semi-circular protrusion mounted on a flat plate, representative of a car wing mirror[36]. Fig. 10.19(U) shows pressure contours on the plate and mirror surfaces with streamtubes indicating the extent of the recirculating flow in the mirror wake. Virtual probes, mounted downstream on

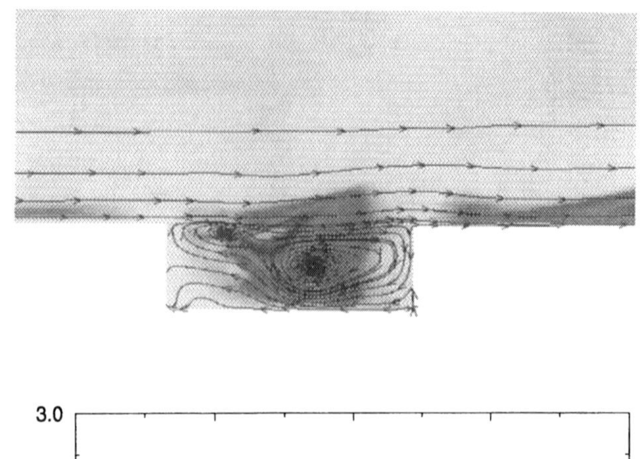

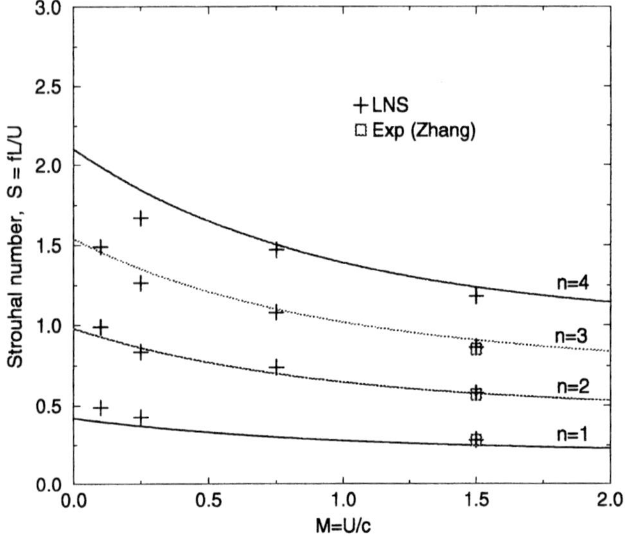

Figure 10.18. Mach 1.5 flow over $L/D = 3$ cavity: streamlines and effective eddy viscosity (upper), predicted Strouhal numbers compared with theory and experiment (lower).

the plate surface were used to collect pressure history data. Predicted sound pressure levels (in decibels) are shown compared with available experimental data[36] in Fig. 10.19(B).

2.8 Turbulent Flow Using Coarse Meshes

The ability to predict turbulent flows on Euler-like grids has long been desired by engineers seeking ever shorter turnaround times for complex flow problems. CFD++'s Y^{++} module achieves this goal by enabling prediction of the salient features of turbulent flows using coarse meshes. The following examples illustrate this capability.

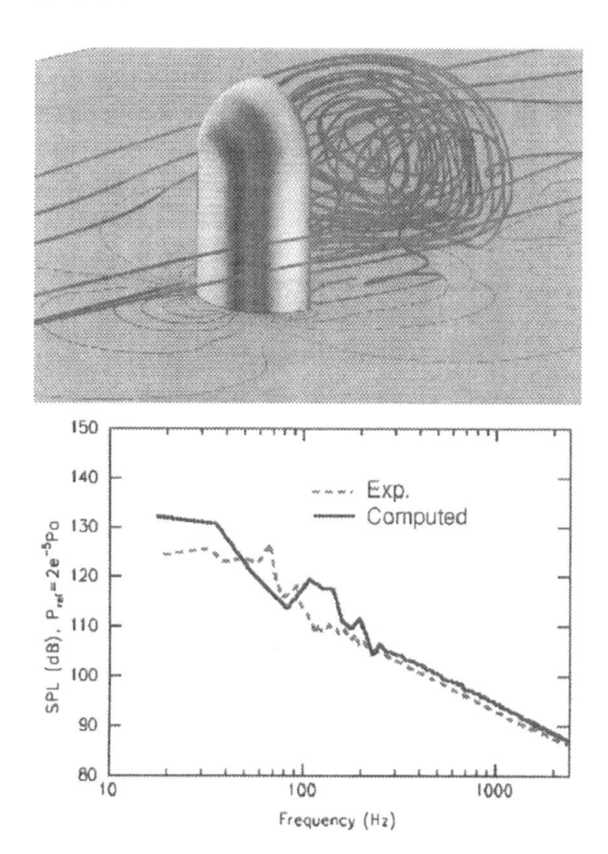

Figure 10.19. Plate/wing-mirror junction: instantaneous snapshot (upper), comparison of sound-pressure levels in the mirror wake (lower).

2.8.1 Transonic Channel Flow Including a Bump. Here a Mach 0.615 flow enters a two-dimensional channel comprising a flat, slightly slanted upper wall and a lower surface which includes a bump-like profile protruding from the otherwise flat wall (see Fig. 10.20(U)). The inflow total pressure is 96KPa. A transonic λ-shock forms toward the bump trailing edge and its interaction with the boundary layer induces a separated flow region, as seen in Fig. 10.20(B). Experimental data were taken by Délery[37]. In the experiment the shock location was controlled by an adjustable throat downstream of the bump. Since no geometrical details of this throat are provided, computors are instructed to adjust the downstream pressure to replicate numerically the experimental shock location on the upper wall. Experience indicates imposing $p/p_{\infty} \approx 0.82$ at the downstream boundary. Adiabatic, non-slip conditions were imposed at the walls and the inflow boundary was set to reservoir conditions. Computations using conventional turbulence models require a 120×120 grid with $y^+ \leq 1$ at the first centroid away from the walls. The resulting pressure

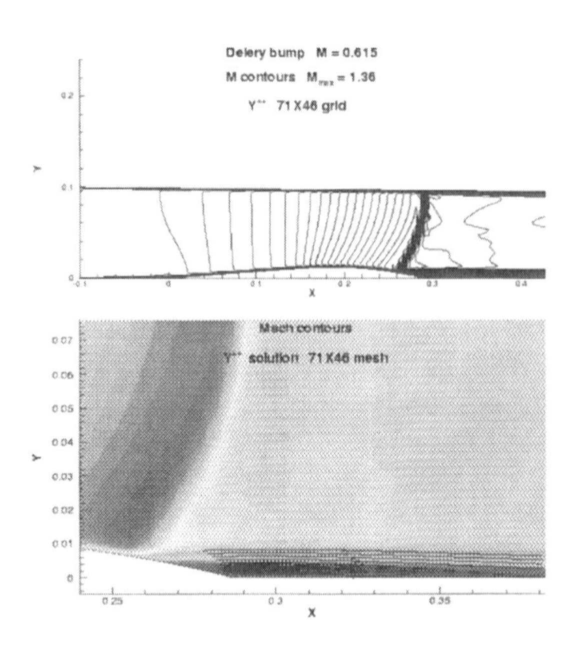

Figure 10.20. Channel with bump: geometry and Mach contours by Y^{++} (upper), shock and separated flow detail (lower).

distribution on the lower wall, produced with the cubic k-ϵ closure[4] is the long-dashed line in Fig. 10.21. The Y^{++} module was tested on a 71×46 mesh with $y^+ \approx 250$ at the first centroids. The figure shows the deterioration in predictive quality when the cubic model was used on this coarse grid (dot-dashed line) whereas using Y^{++} still enabled correct prediction of the shock location (solid line), followed by a pressure plateau indicative of the flow separation region downstream of the shock (see also Fig. 10.20(B)). An inviscid flow solution is also shown in Fig. 10.21 (dashed line labeled Euler). As expected, due to lack of viscous effects the shock is predicted too far downstream and no pressure plateau is captured. This test case indicates that the Y^{++} approach is able, at a small fraction of the time/cost required by a RANS computation, to predict the salient features of complex turbulent flows. This is often all an engineer needs in the preliminary design stage.

2.8.2 Periodic Flow in an Exhaust Manifold. A critical component of an internal combustion engine is the exhaust manifold, which is subject to a complex unsteady flow due to the on/off action of the piston valves. Automotive engineers require the ability to compute this flow repetitively, subject to changes in flow conditions and/or geometrical alterations. A large matrix

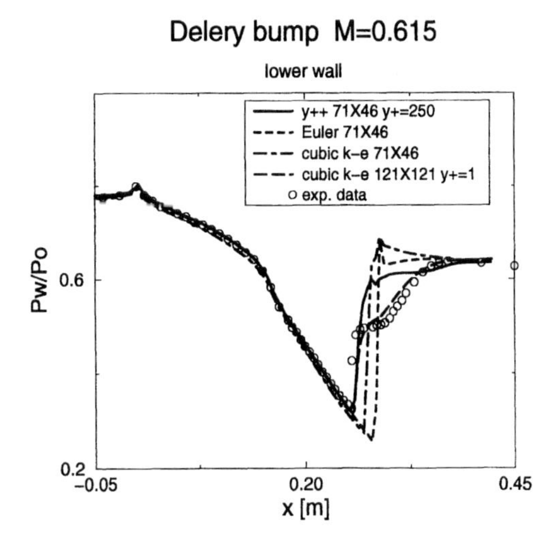

Figure 10.21. Channel with bump: lower wall pressure profiles

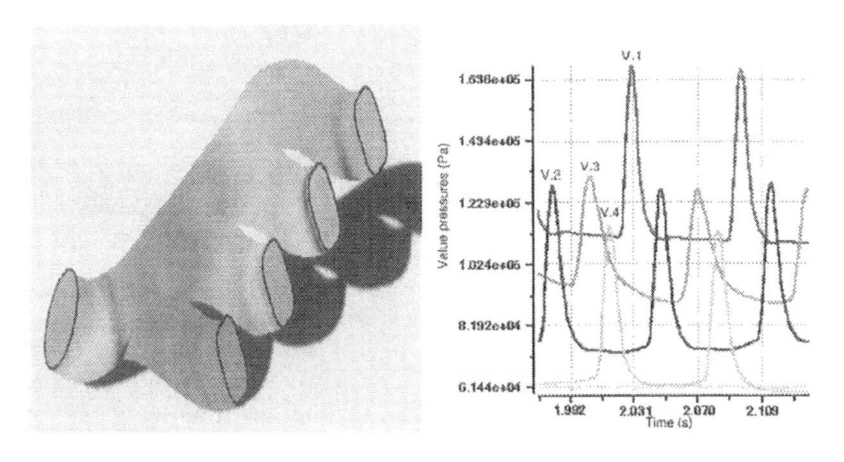

Figure 10.22. Exhaust manifold: surface pressure (left), valves pressure history (right). Valve V.1 is on the top-right and Valve V.4 on the lower-left relative to the common exhaust.

of flow and geometrical parameters must be filled by the engineer in order to optimize the component's design. Such intensive use of CFD in the design cycle currently precludes the use of LES or even hybrid RANS/LES. In fact, the computational meshes are often too coarse even for 3D turbulent flow predictions using traditional RANS approaches. The Y^{++} module in the current solver enables, however, a reasonable prediction of such flows on Euler-like grids. The example below pertains to an exhaust manifold subject to periodic out-of-phase opening and closing of the valves. Fig. 10.22 shows temporal surface pressure distribution and pressure history of each valve. Thanks to the coarse mesh, many such calculations are possible within a short time span to enable the construction of a design matrix.

2.9 Supersonic Reacting Flow

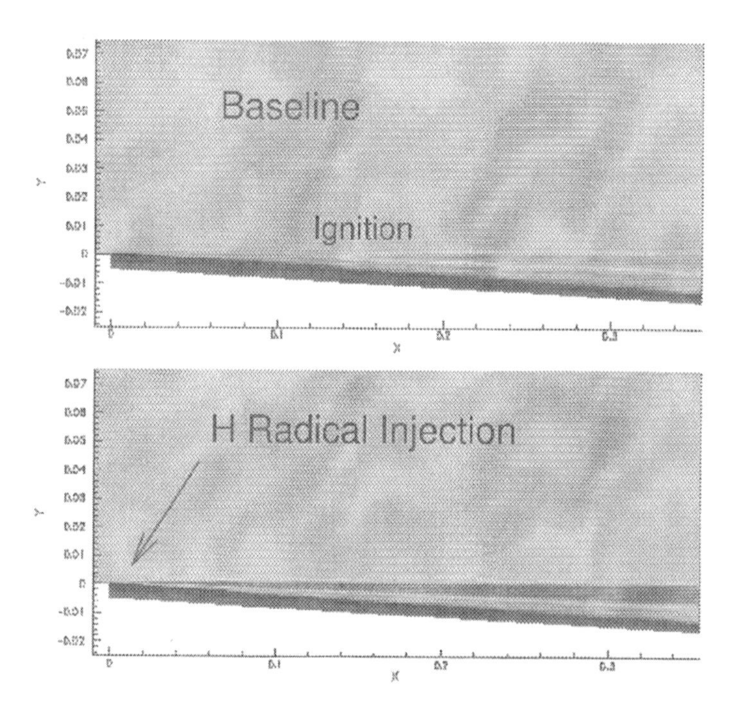

Figure 10.23. Supersonic reacting flow: topology and temperature contours

The design of supersonic combustors for advanced air-breathing engines requires the ability to predict the diffusive mixing and reaction of fuel and air at the elevated temperatures representative of supersonic and hypersonic flight. Experimental data at these Mach numbers are very difficult and expensive to generate with a reasonable level of accuracy. This leaves CFD as an indispensable tool in the analysis and design of supersonic combustors.

In the present example a turbulence/chemistry interaction model was applied to simulate supersonic reacting flow where H_2 fuel is injected into a vitiated air stream at Mach 2.4[38]. The 9 species, 18 reaction step model of Drummond et al.[39] was employed. Fig. 10.23 shows geometry and temperature contours. The upper figure is the baseline case, showing ignition at x $\approx$0.1 m and the lower one shows the effect of H radical injection on propagating the ignition upstream. Fig. 10.24 shows predictions of H_2 and H_2O mole fractions at the

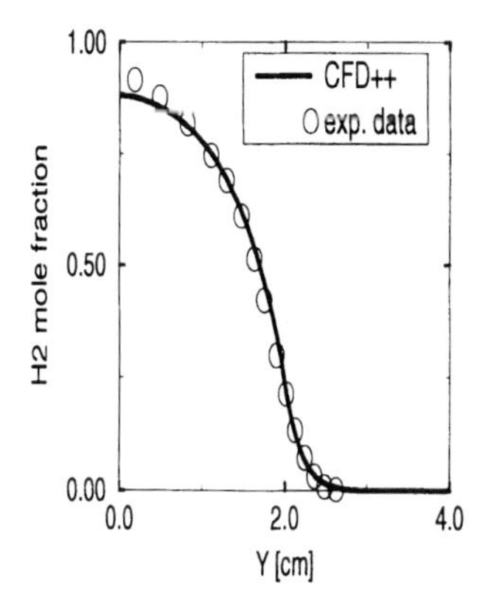

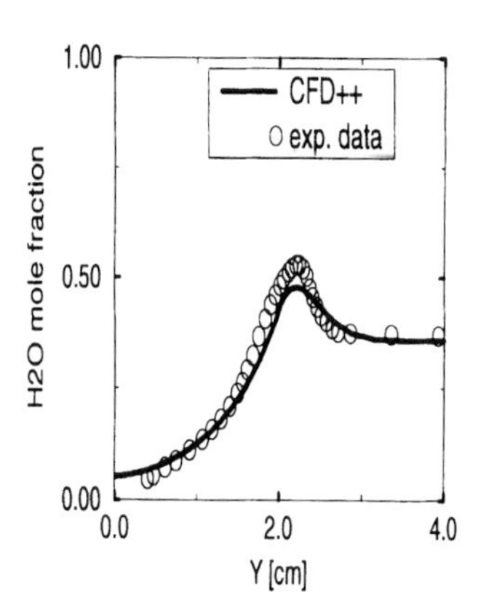

Figure 10.24. Supersonic reacting flow exit plane: hydrogen mole fraction (left), water mole fraction (right).

exit plane. The cubic k-ϵ closure[4] was used in this calculation as part of the turbulence/chemistry interaction model. Comparisons with data are quite encouraging and demonstrate the ability of CFD to predict supersonic reacting flows with a high level of confidence.

3. Concluding remarks

Due to the ever increasing complexity of present day engineering flows, coupled with the attendant increase in cost and reliability of experimental facilities, the role of CFD simulations in the design cycle has recently become prominent. Whereas only a decade ago computations of complex 3D flows in a timely manner were infeasible, today such calculations are commonplace in a wide range of engineering disciplines thanks to the synergy of advanced numerical techniques, reliable turbulence models and powerful computers.

A number of CFD simulations, using CFD++, covering a diverse spectrum of applications, were presented here. These flow cases demonstrated the ability of modern CFD to predict a large variety of complex turbulent flows with a high level of accuracy. The examples included traditional turbulence closures as well as more novel methodologies, namely the hybrid RANS/LES model for unsteady flows and the Y^{++} approach for coarse-grid turbulence prediction. The array of flow cases shown here should convince the reader that modern CFD is a viable engineering design tool.

A major obstacle in rendering CFD a true engineering tool is the need to use good judgement in grid quality decision making. While automatic adaptation exists, the resulting mesh is often too large to enable quick turnaround times necessary in the design cycle. The engineer, therefore, must use his/her experience in selecting a grid which is small enough to permit repetitive calculations in a timely manner yet large enough to enable reliable predictions. Another challenge is the use of CFD as a design tool for unsteady flows. The prohibitively large amount of time/cost needed to accomplish high quality computations of such flows does not permit CFD to be used in the design cycle when unsteadiness is a key feature of the flow in question. In spite of these shortcomings, however, CFD is increasingly used routinely by engineers in diverse disciplines of fluid mechanics, as the examples above indicate.

References

[1] Chakravarthy, S., Peroomian, O. and Sekar, B. (1996) "Some Internal Flow Applications of a Unified-Grid CFD Methodology," AIAA Paper No. 96-2926.

[2] Peroomian, O., Chakravarthy, S. and Goldberg, U. (1997) "A 'grid-transparent' Methodology for CFD," AIAA Paper No. 97-0724.

[3] Goldberg, U., Peroomian, O. and Chakravarthy, S. (1998) "A Wall-Distance-Free k-ϵ Model With Enhanced Near-Wall Treatment," *ASME J. Fluids Eng.*, **120** pp. 457–462.

[4] Goldberg, U., Batten, P., Palaniswamy, S., Chakravarthy, S. and Peroomian ,O. (2000) "Hypersonic Flow Predictions Using Linear and Nonlinear Turbulence Closures," *AIAA J. of Aircraft*, **37** pp. 671–675.

[5] Goldberg, U. and Batten, P. (2001) "Heat Transfer Predictions Using a Dual-Dissipation k-ϵ Turbulence Closure," *AIAA J. Thermophysics and Heat Transfer*, **15** No. 2 pp. 197–204.

[6] Goldberg, U. (2001) "Hypersonic Flow Heat Transfer Predictions Using Single Equation Turbulence Models," *ASME J. Heat Transfer*, **123** pp. 65–69.

[7] Palaniswamy, S., Goldberg, U., Peroomian, O. and Chakravarthy, S. (2001) "Predictions of Axial and Traverse Injection into Supersonic Flow," *Flow, Turbulence and Combustion*, **66** pp. 37–55.

[8] Batten, P., Goldberg, U. and Chakravarthy, S. (2000) "Sub-grid Turbulence Modeling for Unsteady Flow with Acoustic Resonance," AIAA Paper 00-0473.

[9] Harten, A. (1983) "High Resolution Schemes for Hyperbolic Conservation Laws,"*JCP*, **49**.

[10] Harten, A. (1991) "Recent Developments in Shock Capturing Schemes," NASA CR 187502.

[11] Godunov, S.K. (1959) "A Difference Method for the Numerical Calculation of Discontinuous Solutions of Hydrodynamic Equations," *Mat. Sbornik*, **47**

[12] Steger, J.L. and Warming, R.F. (1981) "Flux Vector Splitting of the Inviscid Gasdynamic Equations with Applications to Finite-Difference Methods," *JCP*, **40** pp. 263–293.

[13] van Leer, B., Thomas, J.L., Roe, P.L. and Newsome, R.W. (1987) "A Comparison of Numerical Flux Formulas for the Euler and Navier-Stokes Equations," AIAA Paper 87-1184.

[14] Osher, S. and Solomon, F. (1981) "Upwind Schemes for Hyperbolic Systems of Conservation Laws," *Math. Comp.*, **38** pp. 339–377.

[15] Roe, P.L. (1981) "Approximate Riemann Solvers, Parameter Vectors and Difference Schemes," *JCP*, **43** pp. 357–372.

[16] Einfeldt, B., Munz, C.D., Roe, P.L. and Sjogreen, B. (1991) "On Godunov-Type Methods Near Low Densities," *JCP*, **92** pp. 273–295.

[17] Toro, E.F., Spruce, M. and Speares, W. (1994) "Restoration of the Contact Surface in the HLL Riemann Solver," *Shock Waves*, **4** Springer-Verlag, pp. 25–34.

[18] Batten, P., Clarke, N., Lambert, C. and Causon, D.M. (1997) "On the Choice of Wave Speeds for the HLLC Riemann Solver," *SIAM J. Sci. & Stat. Comp.*, **18(6)** pp. 1553–1570.

[19] Moschetta, J.-M. and Pullin, D. (1997) "A Robust Low Diffusive Kinetic Scheme for the Navier-Stokes/Euler Equations," *JCP*, **133** pp. 193–204.

[20] Liou, M.-S. and Steffen, C.J. Jr. (1993) "A New Flux Splitting Scheme," *JCP*, **107** pp. 23–39.

[21] Harten, A., Lax, P.D. and van Leer, B. (1983) "On Upstream Differencing and Godunov-Type Schemes for Hyperbolic Conservation Laws," *SIAM Review*, **25** No. 1, pp. 35–61.

[22] Allmaras, S.R. (1992) "Contamination of Laminar Boundary Layers by Artificial Dissipation in Navier-Stokes Solutions," *Proc. ICFD Conf.*, University of Reading.

[23] McNeil, C.Y. (1996) "The Effect of Numerical Dissipation on High Reynolds Number Turbulent Flow Solutions," AIAA Paper 96-0891, 34th Aerospace Sciences Meeting, Reno, Nevada.

[24] Perthame, B. and Shu, C.-W. (1996) "On Positivity Preserving Finite Volume Schemes for the Euler Equations," *Numer. Math.*, **73** pp. 119–130.

[25] Peroomian, O., Chakravarthy, S., Palaniswamy, S. and Goldberg, U. (1998) "Convergence Acceleration for Unified-Grid Formulation Using Preconditioned Implicit Relaxation," AIAA Paper No. 98-0116.

[26] Batten, P., Leschziner, M.A. and Goldberg, U.C. (1997) "Average-State Jacobians and Implicit Methods for Compressible Viscous and Turbulent Flows," *Journal of Computational Physics*, **137** pp. 38–78.

[27] "A Selection of Experimental Test Cases for the Validation of CFD Codes," AGARD-AR-303 **2**, 1994.

[28] McDaniel, J., Fletcher, D., Hartfield, R. and Hollo, S. (1991) "Staged Transverse Injection Into Mach 2 Flow Behind a Rearward-Facing Step: A 3-D Compressible Test Case for Hypersonic Combustor Code Validation," AIAA Paper No. 91-5071.

[29] Kussoy, M.I., Horstman, K.C., and Horstman, C.C. (1993) "Hypersonic Crossing Shock-Wave/Turbulent-Boundary-Layer Interactions," *AIAA Journal*, **31** pp. 2197–2203.

[30] White, F.M. (1974) *Viscous Fluid Flow*, 1st ed., McGraw-Hill Book Company.

[31] Launder, B.E. (1988) "On the Computation of Convective Heat Transfer in Complex Turbulent Flows," *ASME Journal of Heat Transfer*, **110** pp. 1112–1128.

[32] Nielsen, P.V., Restivo, A. and Whitelaw, J.H. (1978) "The Velocity Characteristics of Ventilated Room," *ASME J. Fluids Eng.*, **100** pp. 291–298.

[33] Amitay, M., Kibens, V., Parekh, D. and Glezer, A. (1999) "The Dynamics of Flow Reattachment over a Thick Airfoil Controlled by Synthetic Jet Actuators," AIAA Paper 99-1001.

[34] Heller, H.H., Holmes, D.G. and Covert, E.E (1971) "Flow Induced Pressure Oscillations in Shallow Cavities," *Journal of Sound and Vibration*, **18**(4).

[35] Zhang, X. (1987) "An Experimental and Computational Investigation of Supersonic Shear Layer Driven Single and Multiple Cavity Flow Fields," Ph.D. Thesis, Churchill College, Cambridge, UK.

[36] Hold, R., Brenneis, A., Eberle, A., Schwarz, V. and Siegert, R. (1999) "Numerical Simulation of Aeroacoustic Sound Generated by Generic Bodies Placed on a Plate: Part 1 - Prediction of Aeroacoustic Sources," AIAA Paper 99-1896.

[37] Délery, J.M. (1983) "Experimental Investigation of Turbulence Properties in Transonic Shock-Wave/Boundary-Layer Interactions," *AIAA Journal*, **21** pp. 180–185.

[38] Burrows, M.C. and Kurkov, A.P. (1973) " Analytical and Experimental Study of Supersonic Combustion of Hydrogen in a Vitiated Air Stream," NASA TM X-2828.

[39] Drummond, J.P., Rogers, R.C. and Hussaini, B. (1987) "A Numerical Model for Supersonic Reacting Mixing Layers," *Computer Methods in Applied Mechanics and Engineering*, **64** 39.

Mechanics

FLUID MECHANICS AND ITS APPLICATIONS

Series Editor: R. Moreau

Aims and Scope of the Series

The purpose of this series is to focus on subjects in which fluid mechanics plays a fundamental role. As well as the more traditional applications of aeronautics, hydraulics, heat and mass transfer etc., books will be published dealing with topics which are currently in a state of rapid development, such as turbulence, suspensions and multiphase fluids, super and hypersonic flows and numerical modelling techniques. It is a widely held view that it is the interdisciplinary subjects that will receive intense scientific attention, bringing them to the forefront of technological advancement. Fluids have the ability to transport matter and its properties as well as transmit force, therefore fluid mechanics is a subject that is particularly open to cross fertilisation with other sciences and disciplines of engineering. The subject of fluid mechanics will be highly relevant in domains such as chemical, metallurgical, biological and ecological engineering. This series is particularly open to such new multidisciplinary domains.

1. M. Lesieur: *Turbulence in Fluids.* 2nd rev. ed., 1990 ISBN 0-7923-0645-7
2. O. Métais and M. Lesieur (eds.): *Turbulence and Coherent Structures.* 1991 ISBN 0-7923-0646-5
3. R. Moreau: *Magnetohydrodynamics.* 1990 ISBN 0-7923-0937-5
4. E. Coustols (ed.): *Turbulence Control by Passive Means.* 1990 ISBN 0-7923-1020-9
5. A.A. Borissov (ed.): *Dynamic Structure of Detonation in Gaseous and Dispersed Media.* 1991 ISBN 0-7923-1340-2
6. K.-S. Choi (ed.): *Recent Developments in Turbulence Management.* 1991 ISBN 0-7923-1477-8
7. E.P. Evans and B. Coulbeck (eds.): *Pipeline Systems.* 1992 ISBN 0-7923-1668-1
8. B. Nau (ed.): *Fluid Sealing.* 1992 ISBN 0-7923-1669-X
9. T.K.S. Murthy (ed.): *Computational Methods in Hypersonic Aerodynamics.* 1992 ISBN 0-7923-1673-8
10. R. King (ed.): *Fluid Mechanics of Mixing.* Modelling, Operations and Experimental Techniques. 1992 ISBN 0-7923-1720-3
11. Z. Han and X. Yin: *Shock Dynamics.* 1993 ISBN 0-7923-1746-7
12. L. Svarovsky and M.T. Thew (eds.): *Hydroclones.* Analysis and Applications. 1992 ISBN 0-7923-1876-5
13. A. Lichtarowicz (ed.): *Jet Cutting Technology.* 1992 ISBN 0-7923-1979-6
14. F.T.M. Nieuwstadt (ed.): *Flow Visualization and Image Analysis.* 1993 ISBN 0-7923-1994-X
15. A.J. Saul (ed.): *Floods and Flood Management.* 1992 ISBN 0-7923-2078-6
16. D.E. Ashpis, T.B. Gatski and R. Hirsh (eds.): *Instabilities and Turbulence in Engineering Flows.* 1993 ISBN 0-7923-2161-8
17. R.S. Azad: *The Atmospheric Boundary Layer for Engineers.* 1993 ISBN 0-7923-2187-1
18. F.T.M. Nieuwstadt (ed.): *Advances in Turbulence IV.* 1993 ISBN 0-7923-2282-7
19. K.K. Prasad (ed.): *Further Developments in Turbulence Management.* 1993 ISBN 0-7923-2291-6
20. Y.A. Tatarchenko: *Shaped Crystal Growth.* 1993 ISBN 0-7923-2419-6
21. J.P. Bonnet and M.N. Glauser (eds.): *Eddy Structure Identification in Free Turbulent Shear Flows.* 1993 ISBN 0-7923-2449-8
22. R.S. Srivastava: *Interaction of Shock Waves.* 1994 ISBN 0-7923-2920-1
23. J.R. Blake, J.M. Boulton-Stone and N.H. Thomas (eds.): *Bubble Dynamics and Interface Phenomena.* 1994 ISBN 0-7923-3008-0

Mechanics

FLUID MECHANICS AND ITS APPLICATIONS

Series Editor: R. Moreau

24. R. Benzi (ed.): *Advances in Turbulence V.* 1995 ISBN 0-7923-3032-3
25. B.I. Rabinovich, V.G. Lebedev and A.I. Mytarev: *Vortex Processes and Solid Body Dynamics.* The Dynamic Problems of Spacecrafts and Magnetic Levitation Systems. 1994 ISBN 0-7923-3092-7
26. P.R. Voke, L. Kleiser and J.-P. Chollet (eds.): *Direct and Large-Eddy Simulation I.* Selected papers from the First ERCOFTAC Workshop on Direct and Large-Eddy Simulation. 1994 ISBN 0-7923-3106-0
27. J.A. Sparenberg: *Hydrodynamic Propulsion and its Optimization.* Analytic Theory. 1995 ISBN 0-7923-3201-6
28. J.F. Dijksman and G.D.C. Kuiken (eds.): *IUTAM Symposium on Numerical Simulation of Non-Isothermal Flow of Viscoelastic Liquids.* Proceedings of an IUTAM Symposium held in Kerkrade, The Netherlands. 1995 ISBN 0-7923-3262-8
29. B.M. Boubnov and G.S. Golitsyn: *Convection in Rotating Fluids.* 1995 ISBN 0-7923-3371-3
30. S.I. Green (ed.): *Fluid Vortices.* 1995 ISBN 0-7923-3376-4
31. S. Morioka and L. van Wijngaarden (eds.): *IUTAM Symposium on Waves in Liquid/Gas and Liquid/Vapour Two-Phase Systems.* 1995 ISBN 0-7923-3424-8
32. A. Gyr and H.-W. Bewersdorff: *Drag Reduction of Turbulent Flows by Additives.* 1995 ISBN 0-7923-3485-X
33. Y.P. Golovachov: *Numerical Simulation of Viscous Shock Layer Flows.* 1995 ISBN 0-7923-3626-7
34. J. Grue, B. Gjevik and J.E. Weber (eds.): *Waves and Nonlinear Processes in Hydrodynamics.* 1996 ISBN 0-7923-4031-0
35. P.W. Duck and P. Hall (eds.): *IUTAM Symposium on Nonlinear Instability and Transition in Three-Dimensional Boundary Layers.* 1996 ISBN 0-7923-4079-5
36. S. Gavrilakis, L. Machiels and P.A. Monkewitz (eds.): *Advances in Turbulence VI.* Proceedings of the 6th European Turbulence Conference. 1996 ISBN 0-7923-4132-5
37. K. Gersten (ed.): *IUTAM Symposium on Asymptotic Methods for Turbulent Shear Flows at High Reynolds Numbers.* Proceedings of the IUTAM Symposium held in Bochum, Germany. 1996 ISBN 0-7923-4138-4
38. J. Verhás: *Thermodynamics and Rheology.* 1997 ISBN 0-7923-4251-8
39. M. Champion and B. Deshaies (eds.): *IUTAM Symposium on Combustion in Supersonic Flows.* Proceedings of the IUTAM Symposium held in Poitiers, France. 1997 ISBN 0-7923-4313-1
40. M. Lesieur: *Turbulence in Fluids.* Third Revised and Enlarged Edition. 1997 ISBN 0-7923-4415-4; Pb: 0-7923-4416-2
41. L. Fulachier, J.L. Lumley and F. Anselmet (eds.): *IUTAM Symposium on Variable Density Low-Speed Turbulent Flows.* Proceedings of the IUTAM Symposium held in Marseille, France. 1997 ISBN 0-7923-4602-5
42. B.K. Shivamoggi: *Nonlinear Dynamics and Chaotic Phenomena.* An Introduction. 1997 ISBN 0-7923-4772-2
43. H. Ramkissoon, *IUTAM Symposium on Lubricated Transport of Viscous Materials.* Proceedings of the IUTAM Symposium held in Tobago, West Indies. 1998 ISBN 0-7923-4897-4
44. E. Krause and K. Gersten, *IUTAM Symposium on Dynamics of Slender Vortices.* Proceedings of the IUTAM Symposium held in Aachen, Germany. 1998 ISBN 0-7923-5041-3
45. A. Biesheuvel and G.J.F. van Heyst (eds.): *In Fascination of Fluid Dynamics.* A Symposium in honour of Leen van Wijngaarden. 1998 ISBN 0-7923-5078-2

Mechanics

FLUID MECHANICS AND ITS APPLICATIONS

Series Editor: R. Moreau

46. U. Frisch (ed.): *Advances in Turbulence VII.* Proceedings of the Seventh European Turbulence Conference, held in Saint-Jean Cap Ferrat, 30 June–3 July 1998. 1998 ISBN 0-7923-5115-0
47. E.F. Toro and J.F. Clarke: *Numerical Methods for Wave Propagation.* Selected Contributions from the Workshop held in Manchester, UK. 1998 ISBN 0-7923-5125-8
48. A. Yoshizawa: *Hydrodynamic and Magnetohydrodynamic Turbulent Flows.* Modelling and Statistical Theory. 1998 ISBN 0-7923-5225-4
49. T.L. Geers (ed.): *IUTAM Symposium on Computational Methods for Unbounded Domains.* 1998 ISBN 0-7923-5266-1
50. Z. Zapryanov and S. Tabakova: *Dynamics of Bubbles, Drops and Rigid Particles.* 1999 ISBN 0-7923-5347-1
51. A. Alemany, Ph. Marty and J.P. Thibault (eds.): *Transfer Phenomena in Magnetohydrodynamic and Electroconducting Flows.* 1999 ISBN 0-7923-5532-6
52. J.N. Sørensen, E.J. Hopfinger and N. Aubry (eds.): *IUTAM Symposium on Simulation and Identification of Organized Structures in Flows.* 1999 ISBN 0-7923-5603-9
53. G.E.A. Meier and P.R. Viswanath (eds.): *IUTAM Symposium on Mechanics of Passive and Active Flow Control.* 1999 ISBN 0-7923-5928-3
54. D. Knight and L. Sakell (eds.): *Recent Advances in DNS and LES.* 1999 ISBN 0-7923-6004-4
55. P. Orlandi: *Fluid Flow Phenomena.* A Numerical Toolkit. 2000 ISBN 0-7923-6095-8
56. M. Stanislas, J. Kompenhans and J. Westerveel (eds.): *Particle Image Velocimetry.* Progress towards Industrial Application. 2000 ISBN 0-7923-6160-1
57. H.-C. Chang (ed.): *IUTAM Symposium on Nonlinear Waves in Multi-Phase Flow.* 2000 ISBN 0-7923-6454-6
58. R.M. Kerr and Y. Kimura (eds.): *IUTAM Symposium on Developments in Geophysical Turbulence* held at the National Center for Atmospheric Research, (Boulder, CO, June 16–19, 1998) 2000 ISBN 0-7923-6673-5
59. T. Kambe, T. Nakano and T. Miyauchi (eds.): *IUTAM Symposium on Geometry and Statistics of Turbulence.* Proceedings of the IUTAM Symposium held at the Shonan International Village Center, Hayama (Kanagawa-ken, Japan November 2–5, 1999). 2001 ISBN 0-7923-6711-1
60. V.V. Aristov: *Direct Methods for Solving the Boltzmann Equation and Study of Nonequilibrium Flows.* 2001 ISBN 0-7923-6831-2
61. P.F. Hodnett (ed.): *IUTAM Symposium on Advances in Mathematical Modelling of Atmosphere and Ocean Dynamics.* Proceedings of the IUTAM Symposium held in Limerick, Ireland, 2–7 July 2000. 2001 ISBN 0-7923-7075-9
62. A.C. King and Y.D. Shikhmurzaev (eds.): *IUTAM Symposium on Free Surface Flows.* Proceedings of the IUTAM Symposium held in Birmingham, United Kingdom, 10–14 July 2000. 2001 ISBN 0-7923-7085-6
63. A. Tsinober: *An Informal Introduction to Turbulence.* 2001 ISBN 1-4020-0110-X; Pb: 1-4020-0166-5
64. R.Kh. Zeytounian: *Asymptotic Modelling of Fluid Flow Phenomena.* 2002 ISBN 1-4020-0432-X
65. R. Friedrich and W. Rodi (eds.): *Advances in LES of Complex Flows.* Prodeedings of the EUROMECH Colloquium 412, held in Munich, Germany, 4-6 October 2000. 2002 ISBN 1-4020-0486-9
66. D. Drikakis and B.J. Geurts (eds.) *Turbulent Flow Computation.* 2002 ISBN 1-4020-0523-7

Kluwer Academic Publishers – Dordrecht / Boston / London

ERCOFTAC SERIES

1. A. Gyr and F.-S. Rys (eds.): *Diffusion and Transport of Pollutants in Atmospheric Mesoscale Flow Fields.* 1995 ISBN 0-7923-3260-1
2. M. Hallbäck, D.S. Henningson, A.V. Johansson and P.H. Alfredsson (eds.): *Turbulence and Transition Modelling.* Lecture Notes from the ERCOFTAC/IUTAM Summerschool held in Stockholm. 1996 ISBN 0-7923-4060-4
3. P. Wesseling (ed.): *High Performance Computing in Fluid Dynamics.* Proceedings of the Summerschool held in Delft, The Netherlands. 1996 ISBN 0-7923-4063-9
4. Th. Dracos (ed.): *Three-Dimensional Velocity and Vorticity Measuring and Image Analysis Techniques.* Lecture Notes from the Short Course held in Zürich, Switzerland. 1996 ISBN 0-7923-4256-9
5. J.-P. Chollet, P.R. Voke and L. Kleiser (eds.): *Direct and Large-Eddy Simulation II.* Proceedings of the ERCOFTAC Workshop held in Grenoble, France. 1997 ISBN 0-7923-4687-4
6. A. Hanifi, P.H. Alfredson, A.V. Johansson and D.S. Henningson (eds.) : *Transition, Turbulence and Combustion Modelling.* 1999 ISBN 0-7923-5989-5
7. P.R. Voke, N.D. Sandham and L. Kleiser (eds.) : *Direct and Large-Eddy Simulation III.* 1999 ISBN 0-7923-5990-9
8. B.J. Geurts, R. Friedrich and O. Métais (eds.) : *Direct and Large-Eddy Simulation IV.* 2001 ISBN 1-4020-0177-0

KLUWER ACADEMIC PUBLISHERS – DORDRECHT / BOSTON / LONDON

ICASE/LaRC Interdisciplinary Series in Science and Engineering

1. J. Buckmaster, T.L. Jackson and A. Kumar (eds.): *Combustion in High-Speed Flows.* 1994 ISBN 0-7923-2086-X
2. M.Y. Hussaini, T.B. Gatski and T.L. Jackson (eds.): *Transition, Turbulence and Combustion.* Volume I: Transition. 1994 ISBN 0-7923-3084-6; set 0-7923-3086-2
3. M.Y. Hussaini, T.B. Gatski and T.L. Jackson (eds.): *Transition, Turbulence and Combustion.* Volume II: Turbulence and Combustion. 1994 ISBN 0-7923-3085-4; set 0-7923-3086-2
4. D.E. Keyes, A. Sameh and V. Venkatakrishnan (eds): *Parallel Numerical Algorithms.* 1997 ISBN 0-7923-4282-8
5. T.G. Campbell, R.A. Nicolaides and M.D. Salas (eds.): *Computational Electromagnetics and Its Applications.* 1997 ISBN 0-7923-4733-1
6. V. Venkatakrishnan, M.D. Salas and S.R. Chakravarthy (eds.): *Barriers and Challenges in Computational Fluid Dynamics.* 1998 ISBN 0-7923-4855-9
7. M.D. Salas, J.N. Hefner and L. Sakell (eds.): *Modeling Complex Turbulent Flows.* 1999 ISBN 0-7923-5590-3

KLUWER ACADEMIC PUBLISHERS – DORDRECHT / BOSTON / LONDON

Mechanics

SOLID MECHANICS AND ITS APPLICATIONS

Series Editor: G.M.L. Gladwell

90. Y. Ivanov, V. Cheshkov and M. Natova: *Polymer Composite Materials – Interface Phenomena & Processes*. 2001 ISBN 0-7923-7008-2
91. R.C. McPhedran, L.C. Botten and N.A. Nicorovici (eds.): *IUTAM Symposium on Mechanical and Electromagnetic Waves in Structured Media* held in Sydney, NSW, Australia, 18-22 Januari 1999. 2001 ISBN 0-7923-7038-4
92. D.A. Sotiropoulos (ed.): *IUTAM Symposium on Mechanical Waves for Composite Structures Characterization*. Proceedings of the IUTAM Symposium held in Chania, Crete, Greece, June 14-17, 2000. 2001 ISBN 0-7923-7164-X
93. V.M. Alexandrov and D.A. Pozharskii: *Three-Dimensional Contact Problems*. 2001 ISBN 0-7923-7165-8
94. J.P. Dempsey and H.H. Shen (eds.): *IUTAM Symposium on Scaling Laws in Ice Mechanics and Ice Dynamics*. held in Fairbanks, Alaska, U.S.A., 13-16 June 2000. 2001 ISBN 1-4020-0171-1
95. U. Kirsch: *Design-Oriented Analysis of Structures*. A Unified Approach. 2002 ISBN 1-4020-0443-5
96. A. Preumont: *Vibration Control of Active Structures*. An Introduction ($2^n d$ Edition). 2002 ISBN 1-4020-0496-6

Kluwer Academic Publishers – Dordrecht / Boston / London

Printed by Publishers' Graphics LLC